Smart Innovation, Systems and Technologies

Volume 319

Series Editors

Robert J. Howlett, Bournemouth University and KES International, Shoreham-by-Sea, UK

Lakhmi C. Jain, KES International, Shoreham-by-Sea, UK

The Smart Innovation, Systems and Technologies book series encompasses the topics of knowledge, intelligence, innovation and sustainability. The aim of the series is to make available a platform for the publication of books on all aspects of single and multi-disciplinary research on these themes in order to make the latest results available in a readily-accessible form. Volumes on interdisciplinary research combining two or more of these areas is particularly sought.

The series covers systems and paradigms that employ knowledge and intelligence in a broad sense. Its scope is systems having embedded knowledge and intelligence, which may be applied to the solution of world problems in industry, the environment and the community. It also focusses on the knowledge-transfer methodologies and innovation strategies employed to make this happen effectively. The combination of intelligent systems tools and a broad range of applications introduces a need for a synergy of disciplines from science, technology, business and the humanities. The series will include conference proceedings, edited collections, monographs, handbooks, reference books, and other relevant types of book in areas of science and technology where smart systems and technologies can offer innovative solutions.

High quality content is an essential feature for all book proposals accepted for the series. It is expected that editors of all accepted volumes will ensure that contributions are subjected to an appropriate level of reviewing process and adhere to KES quality principles.

Indexed by SCOPUS, EI Compendex, INSPEC, WTI Frankfurt eG, zbMATH, Japanese Science and Technology Agency (JST), SCImago, DBLP.

All books published in the series are submitted for consideration in Web of Science.

Andree Woodcock · Janet Saunders ·
Keelan Fadden-Hopper · Eileen O'Connell
Editors

Capacity Building in Local Authorities for Sustainable Transport Planning

 Springer

Editors
Andree Woodcock
Faculty Research Centre for Arts, Memory
and Communities
Coventry University
Coventry, UK

Janet Saunders
Research Institute of Future Transport
and Cities
Coventry University
Coventry, UK

Keelan Fadden-Hopper
Transport for West Midlands
Birmingham, UK

Eileen O'Connell
Interactions Research
Wicklow, Ireland

ISSN 2190-3018   ISSN 2190-3026 (electronic)
Smart Innovation, Systems and Technologies
ISBN 978-981-19-6961-4   ISBN 978-981-19-6962-1 (eBook)
https://doi.org/10.1007/978-981-19-6962-1

This Springer imprint is published by the registered company Springer Nature Singapore Pte Ltd.
The registered company address is: 152 Beach Road, #21-01/04 Gateway East, Singapore 189721, Singapore

Acknowledgements

This book is based on research conducted by the H2020 CIVITAS SUITS (https://www.suits-project.eu/) consortium funded under the H2020 programme as a Research and Innovation project (MG-5.4-2013, reference 690650) between 2016 and 2020/2021.

It summarises the work of the consortium and colleagues from cities in Europe (Alba Iulia, Coventry/Transport for West Midlands, Stuttgart, Kalamaria, Palanga, Rome, Turin and Valencia) in developing strategies to enhance their capacity to implement sustainable transport measures.

I would like to firstly acknowledge the contribution of all members of the consortium in producing chapters which summarise 4 years of collaboration with our city partners, looking at, for example, organisational change, data collection, business models and procurement. Secondly, I would like to thank city partners and their citizens for contributing to our research. Thirdly, I would like to thank our publishers for their patience and help in producing this volume, the production of which has been very much delayed by COVID-19. Lastly, and most importantly, I would like to thank the EU for their continued support to create more sustainable transport and liveable cities.

Contents

About the Editors

Andree Woodcock is a Professor of Educational Ergonomics and Design at Centre for Arts, Memory and Communities, Faculty of Arts and Humanities, Coventry University UK. She completed her B.Sc. in Psychology and Social Biology from University of London, UK; M.Sc. in Ergonomics from UCL, UK; and Ph.D. in ergonomics from Loughborough University, UK. She has worked as PI on three major EU funded transport projects—FP7 METPEX, H2020 CIVITAS SUITS, and H2020 TInnGO, as well as a number of UK projects. With her background in ergonomics she has always forwarded the needs of transport users, especially those experiencing transport poverty. H2020 TInnGO involved some members of the SUITS consortium focussing on gender and transport.

Janet Saunders is an academic researcher and specialist in user research and user centered design, splitting her time between research roles and consultancy. She has a background in ergonomics, social sciences, business analysis, and user research and has worked extensively in the IT industry, both in the private and public sectors. She has carried out senior research roles with Coventry University, UK: including the DDRI Living Lab project, developing research ethics for co-creation in a Living Lab and the Ergo Work project, an EU project about people with disabilities in the workplace. She was Research Fellow with the SUITS project and more recently was a Researcher for the TInnGO (Transport Innovation Gender Observatory) project, carrying out many aspects of qualitative research and design.

Keelan Fadden-Hopper is a Senior Product Manager at Essex County Council, working as part of a team using digital technology to deliver better public services for residents of Essex. Keelan supports teams across the council to meet the needs of their residents, including through the council's flagship website and other digital services. Keelan previously worked as a Senior Future Mobility Developer at Transport for West Midlands, a transport authority covering a region of 2.9 million people. As part of the team delivering the £22m West Midlands Future Mobility Zone programme, he worked on applications of digital technology to transport services. He led the development of a Mobility as a Service strategy for the West Midlands, planned and delivered a national-first mobility credit trial, and worked on the Midlands Future Mobility CAV Lab and Horizon 2020 SUITS projects. Keelan was awarded the Sampson Young ITS Professional of the Year award in November 2019.

Eileen O'Connell is Managing Director and a founding partner of Interactions Ltd, with over 26 years' experience in qualitative and quantitative research. Eileen's expertise lies in researching the needs, values and motivations of customers, providing insights for behaviour change initiatives, using segmentation techniques to identify potential to change—feeding into development of communications for behaviour change and overcoming resistance. She holds a Bachelor's degree (UCC) and an M.Sc. in Organisation Behaviour from the Trinity College/IMI Graduate School of Management and has completed a Master Class in Personal Construct Psychology. Eileen is a Member of the Institute of Management Ireland and participates in the Going for Growth Community. She is the evaluation manager of H2020 CIVITAS Suits project.

Chapter 1
Introduction

Andree Woodcock

Abstract The H2020 CIVITAS SUITS project was conceived by Professor Andree Woodcock and colleagues in 2014/15. It was scheduled to run between 2016 and 2020 but owing to the COVID-19 pandemic was extended to spring 2021. The aim of the project was to support capacity building of small–medium local authorities developing sustainable transport measures. This volume provides an account of the outputs of the project, in the form of chapters and recommendations for local authorities and consultants who are entrusted in delivering transport services which are inclusive, fit for purpose and enable accessibility for all. Although larger authorities are relatively well equipped to make these changes, smaller, more traditional local authorities may lack the knowledge, capacity and capability to plan, finance and implement sustainable transport measures at a time of great socio economic, technological and cultural change. Such authorities are also required to adopt new ways of working at the same time as designing and planning multimillion-euro transport projects which will support smart city developments and significantly improve the mobility of their citizens. At the heart of H2020 CIVITAS SUITS is a socio-technical approach, which recognises that capacity building is more than just providing training, and it is about empowering members of an organisation to be innovative. This volume has been written to inform designed to inform the daily practices of transport departments and stakeholder groups engaged in commissioning sustainable transport measures of working on Sustainable Urban Mobility Plans.

1.1 Introduction

The project, Supporting Urban Integrated Transport Systems: transferable tools for local authorities (SUITS), grant number 690650 was funded by the European Union (EU) under the H2020 programme as a Research and Innovation Action. It was, in response to the European H2020, 2015 call (MG-5.4–2015) to strengthen the

A. Woodcock (✉)
Coventry University, Coventry, UK
e-mail: A.Woodcock@coventry.ac.uk

© Transport for West Midlands 2023
A. Woodcock et al. (eds.), *Capacity Building in Local Authorities for Sustainable Transport Planning*, Smart Innovation, Systems and Technologies 319,
https://doi.org/10.1007/978-981-19-6962-1_1

knowledge and capacity of local authorities (LAs).[1] It aimed to provide small to medium LAs and associated stakeholders with a sustainable approach to capacity building in the transport arena, helping them to transform into resilient learning organisations able to meet future challenges and lead innovation in their cities.

The consortium (see Appendix 1) was drawn from European wide, academic partners and research establishments, local authorities, consultants and SMEs, many of whom have contributed chapters to this volume.

The project ran from 2016–2021. During this time, the 23 members of the consortium:

- understood the needs, challenges and barriers of LAs responsible for the development of sustainable transport systems and developed a transferable method to enable this understanding;
- developed material to address these needs;
- evaluated the usefulness of the project's outputs in practice.[2]

Due to the COVID-19 crisis, the project was extended by 3 months until 2021 in order for evaluation with city partners to take place.

1.2   Organisation of Volume

This volume covers all aspects of H2020 CIVITAS SUITS project, but its focus is on introducing tools, methods and insights which can be used by practitioners to inform the development of sustainable transport measures.

Section 1.1 sets the scene for the project, describing its context, rationale and ambitions. SUITS had 2 main themes: supporting sustainable transport measures and small–medium local authorities. Although the project was set in Europe, the insights and methods are transferrable, to transport departments outside the EU who share similar problems.

Many LAs have, or are now developing, Sustainable Urban Mobility Plans (SUMP) or Transport Master Plans. The material provided by SUITS does not require the development of a SUMP. Chapter 2 (Andree Woodcock) discusses the context in which these have been developed in Europe and why they may pose problems for a local authority.

---

[1] The terms local authority (LAs) and municipality are used interchangeably throughout this volume, reflecting the different terminologies used across EU.

[2] This evaluation was scheduled to take place in the final year of the project. However, all local authorities have been furloughed of significantly changers their working patterns during this time. Therefore, we have included our transferable evaluation procedure and specimen results which show impact.

In Chap. 3, Stefan Werland and Frederic Rudolph (Wuppertal Institute) set the project in the context of wider EU SUMP 2 initiative [1] and illustrate the strategies the EU has developed to support more integrated sustainable transport planning as a means of addressing climate and other challenges.

Section 1.2 focuses on delivering capacity change in local authorities. This forms the most extensive part of the volume in which we present the outputs of the project, how these were developed, validated and can be used to support capacity building.

Adopting a socio-technical approach, the first year of the project was spent trying to understand how local authorities (LAs) in the project (Coventry City Council/Transport for West Midlands (UK), Torino and Rome (Italy), Municipality of Kalamaria (Greece), Alba Iulia (Romania), Valencia (Spain) with follower cities in Palanga (Lithuania), Stuttgart and Erfurt (Germany)) were addressing the challenges associated with developing sustainable transport measures, the use of real time mobility data and more interdepartmental, and technologically supported ways of working. The larger LAs were taken as representative of larger cities across Europe. With larger departments, they had more time to engage with the project and pilot the tools and methods developed. Their experiences and practices were transferable to the three follower cities who could learn from the larger LAs and benefit directly from our proven outputs.

In Chap. 4, Sofia Kalakou (VTM, with colleagues Miriam Pirra, Ana Diaz and Sebastian Spundflasch) introduce a transferable methodology for discovering capacity gaps in LAs and discuss a set of 'generic gaps' in transport departments which may impede the design and implementation of sustainable transport measures such as gaps in knowledge, ways of working, trust and openness to innovation. This highlights at an operational level the problems which local authorities face when trying to redesign and modernise their transport systems. This is followed by a more in-depth description of key challenges by Sebastian Spundflasch and Heidi Krömker at TUIL (Chap. 5), and how these challenges were further validated.

From the information provided from this analysis of transport departments, Anne Marie Nienaber (Coventry University) adapted Kotter's [2] model of organisational change to transform the somewhat traditional and conservative departments into ones that were willing to be open and innovative. This 6-step process, described in Chap. 6 (Nienaber, Spundflasch and Soares), was followed by SUITS cities during the project. It can now be claimed that this is a proven, sustainable and transferable approach which can be easily implemented by transport and other local authority departments. It is also supportive of the SUMP process [1].

Chapter 7 authored by Anastasia Founta and Olympia Papadopoulou (from Lever Consult) introduces the capacity building, based on their work developing and delivering standalone capacity training modules to LAs. The final topics were ones which emerged from our needs assessment and related to topics not covered by previous material (e.g. on the Eltis website[3]). An introduction to these topics forms the central portion of this volume. Full copies of the training manual and training material

---

[3] The Urban Mobility Observatory. https://www.eltis.org.

are available and downloadable as the SUITS capacity building programme.[4] Each chapter provides an introduction to the subject for those joining transport departments or having to design and implement sustainable transport measures. The authors provide a more detailed discussion of innovation and implementation of sustainable transport measures.

Chapter 9, coauthored by Janet Saunders and colleagues from the local authorities, provides examples of the ways in which sustainable transport measures have been implemented in SUITS cities, with reflections on outcome and impact, challenges and lessons learnt.

This is complemented by Chap. 10 (coauthored by Andree Woodcock, Sebastian Spundflasch, Frederic Rudolph, Kain Glensor, Keelan Fadden-Hopper and Katie Miller-Crolla) which focuses on one particular sustainable transport innovation, Mobility as a Service, which at the time of the project was of especial concern to local authorities.

The following chapters provide sound introductions to a range of topics where gaps in capacity were noted. Chapter 11 looks at ways in which the social impact of transport measures can be measured (Andree Woodcock and Janet Saunders, Coventry University).

Chapter 12 reviews data collection and analysis tools, and describes trials conducted in Kalamaria and Turino to exemplify how mobility data can be used, visualised and integrated to inform decision making. This is authored by the lead of the technical work package, Miriam Calvo (ITENE) and her colleagues Miriam Pirra, Marco Diana, Fotis Liotopoulos and Ferenc Tilesch, from Politecnico di Torino, SBOING and LogDrill, respectively.

The next thee chapters (13–14) focus on financing sustainable transport measures:

- Innovative public procurement processes to implement sustainable mobility policy (Stefan Roseanu, INETCO) (Chap. 13).
- In Chap. 14, Olga Feldman (Arcadis) discusses innovative financing mechanisms for sustainable transport and mobility, along with the challenges and risks associated with each.
- Iana Dulskaia and Franceso Bellini (EUROKLEIS) in Chap. 15 outline new business models and partnerships for sustainable mobility and the transport sector. They provide the business model canvases for contemporary transport measures(such as e-bikes) as a means of increasing the potential longevity of new market entrants.

The final section of this volume documents some of the impacts and reflections by project members.

Chapter 16 by Eileen O'Connell, Director of Interactions and evaluation manager of SUITS discusses in Chap. 8 the approach taken to evaluation of process and impact within the project. Chapter 17 by Ann Marie Nienaber and Katerzyna Gut (Coventry

---

[4] SUITS capacity building programme. https://www.suits-project.eu/capacitybuildingprogram/.

University) provide a snap short of the impact of COVID-19 on our LAs in 2020 and how the work of the project helped them to become more resilient, and some of the issue that have been raised regarding city planning and transport.

Chapter 18 looks at the experiences of transport departments as part of EU projects, based on the consortium's experience of over 50 Framework 7, H2020 and CIVITAS projects. This is essential reading for transport planners and researchers who might be invited to join EU projects. Chapter 20 summarises the impact of the project and addresses ways in which the work needs to be taken forward.

In making the most of this volume, readers are invited to use individual chapters to enhance understanding or implement processes within their own organisation. We have also produced many online free to use resources in the languages of consortium members (e.g. in terms of implementation, data usage and organisational changes). The final section is more reflective in nature and provides essential reading for any city invited to take part in an EU funded project.

The book as a whole is the only detailed account of the SUITS project.

Andree Woodcock
Principal Investigator of SUITS
www.suits-project.eu

This project has received funding from the European Union's Horizon 2020 research and innovation programme under grant agreement no 690650.

References

1. Rupprecht Consult: Guidelines. Developing and Implementing a Sustainable Urban Mobility Plan (2014). http://www.eltis.org/sites/default/files/guidelines-developing-and-implementing-a-sump_final_web_jan2014b.pdf. Accessed 10 Sep 2020
2. Kotter, J.P.: Leading Change. Harvard Business School Press, Boston (1996)

Part I
Contextual Overview: Why is Capacity a Problem?

Chapter 2
Why is There a Need to Develop Capacity in Local Authorities

Andree Woodcock

Abstract SUITS was developed in 2015 and conducted between 2016 and 2021. It was developed specifically to address the climate change crisis through providing resources for local authorities to develop and implement sustainable transport measures. Such measures could take the form of, for example, traffic regulations and enforcement, new lower pollutant vehicles, new transport systems and services, or support for active forms of transport—cycling and walking. The development of such measures may require a strategic, integrated master plan (SUMP), but such an ambitious plan may not be possible for smaller local authorities.

Charles Michel, European Council President, stated in July 2020 that "**Climate neutrality is no longer a question of choice; it is beyond doubt a necessity**".

The transport sector is the only fuel combustion sector which shows an increase in GHG emissions when comparing 1990 with 2018. Between these years, the total GHG emissions increased by 32%, or 231 million tonnes of CO_2-equivalent. The volume of transport, measured as the amount transported times the distance, increased until the economic recession. However, fuel efficiency has not improved substantially enough to offset the increase in transport volume. The energy consumption in transport has increased by 37% from 1990 to 2018, in line with the increase in transport activity. Overall, transport has hardly improved its fuel efficiency. Almost all fuel used in transport consists of petroleum products and there has only been a marginal shift towards renewables.[1]

In 2019, EU leaders endorsed the objective of achieving climate neutrality by 2020, following commitments made by the EU and its member states in the 2015 Paris Agreement. One of the consequences of this was the European Green Deal which included measures such as:

---

[1] https://ec.europa.eu/eurostat/statistics-explained/index.php/Climate_change_-_driving_forces#General_overview.

A. Woodcock (✉)
Coventry University, Coventry, UK
e-mail: A.Woodcock@coventry.ac.uk

© Transport for West Midlands 2023

A. Woodcock et al. (eds.), *Capacity Building in Local Authorities for Sustainable Transport Planning*, Smart Innovation, Systems and Technologies 319, https://doi.org/10.1007/978-981-19-6962-1_2

- investing environmentally friendly technologies
- supporting innovation
- helping the development of cleaner forms of transport
- decarbonising the energy sector
- ensuring buildings become more energy efficient
- working internationally to improve global standards

To achieve climate neutrality means emitting less greenhouse emissions (and reductions in waste inter alia). Transport needs to shift towards more energy efficient, alternative and greener fuels. These need to be inexpensive, so that users can switch behaviour. However, consumers need to reduce their environmental footprint through behavioural change.

Cities are trying to adapt existing systems, services and infrastructures to prioritise the mobility of people and goods. New services have to be seamlessly integrated or replace existing ones, whilst the city continues to function. At the same time, streets and city centres are being reclaimed by people to make them more accessible, satisfying and life enhancing. A more considered understanding of transport has also led to an appreciation of its negative effects on human factors—health, wellbeing, mental health and stress—all of which are effected by the by products of transport—noise, pollution, lack of green spaces, time lost in traffic jams, etc.

Developing new transport services not only enables more sustainable vehicles and operation, but also provides cities with opportunities to assess problems and inequities in current transport provision and make sure that future transport is fit for the needs of smart cities and provide higher quality of service and quality of life for citizens.

H2020 CIVITAS SUITS (https://www.suits-project.eu) fits into this wider picture though its efforts to help cities invest in environmentally friendly technologies (through business models and innovative financing) and helping the implementation of sustainable forms of transport.

Delivering on the EU's sustainable mobility agenda is the responsibility of local authorities, in particular, their transport or mobility departments. Positioned after a period of austerity, the downsizing of local governments through budget cuts, outsourcing, squeezing labour costs and restructuring of public services saw a rise in inequalities between cities. The 'landscape of austerity urbanism' saw an unequal impact of austerity measures on declining and failing cities [15].[2] Therefore, smaller cities have a continuing capacity gap which need to be addressed to ensure that they are able to focus on sustainable mobility planning in the future.

Public and more recently active and sustainable forms of transport are strong policy instruments which local and regional policy makers can use to effect environmental policies, urban development, mobility management and social inclusion. Public transport can support a great number of the public values (Veeneman et al.

---

[2] Peck, J. (2012) Austerity urbanism. American cities under extreme economy, *City*, 16, 6, 626–655.

2006^3) related to these policy fields: from a healthier environment to economic development, from fighting global warming to local upgrading, from reducing congestion to increasing participation.

At the time of writing the initial proposal, the main challenges for the transport sector were 'creating a well-functioning Single European Transport Area, connecting Europe with modern, multimodal and safe transport infrastructure networks and shifting towards low emission mobility, which also involves reducing other negative externalities of transport. From a social perspective, affordability, reliability and accessibility of transport are key' [5].[4]

The EU remains committed to encouraging all sizeable cities to follow the SUMP[5] guidelines, which set out a clear process to develop Sustainable Urban Mobility Plans, which will help to reduce climate emissions related to urban transport. For example, during SUITS, we saw our Italian and Lithuanian partners working on SUMPs.

However, uptake of SUMPs across Europe has been relatively slow, despite multimillion-euro investment, incentives and the added bonus that systematic planning of integrated transport yields more fundable and well thought out proposals. H2020 CIVITAS Prosperity[6] reported that SUMP adoption had risen from 800 in 2013 to 1000 in 2017, with the number in preparation rising from 160 to 350. Countries in SUITS ranged from 'forerunner' to 'engaged' status in terms of SUMP development. The 3 main barriers to starting a SUMP[7] are lack of political will, lack of budget (budgets may be available for planning, but not implementation) and no data. Other potential issues are that it is a new process, many LA employees struggle with the concept of sustainable, integrated transport; the traditional transport planning approach is embedded and remains focused on infrastructure and motorised traffic; small cities, with reduced transport departments are in 'firefighting' mode, addressing current problems so lack the resources to start on this or commission external agents; the length and complexity of the process (with many feedback loops) which may take years to complete. Therefore many cities have started to install sustainable modes of transport in the cities (especially bike and car sharing and micro mobility), without an overall SUMP.

As a project SUITS was designed to, and has been able to provide supportive training and introduce sustainability concepts to smaller local authorities, for example, during webinars and multiplier events. Funded as one of three projects, along with SUMPS-UP and PROSPERITY under the CIVITAS call, the primary

---

[3] Veeneman, W.W., Van de Velde, D.M. and Schipholt, L.L. (2006) The value of bus and train: public values in public transport. In *Proceedings of the European Transport Conference*, Strasbourg, France (pp. 18-20). September.

[4] European Commission (2018) *Transport in the European Union, Current trends and issues*, DGMOVE, Accessed 02/02/2021. Downloaded from https://ec.europa.eu/transport/sites/transport/files/2018-transport-in-the-eu-current-trends-and-issues.pdf.

[5] https://ec.europa.eu/futurium/en/urban-mobility/reinforcing-uptake-sustainable-urban-mobility-plans-sumps.

[6] https://sumps-up.eu/fileadmin/user_upload/Tools_and_Resources/Reports/SUMPs_Up_D5.1_S UMP_in_Member_States_report_28022018_final_doc_without_annexes.pdf.

[7] https://ecf.com/what-we-do/urban-mobility/sump.

focus of SUITS was not on the creation of SUMPs per se, but rather on supporting small to medium local authorities develop sustainable, integrated mobility plans. 44% of EU urban citizens live in 'medium-sized' city regions of less than 500,000. For such cities, the development of a SUMP may not be appropriate. However, these cities still have problems of pollution, congestion and transport inequalities. Their capacity and openness to engage in integrated and sustainable mobility planning has to be supported in ways that work for them.

Firstly, because lack of knowledge and resources place smaller local authorities, their citizens and stakeholders at a significant disadvantage in challenging, leading, promoting or implementing new sustainable transport measures.[8] This is to the detriment of social, economic, health and wellbeing. Congested cities are not attractive to businesses (e.g. in terms of disruption to supply chains, Weisbord and Fitzroy [23]). Cities which are unpleasant to live in will not attract new residents who place an increased emphasis on sustainable, seamless mobility and the quality of urban spaces. This gap can be reduced by building the capacity of local authority employees through training and provision of tools and measures to gather data.

Secondly, organisational issues or cultural gaps in capacity may exist in LAs, which make a SUMP difficult to start. The prevalent silo mentality[9] and hierarchical structure of traditional LAs[10] may significantly affect the speed at which SUMPs or technological innovations are adopted. Taking a socio-technical approach to capacity building was fundamental to the project. This means that one of our starting points was looking at the organisation and culture within which SUMPs or sustainable transport measures were being developed, and assessing what organisational changes needed to be made to make this easier. Organisational culture includes the company vision, values, norms, systems, symbols, language, assumptions, beliefs and habits. The need to consider such factors is not recognised in technology led projects, which may inadvertently result in further widening existing gaps in knowledge and capacity.

Fuchs and Shehadeh [10][11] described the need for Departments of Transportation (in the US) to change, suggesting that '*DoTs that focus on older travel models will be unprepared to serve new kinds of demand. They will also be slower to convert growing data sets into actionable plans and projects that further support these changing trends. Conversely, those that actively embrace change have the opportunity to shape the future of transportation as well as urban, suburban, and rural development*'. They argue that departments dominated by civil engineers may be slow to adapt, and prone to groupthink. Increasing workforce diversity and exposing more senior and traditional departmental managers to best practices and new information is key as

---

[8] http://www.bestufs.net/dowload/BESTUFS_II/key_issuesII/BESTUFS_RecommendationsII.pdf.

[9] https://corporatefinanceinstitute.com/resources/careers/soft-skills/silo-mentality/.

[10] https://www.itproportal.com/features/breaking-down-the-silos-how-local-authorities-benefit-from-more-collaborative-working/.

[11] Fuchs, S. and Shehadeh, R. (2017) The department of transportation of the future, McKinsey and Co, October 19, Accessed 03/02/2021. Downloaded from https://www.mckinsey.com/industries/travel-logistics-and-transport-infrastructure/our-insights/the-department-of-transportation-of-the-future.

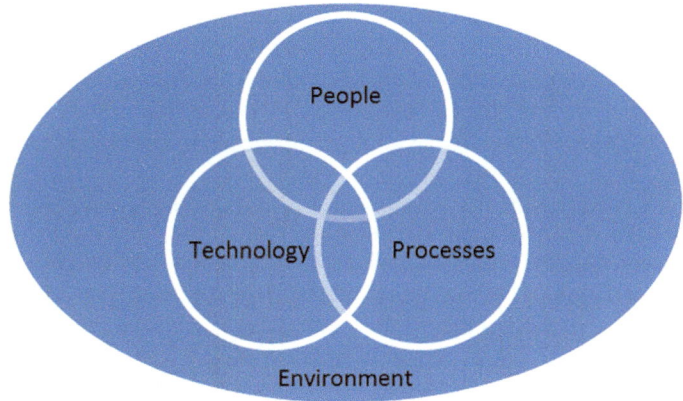

Fig. 2.1 Socio-technical approach

they are organisational gatekeepers. They concluded that '*Radical innovation will be challenging; yet it has been done before, and it can be done again. DoT leaders should acknowledge that the imminent changes to transportation are about to redefine their existing structure and models. They should be visionary, pragmatic, bold and courageous in the years ahead, and stakeholders must coalesce to support them in their transformation. With these measures firmly in place, DoTs can act today to establish a sturdy foundation for the future*'.

To address these concerns, SUITS took a socio-technical approach (e.g. Mumford 2000)[12] to developing capacity in S-M sized local authorities. Through its outputs (documented in Sect. 2.1), it has shared best practice from larger and medium-sized authorities, providing tools and methods which meet the requirements of S-M cities to increase capacity to plan, finance and implement sustainable transport measures and create new business opportunities (Fig. 2.1).

For SUITS, capacity building encompasses the sum of the capabilities of the local authority; the ability of the group or organisation to learn and adapt (here our focus was on the mobility or transport department), and the performance of the organisation in delivering research and having an impact on policies and practice. Organisational culture operates on, within, and is influenced by.

- aspirations,
- strategy,
- organisational skills,
- human resources,
- systems and infrastructure and
- organisational structure.

---

[12] Mumford, E. (2000) Socio-Technical Approach to Systems Design. *Requirements Eng* 5, 125–133 https://doi.org/10.1007/PL00010345.

as such, all areas need to be considered, if the ultimate goal is to create a learning organisation which is resilient, adaptable, innovative and enabling (for employees). The need to develop resilience within LA's was tested during the covid pandemic. During this time the LA's within SUITS were able to draw on, and apply the work of the project to the rapid restructuring needed to address the challenges of the pandemic.

By working closely with city partners over 4 years, insights were gathered about the barriers and enablers to transport planning, where capacity needed to be developed, better tools provided, communication channels enhanced and where integrated freight and citizen data could be used. Relevant training material and tools were provided to bridge existing knowledge gaps supported by tailored organisational programmes to promote sustainable thinking and implementation.

2.1   Brief Overview of the Project

This section provides a brief summary of the H2020 CIVITAS SUITS project as a way of providing a rationale for the following chapters.

SUITS was a four-year Research and Innovation Action, which aimed to increase the capacity of local authorities to develop and implement sustainable, inclusive, integrated and accessible transport strategies, policies, technologies, practices, procedures, tools, measures and intelligent transport systems that recognise the end-to-end travel experiences of all users and freight.

To achieve this, it developed a set of tools relating to planning, financing and implementing sustainable transport measures. In developing and validation these, the project directly supported capacity enhancement in nine local authorities (Valencia (Spain), Rome and Turino (Italy), Transport for West Midlands (UK), Kalamaria (Greece), Alba Iulia (Romania), Palanga (Lithuania), Dachau and Stuttgart (Germany). Each country had variations in approaches to developing and financing sustainable urban mobility and dealing with problems associated with freight and passenger transport. Together they provided a test bed for the development and use of SUITS capacity building material. Cities committed to SUITS varied from small cities and adjunct towns with little capacity or experience in urban mobility planning (e.g. Dachau and Palanga), to acknowledged leaders in sustainable transport systems (e.g. Valencia and Rome).

The project outputs are available freely online, in multiple languages and formats as guidelines, webinars, e-courses, training manuals, decision support tools and interactive software from our project web site,[13] partners site[14] and the ELTIS portal.[15] Many are described in this volume. Together they target the areas shown in Fig. 2.2.

---

[13] https://www.suits-project.eu/.

[14] https://www.mypolislive.net/.

[15] https://www.eltis.org/resources/tools/suits-capacity-building-toolbox.

Fig. 2.2 Specific areas targeted in the project.

This information can help LAs and policy makers make the case for socially and economically sustainable investments in transport. More specifically, SUITS targeted:

1. Organisational change in work practices as a capacity building exercise in its own right, but also to allow/set the stage for individual training and sustainable urban mobility planning (SUMP)
2. The need for integrated urban mobility planning of both freight and passengers based on the capture and use of information relating to the diversity of active, private, public, shared and multimodal forms of transport, journey types and travellers/freight.
3. The need to exploit future transport technologies to improve transport efficiency and enhance quality of life at the same time as humanising technology by creating opportunities for citizen engagement.
4. The need to maximise the effectiveness and sustainability of transport measures through transferable best practice md new funding models and create sustainable opportunities for new business entrants).

Therefore, the primary aim of SUITS was to develop a more holistic approach to capacity building which could support the development of SUMPS and sustainable

Fig. 2.3 Main components
of the project

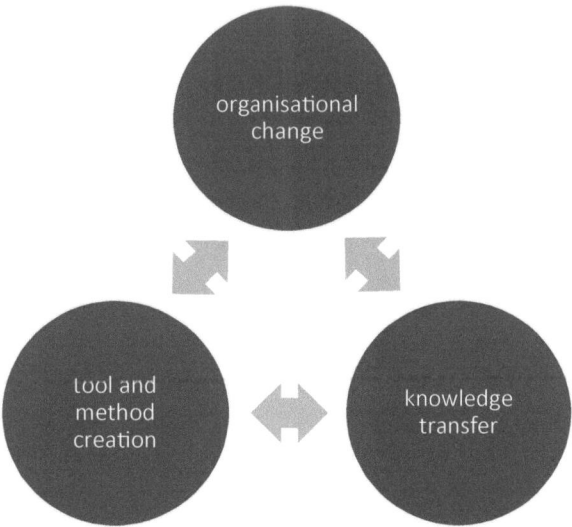

transport measures which recognised the context in which these were developed. This meant that the project had three main foci as shown in Fig. 2.3.

In terms of the structure of the project across the four years, the first 18 months was spent understanding the local authorities and developing tools for self-auditing of capacity gaps and development of KPIs. Each LA audited its organisational capacity, revealing a wide range of issues where they needed development. Not all related specifically to sustainable transport measures, pointing to the need for a more holistic approach. These were rank ordered and performance measures created, producing a list of actions that each LA would address during the project. Additionally, each LA selected sustainable transport measures which would provide opportunities to use project outputs and evaluate impact.

The following 18 months was spent developing and delivering organisation changes and the capacity building programme, which can be transferable to other LAs. In accordance with European Centre for Development Policy Management (ECDPM) best practice, external experts, formed of non-LA partners provided research results to LEVER–Development Consultants SA (Greece) who developed the capacity building programme (CBP). Assisting partners include Coventry University (UK), Ilmenau University of Technology (Germany), Integral Consulting R&D (Romania), EUROKLEIS s.r.l and Politecnico di TORINO (Italy), SmartContinent (Lithuania), ITENE (Spain). The CBP supported:

- Cultural and behaviour change from a socio-technical, perspective at individual, organisational and institutional level.
- The collection and use of real time, open source and legacy data to inform urban mobility plans.

- The ability of local authorities to address the requirements of vulnerable and hard to reach groups and those not using sustainable modes of transport and resisting change.
- The integration of freight and passenger information and measures.
- Increasing the appreciation of the wider benefits of sustainable transport to citizens and cities.
- Sustainable and innovative financing for new transport measures.
- Compliance with new procurement directives.
- Development of appropriate evaluation methods and tools for sustainable transport and mobility needs.

The intention of the final year of the project (2020/2021) was to embed the knowledge and processes within the organisations and to measure impact of these on the development of sustainable transport measures. Many of the chapters in this volume provide examples of the activities undertaken by the local authorities and how these have been affected by the project, with overall impacts being discussed in the concluding chapter.

SUITS met its overall aims and objectives to produce a series of tools, processes and methods which could increase the capacity of small–medium local authorities to implement sustainable transport measures. The global pandemic has revealed new challenges and potential opportunities in the fight for climate neutrality. It is hoped that small–medium LAs, through the work of SUITS and its sister projects are in a better place to deal with these.

Andree Woodcock
PI of H2020 CIVITAS SUITS

Bibliography

1. Bellini, F., Dulskaia, I., Savastano, M., D'Ascenzo, F.: Business Models Innovation for Sustainable Urban Mobility in Small and Medium-Sized European Cities, Management & Marketing. Challenges for the Knowledge Society, 14, 3, Autumn, pp. 266–277 (n/d), ISSN 2069–8887
2. Budhdeo, S., Keelan Fadden-Hopper, K., Drambarean, T., Krinos, I., Woodcock, A.: H2020 CIVITAS SUITS: supporting local authorities' implementing sustainable transport measures, Towards a Humane City Conference, Novi Saad (2019). https://humanecityns.org/conference-proceedings/
3. Diana, M., Pirra, M., Woodcock, A., Matins, S.: Supporting Urban Integrated Transport Systems: Transferable Tools for Local Authorities (SUITS) Proceedings of 7th Transport Research Arena TRA 2018, April 16–19, 2018, Vienna, Austria. Zenodo, 10 p. (2018). https://doi.org/10.5281/zenodo.1441138
4. Diana, M., Pirra, M., Woodcock, A.: Freight distribution in urban areas: a method to select the most important loading and unloading areas and a survey tool to investigate related demand patterns. Eur. Transp. Res. Rev. **12**, 40 (2020). https://doi.org/10.1186/s12544-020-00430-w
5. Directive (EU) 2018/645 of the European Parliament and of the Council of 18 April 2018 amending Directive 2003/59/EC on the initial qualification and periodic training of drivers of

certain road vehicles for the carriage of goods or passengers and Directive 2006/126/EC on driving licences (Text with EEA relevance)

6. Feldman, O., Lugovoi A., Parker, A., Farooq, S.: Financing mechanisms for sustainable transport and mobility. 47th European Transport Conference, Dublin, Ireland, 9–11 October (2019)
7. Feldman, O., Lugovoi, A., Parker, A.: Supporting urban integrated transport systems: guidelines to innovative financing. 17th Annual Transport Practitioners Meeting, Oxford, UK, 10–11 July 2019 (2019)
8. Feldman, O.: Social value at the heart of sustainable planning. 48th European Transport Conference, 9–11 September (2020)
9. Founta, A., Papadopoulou, O., Kalakou, S. and Georgiadis, G.: Building capacity of small-medium cities local authorities to implement MaaS and other innovative transport schemes, Proceedings of the "5th Conference on Sustainable Urban Mobility, 17–19 June 2020, (Virtual CSUM2020), published in the book series "Advances in Intelligent Systems and Computing" by Springer (Building Capacity of Small-Medium Cities' Local Authorities to Implement MaaS and Other Innovative Transport Schemes | SpringerLink) (2020)
10. Fuchs, S., Shehadeh, R.: Sharper estimating tools for getting projects done on time and on budget. Operations Extranet by McKinsey & Company (2017). [online] Operations-extranet.mckinsey.com. https://operations-extranet.mckinsey.com/article/sharper-estimating-tools-for-getting-projects-done-on-time-andon-budget. Accessed 2 Feb 2018
11. Kalakou, S., Spundflasch, S., Martins, S., Díaz, A.: SUMPs implementation: designation of capacity gaps of Local Authorities in the delivery of sustainable mobility projects. 3 rd. International Conference SSPCR Smart and Sustainable Planning for Cities and Regions 2019, Green Energy and Technology "Results of SSPCR 2019—Open Access Contributions" Springer International Publishing (2020)
12. Nienaber, A.M., Spundflasch, S., Soares, A., Woodcock, A.: Employees' Vulnerability—the challenge when introducing new technologies in local authorities. In: Krömker, H. (eds) HCI in Mobility, Transport, and Automotive Systems. Driving Behavior, Urban and Smart Mobility. HCII 2020. Lecture Notes in Computer Science, vol. 12213. Springer, Cham (2020)
13. Nienaber, A.-M., Spundflasch, S., Soares, A., Woodcock, A.: Distrust as a hazard for future sustainable mobility planning. rethinking employees' vulnerability when introducing new information and communication technologies in local authorities. Int. J. Human-Comput. Interaction **37**, 1–12 (2021). https://doi.org/10.1080/10447318.2020.1860547
14. Nienaber, A.-M., Woodcock, A., Liotopoulos, F.K.: Sharing data-not with us! Distrust as decisive obstacle for public authorities to benefit from sharing economy. Front. Psychol. **11**, 576070 (2020). https://doi.org/10.3389/fpsyg.2020.576070
15. Peck, J.: Austerity urbanism. City: Anal. Urban Trends Cult. Theory Policy Action **16**(6), 626–655 (2021)
16. Pirra, M., Diana, M.: Integrating mobility data sources to define and quantify a vehicle-level congestion indicator: an application for the city of Turin. Eur. Transp. Res. Rev. **11**, 41 (2019). https://doi.org/10.1186/s12544-019-0378-0
17. Pirra, M., Deflorio, F., Carboni, A.: Monitoring urban accessibility for freight delivery services from vehicles traces and network modelling. Transp. Res. Procedia **41C**, 410–413 (2019). https://doi.org/10.1016/j.trpro.2019.09.067
18. Roseanu, S., Caraman, D.: Proceduri inovative de achiziții publice pentru sprijinirea investițiilor în infrastructurile urbane de transport pe șină, in Proceedings of the XVIIth Romanian Railway Colloquium, Technical University of Civil Engineering, Bucharest (2019) (under print)
19. Roseanu, S., Caraman, D. (2019) Proceduri inovative de achiziții publice pentru sprijinirea investițiilor în mobilitatea urbană durabilă, iBuy Project 3rd Workshop, Bucharest, November 2019
20. Rudolph, F., Werland, S.: Public procurement of Sustainable Urban Mobility Measures. Urban Mobility Observatory, Brussels (2019)
21. Spundflasch, S., Krömker, H.: Challenges for local authorities in planning and implementing sustainable and user-oriented mobility measures and services. In: Krömker, H. (eds.) HCI in

Mobility, Transport, and Automotive Systems. HCII 2019. Lecture Notes in Computer Science, vol. 11596. Springer, Cham (2019). https://doi.org/10.1007/978-3-030-22666-4_24

22. Spundflasch, S., Rudolph, F., Glensor, K., Fadden-Hopper, K., Miller-Crolla, K., Woodcock, A.: MaaS Implementation; Local Authorities' Perspectives? CIVITAS Forum, Graz (2019)
23. Weisbrod, G., Fitzroy, S.: Traffic congestion effects on supply chains: accounting for behavioral elements in planning and economic impact models. Collection of open chapters of books in transport research, vol. 2011, pp. 119 (2011)
24. Werland, S.: (2020) Diffusing sustainable urban mobility planning in the EU. Sustainability **12**(20), 8436 (2020). https://doi.org/10.3390/su12208436
25. Werland, S., Rudolph, F.: Funding and financing of Sustainable Urban Mobility Measures. Urban Mobility Observatory, Brussels (2019)
26. Woodcock, A.: Developing and implementing sustainable, inclusive, integrated and accessible transport strategies, 1 Oct 2019, In: Open Access Government. 24, pp. 522–523 2 p. (2019)
27. Woodcock, A., Nienaber, A.-M., Olympia Papadopoulou, O., Pirra, M.: Overview of H2020 CIVITAS SUITS: supporting local authorities' delivering sustainable transport measures. Towards a Humane Cities Conference, Novi Saad (2019). https://humanecityns.org/conference-proceedings/

Chapter 3
The Relationship of Capacity Building to SUMPS 2.0: Reflections on Learning Programmes

Stefan Werland and Frederic Rudolph

Abstract This chapter reflects on the ways in which Horizon 2020 capacity building programmes support the diffusion of the sustainable urban mobility planning (SUMP) concept among European cities. It explores the different knowledge requirements of local administrations and how they are addressed through capacity building activities. The chapter builds on the authors' experience of organising, contributing to and evaluating learning events in SUMP-related European research and innovation actions (RIAs), including CIVITAS SUITS and CIVITAS SUMPs-Up. These projects support the spread of the SUMP concept, by providing the necessary capacity building material to help city and regional administrations design, develop and implement more integrated and inclusive transport plans. The chapter introduces the following learning experiences from the capacity building:

- Provision of expert information about the SUMP process.
- Broadening the view: Inspiration and transfer of experiences from other cities (peer-to-peer learning)
- Consolidating the new mobility planning paradigm, internalising the SUMP approach and overcoming outdated planning paradigms
- Motivation, community building and future cooperation
- Development of business cases.

3.1 Introduction: The Concept of SUMPs

SUMP stands for sustainable urban mobility planning[1]. The SUMP concept was introduced in 2009, followed by the first publication of the SUMP guidelines in 2013. Sustainable urban mobility planning aims to establish a new planning approach

---

[1] https://sumps-up.eu.

S. Werland · F. Rudolph (✉)
Wuppertal Institute, Wuppertal, Germany
e-mail: frederic.rudolph@wupperinst.org

© Transport for West Midlands 2023
A. Woodcock et al. (eds.), *Capacity Building in Local Authorities for Sustainable Transport Planning*, Smart Innovation, Systems and Technologies 319,
https://doi.org/10.1007/978-981-19-6962-1_3

for local and regional administrations[2]: moving away from the focus on traffic flow improvement towards a perspective that aims at improving the mobility of the cities' inhabitants. Thus, the primary objective of sustainable urban mobility planning is to improve accessibility and quality of life, including social equity, health and environmental quality and economic viability. This demands the integrated development of all transport modes [1].

A sustainable urban mobility plan (SUMP) is based on the following eight principles:

1. Plan for sustainable mobility in the "functional urban area"
2. Cooperate across institutional boundaries
3. Involve citizens and stakeholders
4. Assess current and future performance
5. Define a long-term vision and a clear implementation plan
6. Develop all transport modes in an integrated manner
7. Arrange for monitoring and evaluation
8. Assure quality

The process of developing and implementing a SUMP has been applied in many urban areas across Europe (and worldwide). It is essentially a staged approach, which is known as the "SUMP cycle" and visualised by a clock face (see Fig. 3.1). This is, of course, an idealised and simplified representation of a complex planning process. In some cases, steps of the SUMP cycle may be executed in parallel (or even revisited), the order of tasks may be adapted to specific needs, or an activity may be partially omitted because its results are already available from another activity.

This need for flexibility is an inherent part of the concept, and planners are encouraged to make reasonable adaptations if required by their specific situation as long as the eight principles of sustainable urban mobility planning are followed. The SUMP cycle is intended as a communication tool to describe in an easily understandable form what urban mobility planning entails. While the SUMP guidelines provide room for flexibility and adaptation to the local context [1], minimum requirements must be met:

- Key milestones must be produced in a factual and participatory manner. These milestones are:

  - a concise analysis of the problems and opportunities of the functional urban area;
  - a vision, objectives and targets agreed upon with stakeholders;
  - a description of actions including their evaluation and financing.

- The implementation process must be closely monitored, and implementation adapted as needed, with citizens and stakeholders actively informed of progress.

---

[2] It is understood that, for the purpose of this volume, this includes local authorities and municipalities.

Fig. 3.1 12 steps of sustainable urban mobility planning. *Source* [1]

3.2 The Relevance of SUMPs

The SUMP approach provides a structured approach to integrated, city and regional wide transport planning. The comprehensive process and guidelines can be used to develop detailed master plans, essential for attracting investment. Indeed, in many EU member states, the transfer of EU funds is contingent on the existence of a SUMP. The widespread use of SUMPs by cities to create a balanced and integrated development of sustainable transport modes is exemplified by the Eltis city database, which in August 2020,[3] reported almost 830 finalised SUMPs and more than 100 plans under preparation for the EU.

Even such a comprehensive process cannot guarantee that the most appropriate mobility solutions have been developed. Here "appropriate" can mean value for money (including whole life cycle costs), inclusivity and accessibility of the transport measures for all citizens, and the extent to which social, cultural and geographic constraints have been recognised. The impact of SUMP has therefore been mixed.

For example, the European Court of Auditors warned that EU cities must shift more traffic to sustainable transport modes. They found that EU-funded projects were not always based on sound urban mobility strategies and were not as effective as intended [2] that projected targets were not reached, and that projects were often delayed.

The reasons for weak implementation of sustainable urban mobility measures included:

- lack of personnel resources and specialist expertise required to identify and implement innovative policy measures, specifically in smaller cities [3].
- lack practical knowledge and experience on how to conduct elements of the SUMP cycle, such as public participation processes, scenario building or expert involvement.

SUITS research also found that the mere existence of a SUMP in itself does not alter the mobility system. Even in cases where cities know which measures will contribute to sustainable transport and contribute to their target reductions, implementation remains a challenge. SUITS conducted a statistical analysis of the modal split of 400 European cities [4] which revealed that developing objectives and planning measures alone does not necessarily lead to actual change. For instance, it appeared that more measures to promote cycling would need to be successfully implemented in order to make a profound contribution to sustainable development in European cities. These statistical analyses are in line with the results of a survey conducted within the SUMPs-Up project [5], a SUITS sister project funded under the same H2020 CIVITAS call. This survey of 441 European cities found that many cities aiming to increase their bicycle share, experienced obstacles implementing bicycle measures: with 140 reporting the need for support in *selecting* bicycle measures, and 264 highlighting the need for support in *implementing* bicycle measures.

---

[3] https://www.eltis.org/mobility-plans/city-database, accessed 26/08/2020.

The European Court of Auditors also found that cities struggle to finance the construction, operation and maintenance of infrastructures and services; or to define coherent policy mixes that combine different interventions [2]. The evaluation of the CIVITAS SUMPs-Up project also found that:

- progressive planning departments often lacked political support, specifically for the implementation of "push-measures" that restrict or discourage private car use and that burden car owners and thus potential voters.
- many urban mobility plans—mostly from cities that were legally obliged to adopt a SUMP by national law—existed merely on paper were provided as one-size-fits all documents by external consultants without proper involvement of the municipal administration and were poorly adapted to local circumstances.

3.3   From the SUMP Concept to Urban Planning Practices

Obviously, the implementation of a novel mobility planning culture in local and regional administrations is a challenging project that requires far-reaching support programmes. It is not enough to merely define a new planning paradigm; a step change towards sustainable transport also requires active support for local and regional authorities and a "hearts and minds approach" to building individual, organisational and institutional capacity.

SUITS started from the premise that local authorities are unlikely to have sufficient resources to develop, support and finance-integrated urban mobility plans without targeted capacity building measures, in order to grow, fund and implement sustainability strategies, appropriate to their future needs and that "special attention has to be paid to the employees in the local authorities, as they are the ones who have to deal with the challenges and have to develop themselves and their organisation further in order to pursue a more sustainable mobility planning in future" [6]. To overcome the implementation gap between the European concept and local planning practices, cities need different kinds of support instruments. This is where capacity building becomes vital. The following section explores the mechanisms through which CIVITAS projects transferred knowledge to European city administrations. It builds on practical experience from the SUITS project and its CIVITAS sister project SUMPs-Up.

3.4   Learning Mechanisms in SUMP-Related Capacity Building Programmes

The promotion of learning and networking among cities has been at the centre of the European Commission's activities for the spread of the SUMP concept: The Commission has instigated a variety of networking initiatives to support sustainable urban mobility initiatives to facilitate organisational learning at city and regional level. Within this, the overall objective of the H2020CIVITAS SUITS project was to enhance the capacity of small and medium local authorities to develop and implement sustainable, inclusive, integrated and accessible transport strategies.[4]

This section provides an overview of the learning mechanisms used in SUMP-related capacity building programmes, including the SUITS project. It explores channels of information provision and good practice and thereby highlights relationships of capacity building with SUMPS 2.0. A more detailed outlook on SUITS capacity building activities is provided in Sect. 3.2.

3.4.1 Provision of Expert Information About the SUMP Process

The most basic learning mechanism is the provision of explicit knowledge about the different steps of the SUMP cycle. The 2019 SUMP 2.0 guidelines provided guidance on the planning process as well as more detailed information on specific aspects, through Topic Guides and Practitioner Briefings. These cover a wide range of topics, including more traditional sustainable urban mobility planning measures and aspects, such as cycling and walking infrastructure, as well as cutting edge topics such as road vehicle automation. Topic Guides and Practitioner Briefings contain good practice examples and ideas for further reading and are meant to be readable for both laypersons and professionals in the cities seeking suggestions and stimulation. EU-funded research projects contribute their research outputs to websites such as Eltis.org or the CIVITAS learning centre.[5]

The provision of information through learning programmes and a facilitated exchange among peer cities may inspire city administrations and politicians, reduce costs and favour the uptake of specific measures proven to be successful in similar settings. In addition, targeted capacity building programmes can bridge the gap between larger and more innovative communities and cities with fewer capacities [9] (p. 52). SUITS and SUMPs-Up both used the following capacity building methods:

- webinars and face-to-face workshops, complemented with e-learning,
- lecture-style teaching/instructions and group work,
- working groups compositions (peer-to-peer, assigned roles),
- development of the SUMP 2.0 guidance material.

The type of knowledge needed and how information is processed is city-specific, depending on a variety of factors including:

- The available capacity and experience with sustainable urban mobility planning;

---

[4] Not necessarily SUMPs. Developing a full SUMP may not be feasible for smaller local authorities, which were the primary audience for SUITS. Nevertheless, such authorities still have capacity needs in selecting and integrating sustainable transport measures.

[5] https://CIVITAS.eu/learning-centre.

- Legal requirements from the national level for sustainable urban mobility planning such as public procurement;
- Company culture and trust and their roles in transitioning towards becoming a learning organisations;
- Planning paradigms and the openness of planners and politicians for sustainable mobility solutions;
- The severity of mobility related problems such as congestion or levels of nitrogen oxide concentration and the targets which need to be met;
- Cultural aspects related to a city's history such as the general openness of the population for active mobility and public transport.

However, despite—or even because of—the plethora of material, many of the resources included in the tool inventories are hard to detect for planning practitioners with limited time budgets. Urban planners may need more targeted guidance, structured practical examples and good practices, and the opportunity to ask direct questions to experts such as mobility researchers or representatives of funding agencies. To meet this need SUITS and similar projects such as SUMPs-Up have delivered information through expert presentations in webinars, and interactive workshops. Such presentations increase the visibility of the SUMP concept for smaller planning departments which may not be aware of it. Such online provision, in native languages, may be most relevant to smaller authorities with limited expertise and capacity to cover personnel and travel costs for participating in European events.

3.4.2 Inspiration and Transfer of Experiences Through Peer-To-Peer Learning

Another mechanism of knowledge transfer being used during workshops, webinars and e-learning courses was peer-to-peer learning. Most European cities that participated in learning programmes face similar challenges, mostly related to congestion and air pollution. In such situations, actors may turn to their peers in search of suitable and proven solutions. Understanding which solutions have worked well in other municipalities reduces the cost and effort needed to identify adequate and effective measures and may avoid potentially costly trial and error learning.

The transferability between cities and the expected impact broadly depends on the selection of examples through the organiser of the learning event. Workshops and other capacity building formats need to adapt to the knowledge of the participants. Naturally, cities are most interested in hearing from cities most similar to them. However, hearing experiences from cities with different characteristics and exploring whether their approaches are transferable to one's own context was also considered a major source of inspiration—even though not all might be replicable within their respective institutional context or legal framework.

In peer-to-peer learning activities, participants cannot only gain direct and first-hand insight into specific transport measures, but also about how to make the most of citizen engagement and stakeholder groups or the added value of interdepartmental cooperation. Usually, low-cost solutions and their impacts on the mobility system are particularly interesting. Moreover, peer-to-peer exchange and hearing from other cities allows participants of learning programmes to understand where their home city or region stands compared to their peers in terms of aspiration level, planning practices and the measures they (intend to) include in their mobility plan.

Formats that support this peer-to-peer mechanism include city presentations and site visits, but also informal exchange and interactive sessions in which groups of participants jointly elaborate on tasks. It is also recommended to include case examples that did not work well in order to discuss potential alternatives, to avoid replication of similar problems and to be aware of potential pitfalls before defining and selecting measures. Fostering learning and knowledge exchange between the nine local authorities during the implementation process was at the core of the SUITS project. A series of 11 workshops took place to guide the different local authorities through necessary organisational changes supported by face-to-face meetings, individual phone calls, emails and discussions [6].

3.4.3 Consolidating a New Mobility Planning Culture and Overcoming Outdated Planning Paradigms

A major obstacle to urban mobility transition is a reliance on outdated, car-oriented planning paradigms in the administration and amongst politicians. Experience from EU research and innovation actions has shown that in many cities, achieving a more sustainable urban mobility system requires an alteration of planning practices and a redefinition of priorities. As Nienaber, Spundflasch and Soares put it, "for transport measures to be successfully implemented, it is not enough to change the technology or the technical aspect. Most change programmes that focus solely on technological and/or technical change, ignoring the importance of social and behavioural aspects, end up by failing" [6]. Or as one participant of the SUMPs-Up learning programme said: "It is the mind-sets that have to change from motorised private transport to a more sustainable system, which is an integrated and multimodal perspective. Not all, including politicians, understand mobility in a right way. Some think that it is only public transport, others see it as a traffic problem".

In order to achieve such social and behavioural change in local planning departments, CIVITAS capacity building activities need to promote a deeper understanding of the SUMP rationale: moving from a transport perspective to a people-based approach, including a strong emphasis on participation, integration, accessibility for all and increased quality of life in cities. The different approach is summarised in Table 3.1.

If successfully employed, the outcome of this mechanism is the internalisation of a new planning culture that follows the SUMP paradigm in mobility planning

Table 3.1 SUMP approach compared to traditional transport planning. *Source* [1]

Traditional transport planning		SustainabLe urban mobility planning
Focus on traffic	→	Focus on **people**
Primary objectives: Traffic flow capacity and speed	→	Primary objectives: **Accessibility** and **quality of life**,including social equity,health and environmental quality, and economic viability
Mode-focussed	→	**Integrated development of all transport modes** and shift towards sustainable mobility
Infrastructure as the main topic	→	**Combination** of infrastructure, market, regulation, information and promotion
Sectoral planning document	→	Planning document **consistent with related policy areas**
Short and medium-term delivery plan	→	Short and medium-term delivery plan embedded in a **long-term vision and strategy**
Covering an administrative area	→	Covering a **functional urban area** based on travel-to-work flows
Domain of traffic engineers	→	**Interdisciplinary**planning teams
Planning by experts	→	Planning with the **involvement of stakeholders and citizens** using a transparent and participatory approach
Limited impact assessment	→	Systematic **evaluation**of impacts to facilitate **learning** and improvement

departments and public authorities. Still, interviews with participants of learning programmes from France—where a similar planning approach was made mandatory for larger cities already in the 1990s—revealed that profound changes in the mind-sets of planners took at least 20 years to materialise [7].

3.4.4   Motivation, Community Building and Future Cooperation

The promotion of networking among progressive cities and mobility departments has been at the centre of the European Commission's activities for the spread of the SUMP concept: The Commission has initialised a variety of networking initiatives to support sustainable urban mobility initiatives. DG MOVE finances the Eltis Mobility Observatory,[6] which is a networking and support platform for cities. Eltis facilitates the exchange of information, knowledge and experience on sustainable

---

[6] https://www.eltis.org/.

urban mobility planning and hosts the SUMP guidelines which were developed by mobility experts in close cooperation with cities, mobility practitioners and city networks.

In 2002, the European Commission launched *the "CIVITAS network of cities for cities"*,[7] which is dedicated to the promotion of cleaner and better urban mobility systems. The network has around 300 members and is co-funded by the EU. CIVITAS supports research and demonstration projects and living labs, offers learning programmes, arranges study visits and provides a tool database for mobility practitioners. The annual Urban Mobility Days provide opportunities for the presentation of on-going research, the dissemination of good practice examples, and exchange of ideas, etc., among cities, decision makers and mobility experts. The CIVITAS award is a highly visible distinction for innovative cities and awarded at the annual Urban Mobility Days. Moreover, the European Commission finances research projects and coordination and support actions on sustainable urban mobility with strong involvement of city partners under the EU's main research programme Horizon 2020 or under the Interreg Programme. Participation in networks and institutionalised exchange forums can intensify learning processes [10] (p. 828).

The degree of knowledge-sharing in learning networks and the acceptance of new insights depends on the quality of the relationships and factors such as reciprocity, trust and face-to-face interaction [10]. Studies on the diffusion of climate policies found that the likeliness of countries to adopt policies increases with their interactions with forerunner countries that have climate policies already in place [11] (p. 479). A number of sustainability related city networks emerged between the mid-1980s and the early 1990s, including ICLEI, Eurocities, the Union of Baltic Cities or Polis, which is a network of cities and regions with a specific focus on sustainable mobility. The existence of city and regional networks is considered to strongly support the diffusion of policies among their members [12, 13] and these city networks were actively involved in the CIVITAS projects.

The CIVITAS network generates a network of cities and experts cooperating in future to solve challenging problems and to provide a sense of being part of a sustainable mobility community. Discussions also have a motivational function, showing the real changes other cities have achieved based on their SUMP. Capacity building should aim to encourage local authorities with their belief that they can really make a change. More specifically, workshops allow for networking and community building, both on a professional and on a personal level. The moderator of the sessions should aim to ensure that all members contribute to group sessions.

3.4.5 Development of Business Cases

A successful transition of mobility systems not only requires changes in local planning practices, but also the support of key stakeholders, including industry and

---

[7] https://civitas.eu.

commercial mobility providers. The SUMP 2.0 concept not only promotes a new mobility paradigm, but it also pushes forward new, climate-friendly mobility technologies and services. Emerging technologies such as big data, automation and electrification need to be included into urban and transport planning. SUMPs have both the potential and the mission to bring together public and private stakeholders to work on the urban future. They can include strategies to incorporate these new technologies into urban planning and thereby to exploit these to frame mobility more sustainably.

Targeted capacity building programmes are a means of bringing together private technology service providers and city officials, i.e. supply and demand. While services such as transport models and data management are explained, workshop participants can discuss their applicability at the urban level. For example, obvious fields of application are the support of multimodal practices—sharing systems, mobility as a service and traffic management through real-time data.

The SUITS capacity building successfully connected cities with technology service providers: For example, in an online seminar on data management and exploitation, a web-based tool for the selection and extraction of global user mobility traces from a database and a web platform (database) for real time and historical urban traffic monitoring were shown and discussed. After the webinar, the provider who had explained the functioning received several new registrations from cities in Europe and worldwide.

3.5 Conclusions

Experience from capacity building programmes revealed a variety of mechanisms through which learning curricula and the provision of targeted information can contribute to learning needs of city administrations. The activities covered the needs of municipalities in different stages of SUMP development, from beginner cities with no experience in sustainable urban mobility planning up to experienced administrations that were currently revising their existing SUMPs. The range of observed mechanisms stretches from instrumental learning, i.e. provision of information by experts to peer-to-peer learning and the exploitation of other cities' experiences up to the deeper internalisation of the SUMP paradigm and the building of a SUMP epistemic community. Capacity building on sustainable urban mobility planning focusses on the process of learning rather than on the contents or methods. City representatives are experts for their local environment—they know more about their city than any trainer or moderator, and they have expertise in certain fields connected to sustainable urban transport and mobility. However, by taking part in the CIVITAS and similar initiatives, these professionals show interest in both learning more and conveying knowledge to others. Capacity building is about networking, about mutual reinforcement, about tips and tricks, about finding funding.

Capacity building can never become dispensable. New and emerging topics demand training, be it the impact of automated driving, last mile delivery or innovative procurement methods. Similarly, the progress and further development of traditional fields of sustainable transport such as push and pull measures or active mobility should continue to be discussed in capacity building.

References

1. Rupprecht, S., Brand, L., Böhler-Baedeker, S., Brunner, L.M.: Guidelines for developing and implementing a Sustainable Urban Mobility Plan, 2nd edn. Eltis, Brussels (2019)
2. European Court of Auditors.: EU cities must shift more traffic to sustainable transport modes, warn Auditors. Press Release, Luxembourg. https://www.eca.europa.eu/Lists/ECADocuments/INSR20_06/INSR_Sustainable_Urban_Mobility_EN.pdf (2020)
3. Chinellato, M., Staelens, P., Wennberg, H., Sundberg, R., Böhler, S., Brand, L., Adams, R., Dragutescu, A.: Users' needs analysis on SUMP take up. Deliverable D1.2 of the SUMPs Up project. http://sumps-up.eu/fileadmin/user_upload/Tools_and_Resources/Reports/SUMPs-Up%20-%20Users%27%20needs%20analysis%20on%20SUMP%20take-up.pdf (2017)
4. Rudolph, F., Damert, M.: Sustainable urban mobility in Europe - from planning to implementation. SUITS Policy Brief 1. Mobility and Transport Research Centre, Coventry University. https://epub.wupperinst.org/frontdoor/deliver/index/docId/6946/file/6946_SUMP.pdf (2017)
5. Staelens, P., Plevnik, A.: SUMP needs assessment: identifying the SUMP status and support needs of European cities. In: Presentation at the 4th European Conference on Sustainable Urban Mobility Plans on 30 March 2017 in Dubrovnik, Croatia (2017)
6. Nienaber, A.-N., Spundflasch, S., Soares, A.: Sustainable urban mobility in Europe–implementation needs behavioural change. SUITS Policy Brief 4. Mobility and Transport Research Centre, Coventry University (2019)
7. Werland, S.: Diffusing sustainable urban mobility planning in the EU. Sustainability, Sustainability (12), 20, 8436 (2020)
8. Rashman, L., Withers, E., Hartley, J.: Organizational learning and knowledge in public service organizations: a systematic review of the literature. Int. J. Manag. Rev. 11(4), 463–494 (2009)
9. Klausen, J.E., Szmigiel-Rawska, K.: The rabbit and the tortoise. climate change policy development on the local level in Norway and Poland. Transylvanian Rev. Adm. Sci. 52E, 38–58 (2017)
10. Knill, C., Tosun, J.: Hierarchy, networks, or markets: how does the EU shape environmental policy adoptions within and beyond its borders? J. Eur. Publ. Policy 16(6), 873–894 (2009)
11. Kammerer, M., Namhata, C.: What drives the adoption of climate change mitigation policy? A dynamic network approach to policy diffusion. Policy Sci. 51(4), 477–513 (2018)
12. Kern, K.: Cities as leaders in EU multilevel climate governance: embedded upscaling of local experiments in Europe. Environ. Polit. 28(1), 125–145 (2019)
13. Kern, K., Kissling-Näf, I.: Politikkonvergenz und Politikdiffusion durch Regierungsund Nichtregierungsorganisationen: Ein internationaler Vergleich von Umweltzeichen. In: Tews, K., Jänicke, M. (eds.) Die Diffusion umweltpolitischer Innovationen im internationalen System, pp. 301–351. VS Verlag für Sozialwissenschaften (2005)

Part II
Delivering Capacity Change in Local Authorities

Chapter 4
Setting Targets for Local Authorities to Increase Their Capacity to Develop and Implement Sustainable Transport Measures

**S. Sofia Kalakou◉, M. Miriam Pirra◉, A. Ana Diaz,
and S. Sebastian Spundflasch**

Abstract In order to assist cities implement their mobility plans, it is essential to analyse which factors might influence their capacity to plan, develop, and implement mobility measures. A four-step process was developed to help cities perform such an audit. In the first step, a comprehensive characterization and contextualisation survey was conducted to understand the context in which cities operated and their situation with respect to mobility planning. Then, each partner city assessed its own capacity to implement mobility plans according to a set of key capacity indicators related to organizational, legal, political, and societal aspects. The results indicated the areas in which each authority needed to focus in order to improve its capacity to implement general mobility plans. In the final step, each city considered their mobility plans and linked these to the capacity indicator. This provided each city with a bespoke set of specific capacity indicator targets which needed to be worked on to successfully apply their mobility plans. The four-step approach presented in this chapter is a transferable tool that can be applied by city authorities to understand and prioritize the capacity needs of transport departments.

S. Sofia Kalakou · A. Ana Diaz
VTM Consultores, Linda-a-Velha, Portugal

S. Sofia Kalakou
Instituto Universitário de Lisboa (ISCTE-IUL), Business Research Unit (BRU-IUL), Lisbon, Portugal

M. Miriam Pirra (✉)
DIATI—Department of Environment, Land and Infrastructure Engineering, Politecnico di Torino, Italy
e-mail: miriam.pirra@polito.it

S. Sebastian Spundflasch
Ilmenau University of Technology (TUIL), Ilmenau, Germany

A. Woodcock et al. (eds.), *Capacity Building in Local Authorities for Sustainable Transport Planning*, Smart Innovation, Systems and Technologies 319,
https://doi.org/10.1007/978-981-19-6962-1_4

4.1 Introduction

Internationally, great efforts are being made to move towards more sustainable cities in line with the United Nations Agenda for Sustainable Development which provides policy goals for various sectors. The move towards greater sustainability has been one of the motivations behind the development of more integrated planning tools and practices. In this context, in the transport sector, Sustainable Urban Mobility Planning (SUMP) has arisen as a policy tool to achieve greater sustainable mobility. Guidelines for the strategic planning of sustainable transport are presented in [1, 2], while operational tools, including criteria and relevant indicators, have been proposed and applied to different geographic contexts [3–8]. However, mobility plans have to be successfully implemented which relies on the understanding and capacity of local governments and stakeholders [9]. Previous work has indicated that the provision of technical support, stakeholder engagement, alignment of investments, and facilitation of collaborations are priorities for SUMP implementation [10] and that monitoring after implementation is also essential [8].

While many studies have focused on the assessment of sustainable urban development and SUMPs, evidence of the capacity of the relevant stakeholders to successfully implement those plans is scarce. However, a survey that aggregated the needs of 328 European cities revealed support is needed in financing, employee training, understanding of legal framework, and monitoring [2, 11]. Other studies have shown that successful implementation of SUMPs in Europe also requires an understanding of legal aspects, the provision of national guidance, participation of citizens and political support [12] and the impact of cooperation among transport and political stakeholders [13], public engagement [14], and data sharing [15] on successful implementation of transport measures.

Given the complexity of integrated planning, the rapid increase in technological innovation in the smart mobility sector, and the need for greater inclusivity and collegiality at all stages of the planning and implementation of sustainable transport measures, a starting point for cities is an assessment of their capacity to implement mobility plans. Such an audit will determine potential barriers (e.g. in knowledge and technical know-how) before they arise, so that corrective actions (e.g. training, IT systems, and key experts) can be put in place prior to commencing the SUMP cycle.

This chapter shows how such an analysis can be made, using the six LAs that participated in the SUITS project as an example. This involved conducting a survey to characterize the cities within the consortium [16], from which an evaluation framework was derived, presented in [17] and used to show the barriers and enablers the city would face in meeting its sustainable mobility challenges [18]. The aim was to understand the gaps and challenges for cities during the planning or implementation of mobility measures, as well as the requirements of cities and mobility planners in terms of support. The chapter presents a short profile of the participating cities, a description of assessment methods, and the results of the assessment for each mobility measure applied by each city.

4.2 Capacity Assessment Method Used in SUITS

The capacity assessment of each city was conducted at the start of the project in 2017 and repeated in part in 2020 to understand the extent to which the outputs of the project had reduced gaps in capacity.

The assessment included four main activities. First, a characterization of city's mobility options and plans was conducted to identify the components of the transport system and address the cities SUMP or other mobility priorities. Then, city stakeholders were given a set of challenges and capacity indicators covering organizational (cooperation/coordination), process, financial, technical, human resources, and working environment), political, legal, and societal aspects was provided to them. An explanation of these challenges is given in Appendix 1, and the capacity indicators are organized and presented with their codes in Appendix 2 and in [17]. Then, the stakeholders were asked to assess their own performance (1–5 scale) on indicators linked to mobility implementation and their level of importance (1–5 scale). To provide a broader and more in-depth understanding ratings, local authorities were joined by relevant transport operators and other stakeholders to represent the views of internal and external stakeholders.

Based on this information, the LA can plot the results and designate areas on which attention and priority should be paid (i.e. those falling within the LH area). This area includes indicators in which the authority underperforms (L) but rates as very important (H). An example from the city of Valencia is illustrated in Fig. 4.1, which shows that there is a need to focus on its financial autonomy (**o**rganizational—**c**oordination aspects), the identification of innovative financing (**o**rganizational—**f**inancial aspects—OF), enhance participatory management (**o**rganizational—**h**uman resources aspects OH), address better regular self-assessment, staff needs, continuous learning (**o**rganizational—**w**orking environment aspects OW), request better cooperation/coordination between political sectors (**p**olitical aspects—P), and understand procurement decisions on fuels, life cycle costs, safety, and security (**l**egal aspects—L).

Following this, the LAs were asked to assess the specific mobility plans on which they most needed support and select which indicators in the LH priority area should be focused on first. This allowed action plans and tKey performance indicators to be drawn up. The application of this process and its results will be discussed after a description of the SUITS cites.

4.3 Description of Cities in SUITS

In this section, the mobility profiles of participating cities are presented, their state on SUMP development is described and the areas in which they expressed they need help are reported.

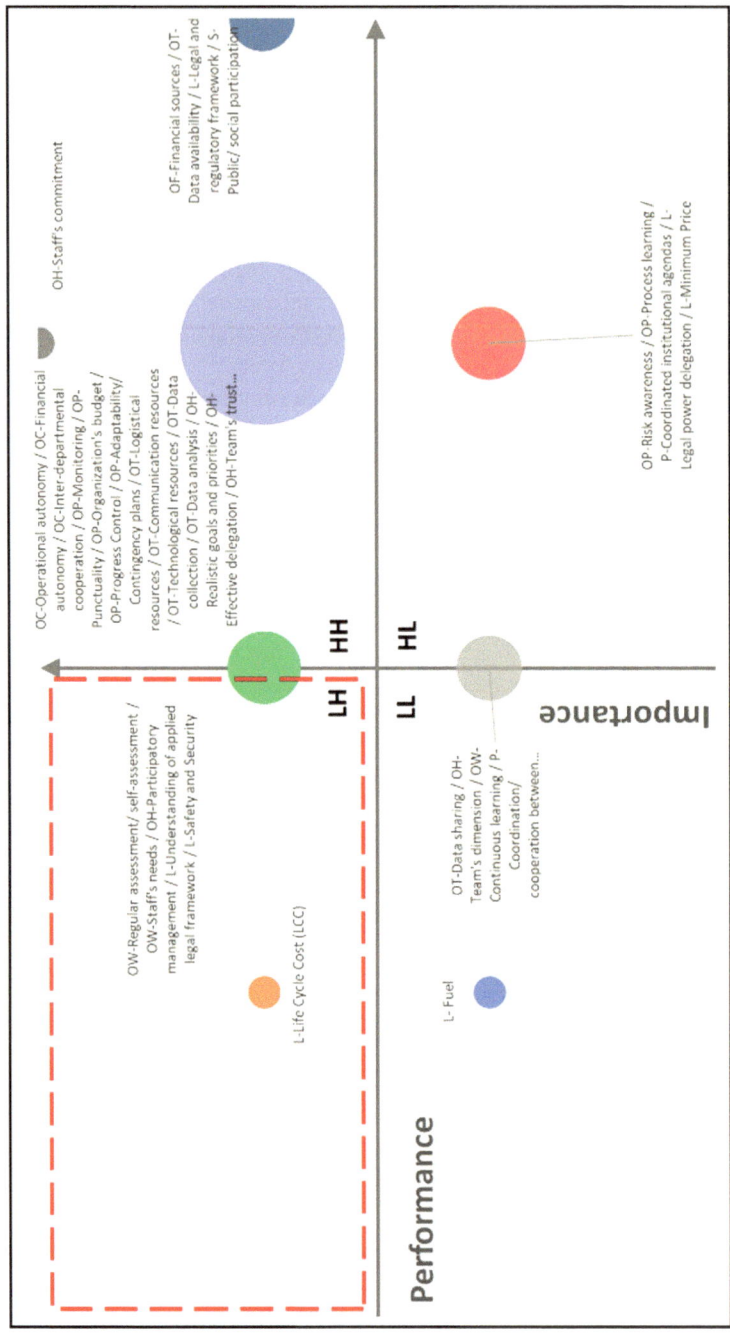

Fig. 4.1 Summary of the outcome of the indicator priority setting process for the city of Valencia

Valencia

Valencia is characterized by a stable population of around 790,000 inhabitants and steady economic growth. Regarding transport, 45% of trips are made using active transport modes. Most of these are made by walking (40.9%) with only 31.9% including the use of private vehicles. Bike sharing is present (3.47 bikes/1000 inhabitants), but car sharing is not available. Public transport provision is wide across the city (1.4 km/1000 inhabitants) and accounts for 23.2% of trips. The road fleet is modern (average age is 6–7 years) with a good percentage of alternative propulsion systems (17.4%). The city is implementing, evaluating, and revising its SUMP. Sectoral plans are available for all considered topics, and specific mobility measures have been already implemented. The city expressed that support is needed to implement projects related to urban traffic safety, urban logistics, and electric mobility.

Kalamaria

Kalamaria is a medium-sized city (of 6.4 km^2) with a growing population of 91,518 (based on 2011 census) but a shrinking employment rate. Private vehicles were the most used mode of transport (71%), with active modes only accounting for 2% of trips. The low level of cycling and pedestrian infrastructure influenced these values. Despite a small number of lines, public transport was used in 27% of trips. At the time of the survey, the city did not have a SUMP or any specific mobility plans but had plans to address this in future. They assessed that they would need support in technical fields and implementation of sustainable transport measures, especially with regard to three mobility policy areas: non-motorized transport, urban traffic safety, and road transport (including parking).

Alba Iulia

Alba Iulia is a medium-sized city (75,000–85,000 inhabitants) with a growing population and an increasing employment rate. It has good public transport but lacks car-sharing and bike-sharing schemes which are expected in a city of this size. Moreover, private motor vehicles are the main transport mode used by 63.5% of the population. The city's commitment to sustainable planning was demonstrated by the mayor's endorsement of the SUMP and its implementation with the help of consultants. Non-motorized transport, urban logistics, and mobility management are the three main areas where technical and implementation support is needed.

Rome

Rome is a metropolis with more than 2.6 million inhabitants, with a stable population and shrinking employment rate. Private vehicles form the main transport mode used for 65% of trips. Public transport accounts for 25% of trips and includes different modes (metro, buses, and tramway) for a rather convenient fare (0.07€ is the ratio of the price of a single ticket over 100 km of PT offer). Regarding propulsion, 8% of the public transport fleet are electric or have alternative propulsion systems. Active modes are used only by 10% of users, probably due to poor cycling infrastructure—lack of cycle lanes (0.09 km/1000 inhabitants), bike parking spaces, and proportionally small pedestrian areas. A car-sharing system is offered, but the fleet size is low compared to similar cities (0.38 cars/1000 inhabitants). A SUMP is being developed and the three mobility policy areas in which most support is needed: public transport, urban logistics, electric mobility, and clean fuels.

Coventry

Coventry is a large city in the U.K. of around 316,900 people, characterized by a growing population and a high number of young people (37% of inhabitants are less than 25 years old). There is no official data on mode shares, but multimodality is promoted with a good public transport offer and the availability of bike and car-sharing systems. The city is preparing a SUMP, measures are planned or have already been implemented for all policy areas proposed. Capacity building is needed in relation to finance and procurement, and the three main mobility policy areas identified as requiring more help were public transport, non-motorized transport, and road transport.

Turin

Turin has a shrinking population of around 870,000 inhabitants with 37.9% over 55 years of age. The modal transport split shows that the majority of trips are made by private vehicles (43%), with similar values for public transport and walking (23% and 25%, respectively) and low numbers for bicycling (only 4%). Both bike sharing and car sharing are present (1.15 bike/1000 inhabitants and 1.15 cars/1000 inhabitants), while high car ownership is observed (619 cars/1000 inhabitants). The city is revising its SUMP. Assistance is needed in planning techniques, and the three main areas of public transport, urban logistics, and mobility management.

These analyses provide a synthetic profile for each city and identify the mobility areas in which help is needed. Such an analysis is necessary as cities have different mobility patterns and levels of understanding/familiarity with sustainability. Considering policy priorities, the most urgent needs for these cities related to non-motorized and public transport, urban logistics, mobility management, and electric mobility. In the next section, the results of the self-assessment and the areas on which the cities needed to focus are discussed.

4.4   Capacity Assessment Results

The indicators that were rated as of high importance but low performance rating during the evaluation process were designated as priority areas> these are summarized in Table 4.1 for each city. As shown in Fig. 4.1, as an example the city of Valencia underperforms in the following capacity indicators that are considered important by them.

As shown in Fig. 4.1, as an example, the city of Valencia underperforms in the following capacity indicators that are considered important by them:

Table 4.1 Aggregated capacity assessment results

List of indicators	Valencia	Kalamaria	Alba Iulia	Rome	Coventry	Turin
Organizational: Cooperation/Coordination (OC)						
Cooperation						
Decision-makers						
Operational autonomy					x	
Financial autonomy	x		x		x	
Inter-departmental cooperation			x			
Organizational: Process (OP)						
Implementation rate			x		x	
Monitoring			x	x	x	x
Punctuality			x	x	x	
Organization's budget				x		
Progress Control		x	x			x
Risk awareness		x	x			
Adaptability/Contingency plans					x	
Process learning		x			x	
Organizational: Financial Resources (OF)						
Financial sources				x	x	
Innovative Financing—Understanding			x		x	
Innovative Financing—Identification	x				x	
Innovative Financing—Training		x	x		x	
Innovative Financing—Use		x	x		x	
Innovative Financing and local economy						
Innovative business model						
Organizational: Technical/Data Resources (OT)						
Logistical resources					x	
Communication resources					x	x
Technological resources					x	
Use of new technologies						
Data availability			x	x	x	
Data collection					x	
Data analysis					x	

(continued)

Table 4.1 (continued)

List of indicators	Valencia	Kalamaria	Alba Iulia	Rome	Coventry	Turin
Data sharing			x		x	
Organizational: Human Resources (OH)						
Staff's commitment				x		x
Realistic goals and priorities						x
Participatory management	x	x				x
Effective delegation						x
Team's trust in processes/tools				x		
Early engagement						
Team's dimension		x		x		
Team's skills						
Support tools/techniques/personnel		x	x	x		x
Organizational: Working Environment (OW)						
Regular assessment/self-assessment	x	x		x	x	
Staff's needs	x	x		x	x	
Continuous learning	x		x	x		
Turnover rate				x		
Political (P)						
Political commitment		x		x	x	
Coordinated institutional agendas				x	x	
Coordination/cooperation between sectors	x	x	x	x	x	
Continuity		x		x		x
Financing		x			x	
Legal (L)						
Legal and regulatory framework		x			x	x
Legal power delegation		x		x	x	x
Understanding of applied legal framework	x	x		x		x
Procurement decision criterions						
Procurement decision—Minimum Price						
Procurement decision—Fuel	x	x	x			x

(continued)

Table 4.1 (continued)

List of indicators	Valencia	Kalamaria	Alba Iulia	Rome	Coventry	Turin
Procurement decision—Life Cycle Cost (LCC)	x	x				
Procurement decision—Safety and Security	x		x			
Procurement decision—Environment			x			
Societal (S)						
Public awareness						x
Public/social participation		x	x			
Public acceptance				x		x
Media reaction						

- Four indicators relating to legal aspects (L): understanding of applied framework, procurement decisions on fuels, life cycle costs, and safety and security.
- Three indicators relating to organizational aspects and specifically to the working environment (OW): regular self-assessment, staff needs, and continuous learning.
- One indicator relating to political aspects (P): cooperation/coordination between sectors.
- One indicator relating to organizational aspects and specifically to cooperation/coordination (OC): financial autonomy.
- One indicator related to organizational aspects and specifically to financial aspect (OF): identification of innovative financing.
- One indicator related to organizational aspects and specifically to human resources (OH): participatory management.
- No societal aspects were identified as priorities.

Indicators relating to the internal working and structure of the LA may be easier to address than those which rely on third parties. At an aggregated level, most cities rated the following indicators as being of high importance/low performance highlighting a common need for support in capacity building in these areas. These related to

- monitoring,
- support tools/techniques,
- innovative financing,
- innovative training,
- regular self-assessment,
- staff needs,
- coordination cooperation between sectors,
- legal power delegation,
- understanding of applied framework,
- procurement strategies for fuel.

Indicators relating to the internal working and structure of the LA may be easier to address than those which rely on third parties. LAs were subsequently supported in making changes to address some of these issues through the organizational change process (see Chap. 6), the capacity building framework (Chaps. 7 and 8) and financial outputs of the project (Chaps. 11, 12 and 13). Many of these indicators do not relate specifically to (sustainable) transport, but have arisen because of a very traditional, conservative, and silo dominated culture. The need for new knowledge, data integration, modernization, and organizational change has been brought to light in part due to the need for greater inclusivity and interdepartmental working to create SUMPs as part of the wider Smart and Resilient City agenda.

4.4.1   Setting Capacity Indicator Targets Per City

In this section, examples are provided of how the cities used the capacity assessment framework to understand the barriers that would arise in addressing their mobility measures. Each mobility measure may give rise to different challenges depending on the nature of the action and the size of the city. At least two representatives from each city (transport planning and implementation) (with some disagreements occurring) and the project team worked to match the results of the capacity assessment to the priorities set by the city partners. In this way, the specific indicators to be improved were set as targets for each city. The results describe the relationship between measures, challenges, and target indicators for each of the cities.

4.4.1.1   Capacity Assessment—Valencia

Two self-assessment results were obtained, one from the municipality and one from the transport authority. The two entities evaluated the following criteria in a *similar* way:

- *Process indicators* and issues related to the working environment, political-legal, and societal aspects. Here, it was felt that the capacity of the city was very good since it responded to monitoring, punctuality, organization's budget, progress control, risk awareness, adaptability/contingency plans, and process learning challenges.

  – In terms of ***human resources,*** both parties felt that there was a correspondence with the required staff's commitment, realistic goals and priorities, participatory management, team's trust in processes and tools, early engagement of staff, team's skills, and use of support tools/techniques and personnel. The only difference was observed in the team's size that was rated lower by the transport authority but still remained at a very good level.

– The ***working environment*** was rated as having a medium performance; regular assessment, staff needs, and continuous learning took place sometimes and the turnover rate was estimated as medium

- *The **implementation rate*** was considered rated highly by both parties.

The municipality and the transport authority *differed* in their views relating to:

- All indicators concerning ***technical and data resources*** except for the assessment of exploitation of communication and technological resources which were rates as average or "sometimes", and the use of new technologies which was not well-rated by either. In the assessment of the rest of the indicators, the transport authority was more optimistic than the municipality.
- For ***political—legal—and societal aspects,*** some variation was noted in the self-assessment of the indicators but there was overall agreement between the two entities.

  – The weakest areas were the coordination/cooperation between sectors, the understanding of the applied legal framework and procurement decisions.
  – *Financial resources* were identified but poorly exploited.

- The ratings also differed in the evaluation of the ***impact of innovative financing and business models***; the municipality rated these two factors much lower than the transport authority, indicating that there is a gap of understanding between the two or that their expectations differed.

Table 4.2 presents the planned mobility measures that Valencia hoped to work on during the lifetime of the SUITS project, the perceived challenges for each measure and the most important capacity indicators the city needed to focus on. The first measure aims to restrict freight traffic through the better organization of delivery operations and the scheduling of deliveries by using new technologies. The second measure aims to reduce car accidents and increase sustainability by applying speed limits (30 km/h), traffic calming, and rearrangement of parking spaces. Sustainable mobility would be enhanced by the third measure looking at the redesign of public spaces through the provision of open space for walking and cycling and traffic calming measures.

Table 4.2 Target indicators for Valencia

Mobility measures	Associated challenges	Capacity indicators to improve
1. Improve freight distribution in the city centre	(1) Institutional cooperation (2) Interaction and cooperation with business partners (3) Citizen participation	(1) Coordination/cooperation between sectors (political) (2) Participatory management (organizational) (3) Understanding of applied legal framework (legal)
2. Progressive pacification of the speed of transit in the centre and other points		
3. Pedestrianization of different areas in the inner city		

Through collation and discussion of results in relation to the mobility measures and respective challenges required to redesign the city, Valencia decided that capacity building measures should relate to **coordination, participatory management, and stakeholder engagement** both inside and outside the organization. Within the organization, objectives related to the transfer of information and the demand for innovative public procurement. Outside the organization, the city needed to work to change the habits and the behaviour of the citizens to increase awareness and commitment to the planned mobility measures. **Political, human, and legal aspects** were considered as **barriers** to implementation.

4.4.1.2 Capacity Assessment—Kalamaria

The self-assessments conducted by the municipality and the transport authority showed clear *differences.*

- The municipality perceived that there was insufficient operational autonomy and cooperation; while the transport authority indicated that cooperation was high.
- The view of the two entities differed in the evaluation of the ***process indicators*** with the highest variation observed in progress control and risk awareness aspects.
- The implementation rate of plans was given a low and medium score. While financial ***resources*** were identified, innovation was not introduced in their exploitation.
- ***Human resources*** were perceived to be well utilized/organized in the case of the municipality, but the transport authority noted there was room for improvement in almost all indicators apart from level of staff commitment.
- In relation to the working ***environment***, the municipality assessed indicators related to self-assessment, continuous learning, and staff's needs more highly than the transport authority, which indicated that actions of self-assessment and staff's needs are rarely considered. Likewise, the turnover rate of employees' occupation and participation was reported as low in the operations of the municipality and high in the case of the transport authority.
- The transport authority rated the exploitation of financing schemes for the transport plans and the coherence among national/regional/local transport plans higher than the authority
- Likewise, the views of the two entities on ***societal aspects*** varied, with views on public awareness widely divergent.

The two entities showed a higher level of agreement on

- ***Technical aspects and data resources***, which were rated as good overall
    - Data collection was given the lowest rating,
    - The use of new technologies was considered high from the municipality's perspective but medium from the authority's perspective.

- Barriers due to the ***legal aspects of the operation*** were considered as not uncontrollable. The same issue that appeared during the assessment of ***political aspects***
- The transport authority rated more highly the exploitation of financing schemes for the transport plans and the coherence among national/regional/local transport plans.
- The views of the two entities on ***societal aspects*** varied, especially public awareness were assessed with both low and high levels. Media reaction to transport plans was mutually rated as average
- Factors related to **financing** were rated very low, while the use of new technological resources and innovating financing received the highest values.

Kalamaria prioritized three mobility measures during the lifetime of the project:

1. The development of an intelligent transport system to improve traffic conditions and parking availability to improve passenger travel quality and air quality by optimizing travel times. The real-time information system will provide drivers with information about current traffic conditions, available parking spots, public transport arrivals, estimated travel time, estimated optimal routes, and combined transport options.
2. Smart pedestrian crossings were implemented near a school to increase road safety. The plan is to extend this measure to more areas.

The implementation of both measures were perceived as facing major challenges relating to citizen participation and employment of innovative technologies. To improve its performance and increase the chances of successful implementation, the following indicators needed to be addressed: public acceptance, use of new technologies, and data collection. The third mobility measure described below, also needed public engagement and support,

3. The installation of 150 space, on-street smart parking system on three roads in the commercial centre, with a sensor-controlled parking management system to optimize the usage of urban parking spaces and reduce the amount of traffic searching for a parking space. The sensors will measure the occupancy of the parking lots which can be used by the drivers to find a vacant parking slot. To successfully implement this, the city needs to improve its public engagement, cooperation and training in innovative financing, political continuity and understanding of applied legal frameworks (Table 4.3).

The mobility measures and respective challenges are linked to improvements needed in the areas of **engagement, financing, innovation management, and participation**. The indicators commonly highlighted through the challenges include **citizen participation** and the **use of innovative technologies and data collection methods**. For Kalamaria, **human resources, organizational, political, legal, social, and financial aspects** are areas/fields where barriers exist that will impede implementation. The municipality considers interventions in human resources and social aspects as opportunities.

Table 4.3 Target indicators for Kalamaria

Mobility measures	Associated challenges	Capacity indicators to improve
Intelligent mobility system information on traffic conditions and parking availability	(1) Citizen participation (2) Use of innovative technologies and data collection methods (3) Application of research knowledge and adaption of good practice examples	(1) Public engagement (societal) (2) Use of new technologies (organizational) (3) Data collection (organizational)
Smart pedestrian crossing	(1) Citizen participation (2) Use of innovative technologies and data collection methods	(1) Public engagement (societal) (2) Use of new technologies (organizational) (3) Data collection (organizational)
Installation of 150 smart parking slots system at three roads (on-street) with sensors	(1) Institutional cooperation (2) Innovative procurement (3) Citizen participation	(1) Cooperation, training in innovative financing (organizational) (2) Continuity (political) (3) Understanding of applied legal framework (legal) (4) Public engagement (societal)

4.4.1.3 Capacity Assessment—Alba Iulia

In the case of Alba Iulia, four stakeholders participated in the capacity assessment of the city: the transport authority, the agency for energy sustainability, the transport operator, and the municipality. Compared to other cities, the self-assessment showed a very good performance, agreed on by most stakeholders. No indicator was awarded a low score. The area with the weakest evaluation was **financial resources**. All agreed that innovative financing—its use and understanding were the two indicators needing improvement. Organizational aspects were also given lower assessment scores and rated as of higher importance (Table 4.4).

The common challenges shared the need for improvement in the common indicators: use of innovative technologies, improved data collection, and understanding of political interests.

4.4.1.4 Capacity Assessment—Rome

In this case, only representatives of the municipality provided feedback on the self-assessment. The self-ratings showed that the operation of the municipality was

- Autonomous at a satisfactory level of ***cooperation*** despite a high number of stakeholders (10).

Table 4.4 Target indicators for Alba Iulia

Mobility measures	Associated challenges	Capacity indicators to improve
Encouragement of cycling	(1) Citizens participation (2) Understanding and applying innovative financing methods (3) Sustainability thinking	(1) Innovative financing use and training, b. data availability and data sharing (organizational) (2) Coordination/cooperation between sectors (political)
Improve public transport	(1) Understanding political interests and affecting political decisions (2) Innovative procurement (3) Sustainability thinking	

- *Financial indicators* were given a low rating, indicating opportunities for improvement.
- *Process indicators* received in general a moderate performance score with risk awareness an exception that was higher rated.
- All factors included in the *technical/data resources* and the *working environment* received average values of performance.

  - The use of new technologies was given a low assessment
  - The turnover rate was medium level.

- Few of the *legal aspects* were rated and the results varied. The *legal* and regulatory framework is explored but resources not always devoted to its efficient comprehension.
- *Political and societal aspects* were given an average rating.

Rome is currently drafting and working on various measures identified in the Mobility Master Plan of 2015. The priority measures chosen by the city to be analysed during the project were the plans for public transport, cycling, and urban logistics. The passenger-related measures aim to reduce car-dependency, promote bike-sharing services, and ultimately increase the quality of life. The city has already endorsed the active participation of citizens through a social media platform, where citizens can make specific suggestions on paths and interconnections among the bike lanes (Table 4.5).

The indicators commonly highlighted through the challenges are **cooperation among LA and their business partners**, **increase in public engagement,** and **better management of legal aspects.** The capacity indicators which need to improve include: staff commitment, political commitment, coordinated institutional agendas, coordination/cooperation between sectors, continuity, and public acceptance in order to observe improvements in its capacity to implement its plans.

Table 4.5 Target indicators for Rome

Mobility measures	Associated challenges	Capacity indicators to improve
Public transport plan	(1) Understanding political interests and affecting political decisions (2) Sustainability thinking	(1) Political commitment, coordinated institutional agendas, continuity, and coordination/cooperation between sectors (political)
Urban logistic plan	(1) Interaction and cooperation with business partners (2) Knowledge management/knowledge transfer	(1) Staff commitment (organizational) (2) Coordination/cooperation between sectors (political)
Cycling plan	(1) Citizen participation (2) Understanding political interests and affecting political decision	(1) Public engagement (societal) (2) Political commitment, coordinated institutional agendas, continuity, and coordination/cooperation between sectors (political)

4.4.1.5 Capacity Assessment—Coventry

During the self-assessment process, Coventry stated that it does not have a SUMP but that one is under development. The self-assessment gave medium to low scores for most indicators. *Legal and societal* aspects received higher scores at the evaluation. No indicator was rated as having an outstanding performance, however, not all aspects are of high importance.

In terms of its planned mobility measures, Coventry aims to contribute to environmental protection by introducing electric vehicle charging points to strengthen sustainable mobility and reduce air pollution. The second mobility measure aims to increase road safety through the employment of a crash-data system to gather collision data from the police in order to better understand collision events (e.g. types of junctions, number of speeding violations, or victim socio-demographic analysis as to who is more likely to be involved in collisions) (Table 4.6).

The common challenges for both measures are citizen participation and cooperation—internally, and with business partners. Anticipated barriers to operation and implementation include the need for better deployment of data resources, specialization of employees, and political engagement. Coventry can improve its capacity to implement mobility measures by focusing on: use of innovative technologies and data collection methods, effective project management and interaction, and cooperation with business partners.

Table 4.6 Target indicators for Coventry

Mobility measures	Associated challenges	Capacity indicators to improve
Clean fuels and low emission vehicles—EV charging points	(1) Citizens' participation (2) Effective project management and monitoring (3) Use of innovative technologies and data collection methods	Continuous learning, data availability, and data resources (organizational)
Safety and security-crash-data analysis	(1) Institutional cooperation (2) Interaction and cooperation with business partners (3) Citizens participation	Continuity (political)

4.4.1.6 Capacity Assessment—Turin

Turin has a SUMP that is currently under revision. Capacity assessment was based on the responses of the local authority.

Indicators that were given a medium rating included

- *cooperation and coordination*.
- *processes* inside the municipality.
- *Political aspects* obtained different results in the sense that most actions were completed with high frequency.
- A moderate score was given to the reaction of media to transport plans (*societal aspects*).

Those rated less highly, and which revealed a need for capacity building interventions included:

- The monitoring process and the implementation rate which were rated as low.
- *Financial resources* showed that there are many opportunities for improvements.
- Data availability, data sharing, and logistical resources were the indicators with the most frequent activity in the *technical/data resources category*. In this context, the utilization of new technologies for data collection was rated as medium.
- *Human resources* were given less importance in LA's operations. The *working environment* was considered insignificant.
- The area of *legal aspects* received, in general, low rates; the only exception was the "minimum price" regulation-related indicator which was referred as a frequently considered aspect.

The analysis showed that Turin's overall assessment was "medium" with no areas of exceptional performance. Human resources and legal aspects were considered to be fields to which attention should be paid for capacity improvement. Many of the other indicators were considered as insignificant (innovative financing, turnover rate, and implementation rate). "Cooperation" and "use of new technologies" had a moderate assessment score and medium level of importance, respectively.

Regarding the transport measures which Turin was working on during SUITS:

- In 2014, Torino signed a memorandum of understanding with all major national and local associations of trade and transport of goods to improve freight management. The objective of this first mobility measure was

  - the reorganization of loading and unloading of goods within the central limited traffic zone,
  - the use of logistic platforms,
  - the progressive substitution of most polluting vehicles,
  - the establishment of relevant tax incentives.

- According to the city's plans, in 2021, a new railway line is expected to connect several areas of the city including the most crucial shopping centres and a large hospital. In order to successfully implement this plan, the challenge is to understand how this innovation will change mobility behaviour and how passengers can be convinced to use it when it includes multimodal choices. To gain this information, the city will analyse data provided by telephone companies for building an origin/destination model. This would lead to a better understanding of how citizens move.

These two mobility measures were linked to the following challenges and the target capacity indicators for this city were identified (Table 4.7).

The results show that **human and legal aspects** may be considered as **barriers** to implementation. Regarding the mobility measures, three common challenges emerged: use of innovative technologies, data collection methods, and innovative procurement. The areas of intervention linked with the most important challenges for Turin are management, innovation, and sustainability. The city needs to improve its performance with respect to the following indicators: support

Table 4.7 Target indicators for Torino

Mobility measures	Associated challenges	Capacity indicators to improve
Urban goods freight distribution with clean vehicles	(1) Interaction and cooperation with business partners (2) Understanding and applying innovative financing methods (3) Sustainability thinking	(1) Procurement decisions criterions (legal)
Intermodal new interchanges for the underground and regional railway	(1) Use of innovative technologies and data collection methods (2) Understanding and applying innovative financing methods (3) Innovative procurement	(1) Support tools/techniques/personnel (organizational) (2) Procurement decisions criterions (legal)

tools/techniques/personnel and procurement decisions criteria. Data are usually acquired to gain information about passenger mobility mainly in public transport and, in a lower measure, to analyse freight transport. Some information would be required on active modes in future.

4.5 Conclusions

This chapter depicted the current state of six cities and the priorities they need to work on in order to improve their capacity to design and implement sustainability plans. Currently, there is no manual for goal setting in the framework of city and transport planning. SUITS project suggests a four-step method for the designation of priority steps for cities to design and implement mobility plans:

1. **Characterization of city's mobility characteristics and identification of mobility plans** by city stakeholders to identify mobility plans and where capacity self-assessment is needed.
2. **Self-assessment of the city to implement mobility plans** using the list of capacity indicators. The indicators are organized into four areas affecting the city's operation: organizational, political, legal, and societal. The capacity assessment areas and related indicators developed by the SUITS team to complete this work are shown in Appendix 2 of this chapter.
3. **Identification of challenges and capacity priority indicators**: The self-assessment process is followed by the identification of perceived challenges (Appendix 1) to the implementation of mobility plans and reveals the capacity indicators the city regards as important and in which it may be underperforming.
4. **Goal-setting**: In the final step, the most important mobility plans identified in Step 1 are linked to the capacity priority indicators, identified in Step 3, to enable cities to prioritize which areas to focus on. Cities were helped to develop key performance indicators to measure progress towards these as part of the organizational change process.

The analysis revealed commonalities in the most substantive problems faced by cities when designing and implementing mobility plans. The most frequent barriers related to internal processes (project monitoring), the working environment (fulfilling staff needs and applying self-assessment), the cooperation with other organizations and the alignment with external aspects, specifically with the legal framework (understanding of legal and regulatory framework and legal power delegation).

When looking at these through the lens of mobility planning, the aspects that appeared more frequently as capacity priority indicators were related to legal aspects (the coordination/cooperation between sectors and understanding of applied legal framework), political aspects (continuity of political agendas), societal aspects (enforcement of public acceptance), and organizational (using innovative financing and training).

The results showed that an improvement in the ability to act effectively can be achieved through improvement of internal organizational and communicational processes and the strengthening of cooperation and participation processes with citizens, business partners, and politicians. The analyses presented so far clearly delineate the need for the activities and the objectives of the SUITS project, the aim of which is to offer to cities the proper tools to reach their sustainable goals.

Appendix 1: Description of Challenges

Challenge area		Challenge description
1	Sustainability thinking	Shaping sustainable mobility requires sustainability thinking among the staff and those who are involved in the process. Anchoring a sustainable mindset is one of the biggest challenges for local authorities, as this cannot be dictated by leadership, rather it is a way of looking at things that needs to develop gradually. The LA must always provide impulses and constantly raise awareness of the issue
2	Institutional cooperation	The challenge illustrates the need to improve the cooperation between local and regional authorities and decision makers who are directly and indirectly involved in the development of sustainable mobility measures. The aim is to motivate the various municipal departments to develop a common vision, to participate and to commit to projects
3	Systematic staff deployment and development	In recent years, the field of mobility has become increasingly broad, complex and difficult to penetrate. Although an incredibly large pool of knowledge and experience is available in general, mobility departments often lack the capacity to develop their own technical know-how in all mobility areas. A major challenge is to develop the needed competencies within the staff systematically, with a view to the long-term, ideally in such a way that synergy effects between the projects can be exploited
4	Project management and monitoring	Effective and efficient project management forms the basis for successful projects. This aspect is still a big barrier and often leads to serious delays or even the failure of mobility projects. The challenge is to critically support and optimize project management and monitoring processes
5	Knowledge management and transfer	Shaping mobility depends to a large extent on experience. The challenge is to enhance and establish a sustainable process for knowledge management/knowledge transfer among mobility departments and stakeholders. The aim is to apply and try out established methods in order to learn from experience and from that of others. It is about applying these findings to new projects and transmitting them to new employees
6	Understanding and applying innovative financing	The challenge is to increase the ability to identify funding sources and to use innovative financing methods. This requires capacity to identify, evaluate, adapt and apply financing methods to projects for which there is no funding available or urban funds are insufficient

(continued)

(continued)

Challenge area		Challenge description
7	Innovative procurement	The challenge is to integrate sustainability criteria and requirements to procurement processes and sensitize procurement agents to sustainability aspects and opportunities arising from the procurement reform
8	Understanding political interests and decision making	No matter how well planned a measure may be, without political backing, it will not be implemented. The challenge is to increase the capacity to assess political moods and to affect political bodies through evidence and argument
9	Understanding legal and regulatory framework	As many policy areas are directly or indirectly affected by the development of mobility measures, various legal and regulatory frameworks need to be considered. Some of these regulations also may change over time. The challenge is to further develop strategies and skills, to access the legal framework conditions and to take them into account for planning and implementation of mobility measures
10	Citizen participation	The challenge is to increase the capacity to identify and actively involve citizens in the development process of measures and strategies. This requires a precise understanding of benefits and concrete methods of citizen participation. Citizens need to be informed about measures, goals and backgrounds in order to engage with the measures
11	Estimating the feasibility and acceptance of measures	It is particularly difficult to obtain the necessary political support for innovative measures when there is a lack of experience and a high degree of uncertainty in terms of feasibility and acceptance. The challenge is to use methods to try out innovative measures in a scaled version, in a closed system beforehand, to gain a better understanding for upcoming problems and to be able to make predictions for workability and acceptance
12	Interaction and cooperation with business partners	The interaction and cooperation with business partners has become increasingly important in order to implement new mobility services (e.g. sharing services). The challenge is to combine new offers with existing services, adapt them to the local characteristics and make them attractive to citizens. The conditions must be attractive for providers to offer such services in the city. Close cooperation with business partners is a key factor
13	Identification and utilization of synergy effects	The challenge is to identify early connections and dependencies between mobility strategies and measures or between different mobility services
14	Use of innovative technologies and data-collection methods	The challenge for the cities and the mobility departments is to raise awareness of technologies, tools and methods for the effective and efficient collection and evaluation of data and it's use for the planning, implementation and evaluation of mobility measures. It is also a matter of looking across other departments to see who is already collecting certain data, or who might still be interested in certain data. Multiple use of the data and the exploitation of synergy effects is particularly important

(continued)

(continued)

Challenge area		Challenge description
15	Application of research knowledge and adaption of good-practice examples	The challenge is about a greater application of research findings and knowledge. It is also about a better understanding of the transferability of good-practice examples. The identification and understanding of contextual factors that are relevant to the success or failure of measures is challenging and that must be taken into account when trying to adapt measures to the specific conditions of a city

Appendix 2: Description of Self-assessment Indicators

Organizational (O)	
Indicator's name	**Indicator's description**
Subcategory: Coordination/Cooperation (OC)	
Cooperation	Level of collaboration among the LA and the organizations that participate in all stages of planning and implementation of a plan (financing, procurement of products and services, public–private partnerships)
Decision makers	Number of policy makers involved in planning and implementation
Operational autonomy	Organization's autonomy to implement plans independently of other stakeholders' approval
Financial autonomy	Financial independence from central government and other financial agents
Interdepartmental cooperation	Level and frequency of cooperation and networking between the involved departments inside the same organization
Subcategory: Process (OP)	
Implementation rate	Number of implemented or planned measures
Monitoring	Project-management activities to control technical and processual issues
Punctuality	Rate of compliance with deadlines with clear milestones' identification
Budget management	Ability to realistically include plans/measures in the organization's budget
Progress control	Regular process evaluations to determine gaps and flaws in the plan's workflow execution, avoiding delays and redundant work
Risk awareness	Frequency of identification and assessment of possible risks that may appear during all the project's lifetime
Adaptability/contingency plans	Capacity to adjust plans/measures in reaction to an extraordinary event. Existence of risk-control measures defined to control the impact of the risks that affect the project

(continued)

(continued)

Organizational (O)	
Indicator's name	**Indicator's description**
Process learning	Organization's acknowledgement of internalizing past experiences, both positive and negative, to solve present/future issues that may arise
Subcategory: Financial sources (OF)	
Financial sources	Efficient use of national/international, public/private investment sources
Understanding (IF) innovative financing	An understanding of the benefits that innovative financing methods have on the financial capacity of the organization
Identification of IF	Ability to identify innovative financing opportunities
Training of IF	Number of people in the organization who are trained in innovative financing
Use of IF	Organization's employment of innovative financing resources
IF and local economy	Economic status of city increased through projects funded by innovative finance
Innovative business model	Organization's development of Innovative Business Models in the projects developed/implemented
Subcategory: Technical/Data resource (OT)	
Logistical resources	Available resources' quantity/quality needed to properly complete the tasks required for planning and implementation. Easy access to logistical tools
Communication resources	Available resources' quantity/quality needed to properly complete the tasks required for planning and implementation. Easy access to communication tools
Technological resources	Available resources' quantity/quality needed to properly complete the tasks required for planning and implementation. Easy access to technological tools
Use of new technologies	Willingness to use new technologies and familiarity with their application for data collection
Data availability	Availability of the necessary data required to complete all project's tasks
Data collection	Availability of necessary tools, networks and resources to efficiently collect data from diverse sources and in different formats
Data analysis	Availability of the necessary tools, networks and capabilities needed to efficiently analyse data collected of diverse sources and formats
Data sharing	Being able to retrieve valuable information as an output from the data analysis. Quantity and quality of data shared among departments (paper-form, electronic, etc.)
Subcategory: Human resources (OH)	
Staff's commitment	Staff's alignment, in attitude and performance, with the goals of the organization

(continued)

(continued)

Organizational (O)

Indicator's name	Indicator's description
Realistic goals and priorities	Link between managers' notion of the team's capacity, and the real team's capacity to deliver the expected outputs
Participatory management	Level of bidirectional communication between various management levels of the organization. Global knowledge increment
Effective delegation	Each member of the organization has a clear vision of her participation and responsibilities for the successful completion of the Plans. Clear understanding of one's role and participatory timeline
Team's trust in processes/tools	All staffers involved in the plans' planning and implementation phases feel completely comfortable with the tools and methodologies needed to successfully carry out all projects' tasks
Early engagement	Everyone participating in the project is involved from the beginning enabling all stakeholders to have a full view of the entire process
Team's dimension	Human resources available to complete all the project's tasks
Team's skills	Knowledge, competences and abilities of the team to meet project's needs
Supporting resources	Responsiveness to operational/process inefficiencies
Subcategory: Working environment (OW)	
Regular assessment/self-assessment	Identification of strengths and weaknesses of each member of the team
Staff's needs	Team members' needs are encouraged to be transparent inside the organization
Continuous learning	Permanent effort in keeping the staff updated regarding tools and techniques that would enable the project. Includes the level of evolvement in workshops, seminars, conferences, etc
Turnover rate	Reflects the stability in the composition of the team
Political (P)	
Political commitment	Defines how the project will be led and if it is a priority in the political agenda
Coordinated institutional agendas	Consistency in national/regional/local priorities. Correspondence between the plan and the national political agenda
Coordination/cooperation	Effective networking between the national departments of transport, land use, mobility, energy, etc
Continuity	Commitment to the continuation of the project independently of the authorities elected; the plan's progress is maintained unimpeded when moving from one political framework to the next one elected
Financing	Existence of financial programs within the national general budget to undertake the implementation of the Plan
Legal (L)	

(continued)

(continued)

Organizational (O)

Indicator's name	Indicator's description
Legal and regulatory framework	Contribution of legal and regulatory frameworks to efficient decision-making processes
Legal power delegation	Organization's autonomy to solve its own legal issues regarding the planning and implementation of the projects
Understanding of applied legal framework	All applicable legal framework should be clearly understood by all the involved stakeholders
Procurement decision criterions	Way of using decision criteria in the public procurement procedure (price, fuel etc.)

Societal (S)

Public awareness	Use of communication channels related to the project, its design, implementation and impact included
Public participation	Actions taken to engage citizens in the development of the project
Public acceptance	Level of willingness to support and engage with the implementation
Media reaction	Responsiveness of social media

References

1. Sustainable NI.: Sustainability assessment toolkit: an introduction, Version 4:0. Retrieved from: https://www.sustainableni.org/sustainability-reporting (2016). Accessed 30 July 2020
2. Plevnik, A., Balant, M., Rye, T.: National support frameworks for Sustainable Urban Mobility Planning. National SUMP Supporting Programs. European Platform on Sustainable Urban Mobility Plans. https://www.eltis.org/sites/default/files/national_support_frameworks_for_sustainable_urban_mobility_planning.pdf (2019). Accessed 30 July 2020
3. Zheng, J., Garrick, N., Atkinson-Palombo, C., McCahill, C., Marshall, W.: Guidelines on developing performance metrics for evaluating transportation sustainability. Res. Transp. Bus. Manag. **7**, 4–13 (2013)
4. DSDG-UNHQ-Division for Sustainable Development Goals.: Report of the Inter-Agency and Expert Group on Sustainable Development Goal Indicators (E/CN.3/2016/2/Rev.1). https://sustainabledevelopment.un.org/about/dsd (2016). Accessed 30 July 2020
5. Perra, V.M., Sdoukopoulos, E., Pitsiava-Latinopoulou, M.: Evaluation of sustainable urban mobility in the city of Thessaloniki. Transport. Res. Proced. **24**, 329–336 (2017)
6. Alonso, A., Monzon, A., Cascajo, R.: Comparative analysis of passenger transport sustainability in European cities. Ecol. Ind. **48**, 578–592 (2016)
7. Mozos-Blanco, M.A., Pozo-Menendez, E., Arce-Ruiz, R., Baucells-Aleta, N.: The way to sustainable mobility. A comparative analysis of sustainable mobility plans in Spain. Transport Policy **72**, 45–54 (2018)
8. Ali-Toudert, F.: Comprehensive assessment method for sustainable urban development (CAMSUD) - A new multi-criteria system for planning, evaluation and decision-making, Progress *in Planning*. In Press, Corrected Proof (2019). Available online 26 June 2019
9. Zoeteman, B.C.J.: What's behind the leadership sustainable development from politicians to CEOs? Environ. Dev. **8**, 113–130 (2013)

10. Skoudopoulos, E., Kose, P., Gal-Tzur, A., Mezghani, M., Boile, M., Sheety, E., Mitropoulos, L.: Assessment of urban mobility needs, gaps and priorities in Mediterranean partner countries. 6th Transport Research Arena April 18–21, 2016. Transport. Res. Proced. **14**, 1211–1220 (2016)
11. May, A.D.: Developing sustainable urban land use and transport strategies: a decision-makers' guidebook, 2nd edn. European Commission DGRTD, Brussels (2005)
12. ELTISplus.: The state-of-the-art of sustainable urban mobility plans in Europe (2012), p 48
13. Steurer, N., Bonilla, D.: Building sustainable transport futures for the Mexico City Metropolitan Area. Transp. Policy **52**, 121–133 (2016)
14. Gil, A., Calado, H., Bentz, J.: Public participation in municipal transport planning processes– the case of the sustainable mobility plan of Ponta Delgada Azores, Portugal. J. Transport Geography **19**, 1309–1319 (2011)
15. Tafidis, P., Skoudopoulos, E., Pitsiava-Latinopoulou, M.: Sustainable urban mobility indicators: policy versus practice in the case of Greek cities. Transport. Res. Procedia **24**, 304–312 (2017)
16. Diana, M., Pirra, M., Woodcock, A., Martins, S.: Supporting urban integrated transport systems: transferable tools for local authorities (SUITS). Zenodo (2018). https://doi.org/10.5281/zen odo.1441138
17. Kalakou, S., Spundflasch, S., Martins, S., Diaz, A.: SUMPs implementation: designation of capacity gaps of local authorities in the delivery of sustainable mobility projects. In: Bisello, A., Vettorato, D., Ludlow, D., Baranzelli, C. (eds) Smart and Sustainable Planning for Cities and Regions. SSPCR 2019. Green Energy and Technology. Springer, Cham (2021). https://doi.org/10.1007/978-3-030-57764-3_16
18. Spundflasch, S., Krömker, H.: Challenges for local authorities in planning and implementing sustainable and user-oriented mobility measures and services. In: Krömker H. (eds) HCI in Mobility, Transport, and Automotive Systems. HCII 2019. Lecture Notes in Computer Science, vol 11596. Springer, Cham (2019). https://doi.org/10.1007/978-3-030-22666-4_24

Chapter 5
Emergent Capacity Needs

Sebastian Spundflasch and Heidi Krömker

Abstract Sustainable mobility planning is associated with a number of challenges for mobility planning actors and especially for mobility planners in local authorities. Understanding these challenges is of great importance, on the one hand for the work in the SUITS project, where the aim is to improve the capacity of local authorities with regard to sustainable mobility planning, and on the other hand for future support initiatives, which can thus be better tailored to the needs and requirements of cities, especially small and medium-sized ones. Through collaboration with the nine city partners in the SUITS project, 15 challenges were identified. Some of these can be attributed to internal and organisational aspects and may be considered to be basic prerequisites for sustainable and innovative planning, whereas other challenges relate to the implementation of concrete mobility projects. This chapter describes the challenges identified and helps to further understand requirements of the mobility planners in the local authorities. Subsequently, a survey was conducted, where actors in the field of transport were asked to assess the relevance of the identified challenges.

5.1 Introduction

The topic of mobility is of particular importance in municipal planning. In the public perception, transport and mobility are among the main causes of pollution and climate change. Accordingly, special attention is paid to this area and developments are critically scrutinised by the public. The pressure on local authorities and their mobility planners is considerable. In the context of constantly shrinking budgets, they are called upon to develop sustainable mobility solutions that help to

S. Spundflasch (✉) · H. Krömker
Technische Universität Ilmenau, Ilmenau, Germany
e-mail: sebastian.spundflasch@tu-ilmenau.de

H. Krömker
e-mail: heidi.kroemker@tu-ilmenau.de

© Transport for West Midlands 2023   61
A. Woodcock et al. (eds.), *Capacity Building in Local Authorities for Sustainable Transport Planning*, Smart Innovation, Systems and Technologies 319,
https://doi.org/10.1007/978-981-19-6962-1_5

minimise traffic congestion, reduce environmental pollution, meet citizens' mobility needs and ultimately contribute to improving the quality of life in urban and rural areas.

The mobility sector has also undergone considerable changes in recent years. A greater public awareness of issues relating to environmental protection, sustainability and economic efficiency has led to a change in people's mobility behaviour and corresponding mobility needs, especially those of younger people. And although motorised private transport is still the number one means of transport, a change in people's thinking is occurring. Especially in the larger cities, with well-developed public transport systems, many people realise that owning a car no longer brings significant benefits. According to the motto "using instead of owning", alternative mobility offers are becoming increasingly important. Numerous new mobility services and service providers are entering the market, exploiting the possibilities offered by innovative information and communication technologies, and thus making mobility easier for the users and at the same time, contributing to reducing environmental effects.

However in the past, the focus was on infrastructure development for traffic, today the focus is on mobility, with a more efficient use of the existing infrastructure, on sustainable modes of transport and on generating added value to services, like providing real-time travel information and efficient ticketing in public transport or the seamless linking of different mobility offers.

However, the dynamics of the rapidly changing mobility sector present major challenges for cities and their mobility planners. They are required to constantly expand their knowledge and their capacity and continuously improve their strategies.

The aim of the SUITS project is to develop methods and tools to increase the capacity of local authorities, especially in small to medium-sized cities, to plan and implement mobility measures with a higher focus on sustainability. The aim of increasing the capacity, first requires a clear understanding of what capacity means in the present context. The general capacity of a mobility department depends on numerous parameters like staffing numbers and levels, leadership and political mood, financial resources, etc., things that cannot easily be changed. However, this is not where SUITS is aiming at. Rather, it is about identifying and considering capacity aspects to which science and research really can make a contribution, by providing support programmes, information and training materials and thus increase the awareness and the necessary capacities of the cities towards a more sustainable mobility planning.

During the project and while working with different partner cities, 15 challenges emerged, which local authorities and mobility planners have to cope with in the context of sustainable mobility planning. These are crucial starting points for future capacity building. In order to support cities more effectively, it is important to understand these "emergent" challenges to tailor appropriate support programmes and materials to their needs.

5.2 Challenges in Sustainable Mobility Planning

Cities face numerous challenges in transport and mobility planning. The SUITS project investigated which challenges cities have to cope with when developing mobility measures and strategies focusing on sustainability. In close cooperation with representatives of the project's partner cities, gaps and barriers as well as drivers and enablers to sustainable mobility planning were identified through several workshops and meetings, and ultimately, 15 challenges were derived from the results (as given in Table 5.1).

These challenges reflect the issues and problems that cities and mobility planners inevitably have to deal with in sustainable mobility planning (many of which were raised in Chap. 4). The identified challenges are not independent of each other, there are overlaps and parallels, but nevertheless they are clearly distinguishable. Depending on the project, individual challenges can have a higher or lower importance. The challenges are relevant to all cities, but of course some cities face certain challenges more successfully than others, depending on their respective capacity. In any case, they represent important starting points when it comes to capacity building on sustainable mobility topics.

Talking about capacity includes, on the one hand, the technical knowledge and expertise of various mobility topics and on the other, the willingness to engage with new topics and develop continuously in a constantly changing field. This applies to

Table 5.1 Challenges in sustainable mobility planning

No.	Name of the challenge
Challenges with regard to internal organisation	
1	Sustainability thinking
2	Institutional cooperation
3	Systematic staff deployment and development
4	Effective project management and monitoring
5	Knowledge management/knowledge transfer
6	Understanding and applying innovative financing methods
7	Innovative procurement
Challenges about concrete mobility projects	
8	Understanding political interests and affecting political decisions
9	Understanding legal and regulatory framework
10	Citizen participation
11	Estimating the feasibility and acceptance of measures
12	Interaction and cooperation with business partners
13	Identification and utilisation of synergy effects
14	Use of innovative technologies and data collection methods
15	Application of research knowledge and adaption of good practice examples

entire departments as well as to individual employees. The capacity depends to a large extent on the staff available in an organisation. Overall mobility departments in larger cities are better situated than smaller ones. In most cases, their larger number of staff makes it possible to build up a wider range of knowledge and expertise. In smaller cities, there are usually very few people, sometimes only one person, who have responsibility for mobility issues.

5.2.1 Challenges Regarding Internal Organisation

A certain number of challenges concern the internal organisation of the local authority. These challenges are not linked to concrete mobility projects. Rather, they form the organisational prerequisites and the necessary attitudes local authorities need when aiming at the development of innovative and sustainable mobility.

Figure 5.1 shows, by way of example, that the development of mobility measures typically requires the interaction and cooperation of numerous departments within the local authority. The challenges described in the following concern the need for an efficient interplay between these different units.

Fig. 5.1 Mobility planning requires the cooperation of various departments within the local authority

5.2.2 Sustainability Thinking

Developing sustainable mobility requires sustainable thinking among the staff working on mobility issues, but also ideally across the entire authority. This represents one of the biggest challenges as sustainability thinking is not something that can be dictated by leadership. Sustainability thinking is an attitude that needs to develop and thus, employees must be continuously sensitised for the topic and its importance. Sustainability has to be anchored in the local authorities' agenda as an elementary part of the development of mobility strategies and measures.

Sustainable mobility planning is not characterised by simply copying good practices from other cities. Of course, good practices can serve as a source of inspiration, but to ensure that they work properly in the context of another city, they need to be adapted and tailored to the specific conditions. In addition to the understanding of these conditions, this task requires an awareness of the employees for the basic principles of sustainability. Only in this way can they customise the measures to achieve the maximum effect.

In the mobility sector, the term sustainability is closely related to the shift towards clean and environmentally friendly transport. But there's a lot more to it than that. The outcome of the United Nations General Assembly [1] names three pillars for sustainable development: Social, Environment and Economic. The SUMP Guideline [2] specifies these principles for the planning of sustainable mobility and mentions, e.g. focus on the people, accessibility and quality of life, participative planning involving all relevant actors, development of cost-effective solutions, evaluation of (social, environmental and economic) effects and establishment of a learning and improvement process. The basic ideas and the different principles must be internalised by the staff in order to apply them in the different mobility projects.

To face this challenge, the West Midlands Combined Authority in UK, for example organises in-house workshops with external trainers, but also workshops in which staff jointly try to develop ideas for concrete activities. Furthermore, employees are encouraged to share and discuss issues, and they have found on specific topics of sustainable mobility.

5.2.3 Institutional Cooperation

The development of mobility measures and services is a complex undertaking and touches different topics and areas of responsibility within the local authority. Thus, an intensive cooperation between different administrative units, as shown in Fig. 5.1 is required. A key success factor is the willingness of the different departments to cooperate and to jointly plan and implement mobility strategies and specific measures.

A systematic and efficient collaboration requires the:

- development of a common vision
- bundling of competencies
- recognition and exploitation of synergy effects
- allocation of capacities
- definition of roles and responsibilities

The local authority of Stuttgart, for example has addressed this challenge through the creation of a strategic steering committee that promotes cooperation between the departments. In regular meetings, ongoing and planned projects are discussed, and a common understanding of goals is developed which makes it easier to identify collaboration potentials. In the West Midlands Combined Authority, monthly Newsletters are created and sent to employees, which contain, for example news about ongoing projects, concrete team activities, current problems and results. This leads to a higher transparency and general awareness about the projects of other departments and also reveals potential for cooperation.

Since mobility services may not necessarily end at the city borders, inter-municipal cooperation plays an increasingly important role. The joint development and management of mobility services can greatly increase efficiency and lead to more sustainable concepts [3].

5.2.4   Systematic Staff Deployment and Development

The field of transport and mobility has become increasingly broad, complex and difficult to penetrate. Accordingly, it is a big challenge, especially for smaller cities, to further develop the technical know-how and build up a broad range of expertise which is needed to plan and implement innovative services and sustainable mobility measures. Larger cities have an obvious strength in that they can more easily pull together horizontal teams with the necessary skills. Smaller cities, where the mobility departments are by nature smaller, need to think carefully about which competencies they develop within the organisation and which expertise they buy from external consultants. To date, projects that require expertise in innovative subject areas are often "outsourced", at least partly. This is not necessarily wrong, as the departments cannot know everything by themselves and sometimes it is better to involve experts. However, local mobility planners are the experts on the city. They need to complement this a basic understanding of latest transport innovations to ensure efficient cooperation with external planning agencies.

Cities must look at their long-term visions and consider the direction in which they want to develop their staff, especially in view of the demands that innovative and sustainable development of mobility measures and services entail.

5.2.5 Effective Project Management and Monitoring

Although the importance of project management is obvious, insufficient project management is still seen as a big barrier often leading to serious delays or even failure of projects. This applies equally to the subject of sustainable mobility planning. During the planning phase, the biggest problems relate to over-ambitious planning combined with a lack of experience on innovative topics which leads to unrealistic plans or higher total costs. Another major challenge is the systematic monitoring and early detection of problems and deviations, often caused by the fact that when different departments work together on specific projects, they may have different views and priorities.

In the MAX project "Successful Travel Awareness Campaigns and Mobility Management Strategies", the guideline "MAX-Sumo—Guidance on how to plan, monitor and evaluate mobility projects" [4] was developed to support project management in mobility projects. Another interesting guideline developed in the GUIDEMAPS European Union project is "Successful transport decision-making— A project management and stakeholder engagement handbook" [5] which provides useful models and approaches to improve project management in the mobility sector.

5.2.6 Knowledge Management/Knowledge Transfer

Knowledge management is the "process of creating, sharing, using and managing the knowledge and information of an organisation" [6]. Well-functioning knowledge management results in greater efficiency, less duplication and significantly better job performance. It ensures that new employees can be trained more quickly, influences the quality of decisions and can improve communication within and across project teams [7].

The challenge is to set in place structures for the exchange of experience and knowledge within the organisation and between different units. This includes the exchange of knowledge between employees as well as the documentation of findings and experiences in a knowledge management system.

Improving knowledge management requires, for example:

- Knowing and using the technological possibilities to support this process
- Setting the basic structure and ensuring systematic data preparation in a way that the information can be easily found and used in daily operation
- Motivating employees to actively participate in enriching and maintaining the knowledge management system

To establish well-functioning knowledge management within an organisation, it is not enough just to purchase a software product. Knowledge management is a process that has to be supported by all employees, permanently maintained and optimised. This means a lot of effort, but an effective knowledge management process greatly improves the capacity of an organisation and enhances its ability to innovate [8]. Especially, since mobility planning depends to a large extent on experience, an effective knowledge management plays an important role on the way to becoming a learning organisation and in retaining knowledge within an organisation when employees leave. It has been our experience that organisational churn is quite high in local authorities.

As part of a knowledge management system, knowledge maps can help make the knowledge and competences of individual employees visible and to use them more effectively, especially in larger departments and across different departments. This also makes it easier to decide in which areas further expertise still needs to be built up.

5.2.7 Understanding and Applying Innovative Financing Methods

The matter of financing plays one of the most important roles in mobility planning as it ultimately decides whether a measure is implemented or not. There are significant differences between cities located in structurally and economically strong regions where sufficient municipal funds are available to finance mobility projects and weak regions where only small local budgets are available.

However, there is limited awareness of innovative or alternative financing options that go beyond the pure acquisition of municipal or federal funds. Insights from working with project cities made clear that financing is one of the most important issues, but at the same time, the area with most conservative action. Not understanding financing mechanisms or funding that may be available can be a major impediment to the implementation of new transport measures, as many require sizable investments and partnerships.

Therefore, the main challenge, especially for cities with limited financial resources is to increase their capacity to identify, adapt and apply alternative/innovative financing methods for projects for which there is no funding available or urban funds are insufficient. One of the reasons for the poor use of innovative financing methods such as congestion charges, crowdfunding or municipal green bonds is the lack of expertise and experience and a lack of political support to try things out. In this context, a series of guidelines on innovative financing were developed within the SUITS project, informing about alternative ways to obtain the necessary financial resources (please see Chaps. 11–13).

It is also striking that the topic of financing is usually not discussed openly. Looking at good practice examples available on the Internet, the aspect of financing is mostly not addressed. The Romanian city of Alba Iulia, for example receives only little municipal and national funding for its transport projects, but has successfully built up the necessary expertise to attract European funding and move to more innovative financing and procurement based on SUITS outputs.

5.2.8 Innovative Procurement

Public procurement describes the process of acquiring goods, works or services by the public sector. Procurement is a complex activity and often seen as a necessary evil, but it offers great potential in terms of sustainability. The EU's Procurement reform [9] enables goods and services to be procured in a more sustainable manner. However until recently, the lowest price criterion (see Chap. 11) was the most important award criterion, nowadays, as a result of the reform, criteria such as life cycle costs, pollution reduction, energy consumption or the allocation of external transport costs play an increasingly important role in procurement. This makes it possible for cities to encourage bidders to pay more attention to issues of sustainability and wider impacts. As not only the lowest price, but also the most sustainable ideas may win the bid, it is worthwhile for the bidders to become creative and think of more sustainable solutions.

However, the mobility departments have only a limited influence, as procurement is usually carried out by central administrative units, who may lack specific knowledge on mobility. In order to exploit the full potential of the new opportunities raised by the procurement reform, close cooperation between mobility planning departments and procurement departments is required.

In the SUITS project, a guideline and a decision support tool for innovative procurement was developed to help cities to understand and take advantage of the new opportunities (see relevant chapters and https://www.suits-project.eu/ids-tool/).

The Mobility Department of Alba Iulia, a partner in the SUITS project, has succeeded in raising the awareness within the procurement department of the application of innovative award criteria for the procurement of new buses. This was only possible through intensive cooperation and the use of the SUITS guidelines.

5.2.9 Challenges Regarding the Development of Sustainable Mobility Measures

The challenges described above, mainly concern the internal organisation of the mobility departments and the Local Authorities and prepare the ground for sustainable action. In addition, there are a number of challenges that play an important role when it comes to the development of concrete mobility measures. When shaping mobility, local authorities operate in an area of conflicting interests. Different stakeholders such as citizens, politicians or business partners have certain interests, often contrary to each other, which must be taken into account.

Furthermore, the development of mobility measures and strategies depends to a large extent on subject specific knowledge available not only to mobility planners and planning departments, but also to other stakeholders like mayors or members of a city council. The sources from which this knowledge is taken are very different. Working with the cities in the project, it became clear that findings from various EU

Fig. 5.2 Actors and conditions influencing the planning and implementation of mobility projects of local authorities

research projects, available on web resources like CIVITAS or ELTIS[1] in the shape of guidelines, project reports and good practice examples are used relatively little. Inspiration and exchange of experiences arise primarily from personal exchanges with other mobility planners, through content presented at meetings and congresses and through media reports.

The potential behind this large pool of knowledge is still not fully exploited and the reasons for this are manifold. Despite best efforts, the aforementioned sources of knowledge are simply not known. Furthermore, documents may not be available in the respective national language. Another reason lies in the presentation of the material, for example important background information may be missing such as details about the context in which a measure was implemented, financing and legal aspects; information which sounds potentially relevant may not be so or takes a lot of tracing through different websites (Fig. 5.2).

When it comes to adapting good practice examples in their own city, understanding local context is vital. There are no generic solutions to mobility problems. What works in one city may not help with solving problems in another city. Therefore, in addition to technical knowledge, it is important to understand the wider factors that may influence the extent to which good practices can be transferred. Here a systematic classification of contextual factors, which could be used when describing examples of good practice, would make a useful contribution.

The challenges related to the planning and implementation of concrete mobility measures and strategies will be described in more detail in the following section.

---

[1] The webpages www.civitas.eu and www.eltis.org contain knowledge and numerous materials that have been developed in the framework of various EU projects on sustainable mobility.

5.2.9.1   Understanding Political Interests and Affecting Political Decisions

Receiving strong political backing is a prerequisite for the successful implementation of mobility projects, but at the same time is one of the biggest challenges for mobility planners in local authorities. No matter how well planned a project is, without political support, it will not be implemented. Complex projects and the absence of appropriate experience are considered as risky by decision-makers. Accordingly, it is difficult to receive support for such projects. Moreover, political moods are often unstable and depend on numerous factors difficult to calculate. For example, the closeness of local elections seems to significantly affect decision-making and the implementation of transport measures which may favour certain parts of the electorate. Mobility planners need to take this into account and propose the right measure at the right time.

Decision-makers often have to make decisions, on the basis of extensive information or expert reports which are typically not easy to access or to understand. This can lead to cognitive overload. Particularly in the case of sustainable mobility projects, which are often associated with additional expenditure, the benefits tend to be indirect in nature become manifest over a longer time frame and sometimes appear diffuse and difficult to measure. Requiring large levels of investment and committing a city to a particular technology may mean that decision-makers are reluctant to make decisions.

An opportunity exists to prepare information for decision-makers in a user-friendly and evidence-based manner. This can significantly influence decisions. For example, SUITS has made use of this in its 4 policy briefs, for example on "Social Impact Assessment of transport measures and systems" [10].

5.2.9.2   Understanding Legal and Regulatory Framework

The development and implementation of mobility projects often affects different policy areas. Accordingly, a lot of legal and regulatory frameworks must be considered. Depending on the mobility measure, this can be very challenging, for example when it comes to data protection issues or the use of innovative financing methods. Special challenges arise when changes in the legal framework become necessary in order to implement certain measures. This can lead to major delays or project failure.

5.2.9.3   Citizen Participation

When developing innovative mobility measures and services, it is important to tailor them to the local conditions and to the user needs. The active and early involvement of citizens in the planning process can have positive effects on the later acceptance and is now mandatory in the development of SUMPS.

Citizen participation includes both the early and active involvement of users in planning processes in order to understand their requirements and possible concerns, as well as informing citizens about contents and backgrounds of certain mobility projects and the "vision for the city". This is particularly important when certain measures aim to influence or change the mobility behaviour of citizens.

Although the importance of involving citizens is widely known, effort to engage in meaningful activities is scarce. Reasons include:

- Lack of experience and knowledge on participation methods;
- Difficulty in estimating the required effort and the potential benefit in advance;
- The difficulty of considering citizens´ problems and requirements, which are often contradictory, in projects that are complex anyway;
- Lack of ability to recruit diverse groups into engagement activities.

Moreover, citizens who do take part in participation processes may not be representative or reflect the opinion of the broad general public. In addition, citizens may only look at single measures and issues that affect them personally. They may not have been provided with sufficient details of the big picture, surrounding individual projects, to make informed decisions.

With regard to a sustainable planning, mobility planners must consider different user groups, especially those who may be considered as "hard-to-reach" like older people or people without mobile phones or access to the Internet in order to meet the EU's demand for a transport system that is truly accessible and usable by **all** citizens [11].

The main challenge is to find ways of involving citizens early in the planning processes, of raising awareness and showing benefits of new measures, as well as to provide background information to ensure comprehensibility and understanding, because "only when there is sufficient public support for change, will the action take place" [12]. Other issues relate to how to use the information which has been provided to shape planning decisions and communicating back to citizens how their participation has made a change.

In recent years, numerous guidelines and good practice examples on the topic of public participation have been developed by EU projects, for example in the EU DYN@MO [13] project or the CH4LLENGE project [14]. This has, in part, led to the topic enjoying increasingly high priority, especially in the context of mobility planning. Unfortunately, there is still a shortage of staff capacity to undertake extensive participation processes, especially in the administration of smaller cities.

5.2.9.4 Estimating the Feasibility and Acceptance of Measures

Another challenge is estimating the feasibility and acceptance of mobility measures in advance. In the case of innovative mobility measures or services, little experience may be available. Here the piloting of small-scale versions may be an appropriate means to gain a better understanding of upcoming problems and thus predict feasibility and acceptance. An example could be a public transport smartcard system, offering automated fare collection (a pay-as-you-go) by making use of innovative check-in/check-out systems in the vehicles. This could be tested on a specific line with a specific group of passengers, e.g. commuters. Another example may be temporary installation of shared spaces. Several cities with the same interests can join together to test projects, e.g. piloting of smart travel cards. Demonstrable outcomes may facilitate political support.

The involvement of stakeholders during feasibility studies is essential for success [15]. This is especially the case in the development of mobility services in which external business partners play a major role, where particular challenges may occur, such as, for example these service providers may be unwilling to share relevant data about their business models, usage numbers, etc.

5.2.9.5 Interaction and Cooperation with Business Partners

The interaction and cooperation with business partners are increasingly important, especially with regard to new mobility schemes like sharing services, offered by external providers. From the city side, conditions must be created that make it attractive for providers to offer their services in the city. On the other hand, cities must communicate to suppliers precisely what they need and what effects expect to achieve with the new service. Local authorities and business partners need to understand each other's interests, define common goals and if necessary, find compromises. The challenge here is to jointly tailor services according to the user needs, while also taking into account the city's local conditions. This is often associated with major challenges, as private service providers naturally place economic interests first. However, the intensive exchange must also be continued after implementation, in the running phase of the project in order to be able to assess its effectiveness, identify upcoming problems and continuously develop the service. Some of the concerns relating to dealing with multiple service providers is discussed in Chap. 15.

5.2.9.6 Identification and Utilisation of Synergy Effects

Synergy effects between different mobility measures and services are a very important issue in the context of sustainable mobility planning, especially as the aim should be to develop holistic solutions. The SUMP guidelines [2] refer several times to the identification and consideration of synergy effects. Synergies can be positive, when for example different mobility measures contribute to the same objective and enhance each other, but negative synergies can also occur, if measures torpedo each other or work in opposite directions. The challenge is to identify these synergies, to exploit multiplier effects and to eliminate mutually distracting effects. In practice, this represents a major challenge for the cities. The field of mobility and transport is complex, there are a lot of dependencies between the different services and measures and certain dynamics are difficult to foresee. This important issue could be addressed by future support projects, since so far only a little information is available, and the topic becomes more important with the increasing complexity of the mobility sector.

5.2.9.7   Use of Innovative Technologies and Data Collection Methods

Information and communications technology (ICT) is seen as one of the biggest enablers towards sustainable mobility planning. However, data collection by means of innovative technologies and the use of this data for mobility planning is still a big challenge. Many mobility departments, especially in smaller cities, lack detailed knowledge of which technologies, which data is worth collecting and how this data can be visualised.

During the SUITS project, cities demonstrated great interest in the use of innovative technologies and data collection methods. But the potentials seem to be far from being fully exploited. Reasons for this are a lack of technical know-how on the part of the mobility planners, non-transparency regarding suitable technologies and high implementation costs. In addition, the cost–benefit ratio in many cases is difficult to justify to decision-makers.

However, it is not always necessary to collect one's own data. Often data is already available from other bodies or institutions and can be obtained through cooperation. The mobility department of Rome for example, established cooperation with the local traffic management centre and a telephone operator to get access to movement flow data. Torino acquired mobility data from freight operators as part of an agreement to operate in certain regions of the city and is exploring mobile phone usage to understand the impact of its new rail line.

5.2.9.8   Application of Research Knowledge and Adaption of Good Practice Examples

Numerous projects on the subject of sustainable mobility have generated many findings and research knowledge, available in large volumes of guidelines, reports, good practice examples, etc. A major challenge for all cities is to put this knowledge into practice. Talking to mobility planners during the SUITS project became clear that guidelines are often too extensive, written too academically, are difficult to grasp and accordingly hard to implement. The impression could easily be gained that they are written primarily for an academic audience. The main interest of the mobility planners is in good practice examples from other cities. However, available examples do not necessarily address the problems that arise when trying to implement measures in one's own city. What proves to be a good practice in one city does not necessarily mean that it will succeed under different conditions. A sufficient consideration of the contextual factors is usually missing. Good practice examples have great support potential but must be better tailored to the knowledge interests of the mobility planners.

5.3 Evaluation of the Challenges

In order to validate the identified challenges, mobility experts, including mobility planners, mobility consultants and academics, were asked in an online survey to assess the relevance of the identified challenges in the planning and implementation of sustainable mobility measures. Furthermore, where they had concrete experience working with cities, they were asked to assess how the cities already cope with the different challenges in the frame of their actions. A total of 46 people took part in the survey [16]. The results are briefly presented in the following section.

5.3.1 Importance of Challenges

The experts were asked to draw on their experiences to rate (on a five-point Likert scale), the general importance of challenges for cities when planning and implementing sustainable mobility measures and indicate any new challenges. No additions were made on this by the experts. Figure 5.3 provides an overview of the results.

The ratings show that challenges we identified are important for the planning and implementation of sustainable mobility measures and services. Only the challenge of *Innovative procurement* had a slightly lower significance. This confirms to some extent that the issue has not yet been given much importance and the potentials and opportunities raised by the procurement reform are not yet widely known. *Institutional cooperation, citizen participation* and *Effective and efficient project management* are particularly important, which also coincides with the experiences of the discussions with the mobility planners. Overall, it was hoped that the survey would bring to light further challenges that had been missed in the project's cooperation with the cities. Unfortunately, this was not the case.

5.3.2 Estimation of How Cities Cope with the Challenges

In addition, the experts were asked whether they work for a city or whether they had gained concrete experience from working with a particular city. In this case, the experts were asked to assess the following question: *How well do you think, the city you indicated is coping with the individual challenges in their mobility planning activities?* A five-point Likert scale was also used here. Figure 5.4 shows the results.

The results show that in terms of the attested performance of the cities, and there is room for improvement on all challenges. The performance of the challenges *Institutional cooperation, understanding political interests and affecting political decisions* as well as the *understanding of the legal and regulatory framework* are rated as comparatively good. It is interesting to note that the challenge *Innovative*

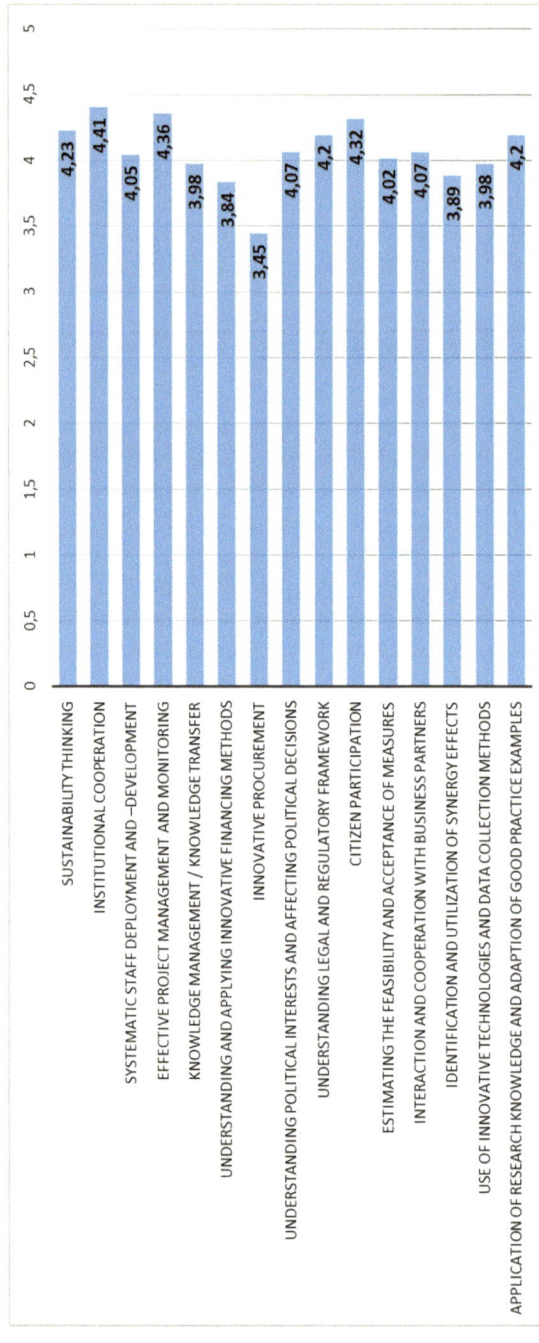

Fig. 5.3 Mean values of the importance rating of the identified challenges, using a five-point Likert scale (5 = very important 1 = not important at all)

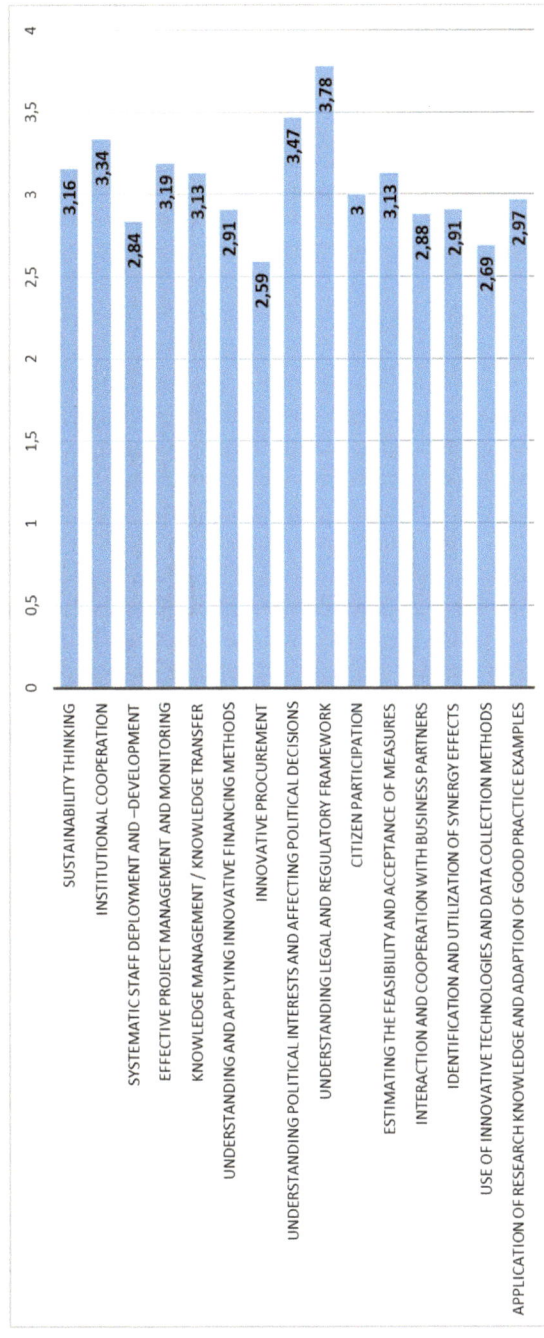

Fig. 5.4 Mean values of the performance rating using a five-point Likert scale (5 = strong performance 1 = poor performance)

procurement, for which the importance was rated relatively low in the first question, was also rated to be low in terms of cities' performance. This again clearly shows that the potential of innovative procurement must be made even clearer in future. The challenge *use of innovative technologies and data collection methods* scores relatively poorly, too. This is consistent with the experience of working with the project cities. They confirmed that there is a great interest in the use of innovative technologies and mobility data for mobility planning, but very often the appropriate know-how and the necessary funds are insufficient.

5.4 Conclusion

Identifying and understanding the challenges cities face in the context of sustainable mobility planning has made an important contribution to defining possible starting points for capacity building. Of course, not all challenges could be supported within the SUITS project. However, the results provide an important starting point also for future research activities and will allow to tailor support offers and materials even more precisely to the knowledge interests of mobility planners.

Existing support initiatives usually aim to provide knowledge on specific areas such as clean vehicles, mobility as a service or walking and cycling. But the challenges identified in the SUITS project mirror the questions coming up in every project, for example: How can we get political support for our project? How can we get administrative departments to work more closely together? How can we test solutions in advance to ensure feasibility? How can our project be financed even if no municipal funds are available? Which technologies and data do we really need to improve our mobility planning? These questions and challenges are not sufficiently addressed in existing guidelines or in good practice examples which have been developed in different research initiatives. In this respect, there is still potential to achieve a better acceptance of the offerings and materials developed by science and research in future and to achieve a higher impact. The challenges identified can contribute to this. At this point, it must also be noted that the challenges represent a result which does not claim to be complete. They are intended to be expanded and refined. But ideally, however, they are a step in a direction towards a higher consideration of the requirements of the later users, the mobility planners in local authorities.

In the SUITS project, the challenges were used to inform the organisational change process implemented by the SUITS partner cities. Here they have helped to find starting points for change initiatives and to structure the cities' plans for change more clearly. Each city had selected three challenges on which they wanted to focus and improve their capabilities (see previous chapter). In addition, the Transport Innovation Team of the West Midlands Combined Authority uses the challenges as a kind of risk register. When new projects are planned, they considered which activities they needed to undertake to address each potential challenge.

Overall, working with the cities in the SUITS project has shown that the cities are already sensitised to the importance of sustainability. Many exciting mobility projects with a focus on sustainability are already being implemented. Unfortunately, there is often a lack of the necessary resources, political support and financial means to bring the topic even more into focus.

References

1. United Nations General Assembly: World Summit Outcome, Resolution A/60/15 (2005). http://www.un.org/en/development/desa/population/migration/generalassembly/docs/globalcompact/A_RES_60_1.pdf. Accessed 10 Sept 2020
2. Rupprecht Consult: Guidelines. Developing and implementing a sustainable urban mobility plan (2014). http://www.eltis.org/sites/default/files/guidelines-developing-and-implementing-a-sump_final_web_jan2014b.pdf. Accessed 10 Sept 2020
3. VNG International: Inter-municipal cooperation Introduction guide-approach to a successful IMC (2010). http://www.vng-international.nl/wp-content/uploads/2015/06/IMC_EN.pdf. Accessed 10 Sept 2020
4. MaxSumo: Guidance on how to plan, monitor and evaluate mobility projects (2009). http://www.epomm.eu/index.php?id=2602. Accessed 10 Sept 2020
5. Guidemaps: Successful transport decision-making. A project management and stake-holder engagement handbook (2012). https://civitas.eu/content/guidemaps-successful-transport-decision-making-project-management-and-stakeholder-engagement. Accessed 10 Sept 2020
6. Girard, J.P., Girard, J.L.: Defining knowledge management: toward an applied compendium. Online J. Appl. Knowl. Manag. **3**, 1, 1–10 (2015). http://www.iiakm.org/ojakm/articles/2015/volume3_1/OJAKM_Volume3_1pp1-20.pdf. Accessed 10 Sept 2020
7. Mohajan, H.: The roles of knowledge management for the development of organizations. J. Sci. Achiev. **2**(2), 1–27 (2017). https://mpra.ub.uni-muenchen.de/83038/1/MPRA_paper_83038.pdf. Accessed 10 Sept 2020
8. Bornemann, M., et al.: An illustrated guide to knowledge management (2003). https://www.wm-forum.org/wp-content/blogs.dir/2/files/2014/01/An_Illustrated_Guide_to_Knowledge_Management.pdf. Accessed 10 Sept 2020
9. European Parliament, Council of the European Union: Directive 2014/24/EU on public procurement and repealing Directive 2004/18/EC (2014). https://eur-lex.europa.eu/legal-content/EN/TXT/PDF/?uri=CELEX:32014L0024. Accessed 10 Sept 2020
10. SUITS Policy Briefs. Civitas SUITS Project (2015–2021). https://www.suits-project.eu/capacitybuildingprogram/policy-briefs/. Accessed 10 Sept 2020
11. European Commission COM: 283 final—an agenda for a socially fair transition towards clean, competitive and connected mobility for all (2017). https://eur-lex.europa.eu/legalcontent/EN/TXT/PDF/?uri=CELEX:52017DC0283&from=EN. Accessed 10 Sept 2020
12. Banister, D.: The sustainable mobility paradigm. Transp. Policy **15**(2), 73–80 (2008)
13. EU CIVITAS DYN@MO Project (2012–2016). https://civitas.eu/sites/default/files/participation_2.0_in_the_sump_process_dynamo_web.pdf
14. EU CH4LLENGE Project—Addressing key challenges of sustainable urban mobility planning (2013–2016). http://www.sump-challenges.eu/
15. Wesley, F., Seaton, S.: Determining stakeholders for feasibility analysis. Ann. Tour. Res. **36**(1), 41–63 (2009)
16. Spundflasch, S., Krömker, H.: Challenges for local authorities in planning and implementing sustainable and user-oriented mobility measures and services. In: 1st International Conference, Mob*iTAS* 2019, Held as Part of the 21st HCI International Conference, HCII 2019, Orlando, FL, USA, July 26–31, 2019, Proceedings (2019)

Chapter 6
Behavioural Change in Local Authorities to Increase Organisational Capacity

Ann-Marie Nienaber⬩, **Sebastian Spundflasch, and André Escórcio Soares**

Abstract Local authorities' transport departments face extraordinary requirements regarding future mobility planning that affects and disrupts their internal business models and institutional logic in substantive ways. In this chapter, we highlight how organisational change can be implemented in local authorities to allow organisational capacity to increase and to enable employees to cope with the increasing expectations and requirements of future mobility planning. Our bottom-up approach is based on a socio-technical approach, taking into account both social (e.g., changing social behaviours or values) and technical aspects (e.g., new technologies). Applying Kotter's Eight Stage Process, we outline the implementation process of organisational change followed in seven local authorities in Europe as part of the SUITS project. The multiple case study approach allows us to indicate the crucial points along the path towards organisational change and to provide a step-by-step guide for successful implementation of organisational change in local authorities. We provide best practice examples from our work that may help other European cities increase their organisational capacity and be prepared to cope with the extraordinary requirements in relation to future mobility planning.

A.-M. Nienaber (✉)
Centre for Trust, Peace and Social Relations, Coventry University, Coventry, UK
e-mail: ann-marie.nienaber@coventry.ac.uk

S. Spundflasch
Institute for Media Technology, Technische Universität Ilmenau, Ilmenau, Germany

A. E. Soares
Institute for Advanced Manufacturing and Engineering (AME), Coventry University, Coventry, UK

6.1   Introduction

Mobility planning has become one of the key topics for local authorities over recent years [1]. Local authorities' (LAs) mobility or transport departments are faced with a variety of political and societal challenges in relation to future sustainable mobility planning.

On the one hand, the mobility field has undergone significant changes in recent years and is becoming increasingly complex, with a paradigm shift to Mobility as a Service (MaaS) and an ever-increasing amount of innovative and service providers entering the market [2]. On the other hand, citizens' mobility requirements have changed considerably. Citizens are more aware of health and sustainability issues associated with transport modes, demand resource efficient travelling and wider accessibility and inclusivity. Accordingly, many LAs, especially in small-to-medium cities are facing challenging situations, affecting and disrupting their business model and their institutional logic. This threatens public sector employees in a number of substantive ways. The changes require new organisational structures and processes which challenge employees to work in new ways, including using techniques and tools they have not used before. Further, new roles and responsibilities are needed (especially if a city commits to producing a SUMP), which may make former routines obsolete and require increased flexibility and motivation. In such situations, parallel and transparent organizational change is needed [3, 4].

In this chapter, we adopt a socio-technical approach to change. This approach is underpinned by socio-technical systems theory that recognizes the importance of cognitive and behavioural change when implementing technological innovation [5–7]. Along with Kotter's Eight Stage Process for organisational change published in 1996 [8], we want to outline how organisational change was implemented in different LAs in Europe in the frame of the SUITS project.

Kotter's Eight Stage Process is one of the key models in the literature of change management [8], but there is a lack of real-life case studies showing clear implications for practice (e.g., [9, 10]) with few exceptions (e.g., [11–17]). Furthermore, there is a scarcity of documented experience in the field of public administration about implementing organisational change using this process [10]. Therefore, SUITS contributes significantly to research and management practice in the field of public administration.

The structure of the chapter is as follows. After showing the need for organizational change, we describe Kotter's model of organisational change. After a short theoretical description of the different steps, we focus on the lessons learnt when we started to implement this process with the LAs in SUITS. We provide several best practice examples and show practical guidelines about how to achieve successful organisational change to increase local authorities' capacity to plan and implement sustainable mobility.

6.2 Theoretical Background

6.2.1 Organisational Change in Local Authorities

Implementing organisational change is a challenging undertaking, as the financial and personnel costs are relatively high, and success rates alarmingly low [18], despite best efforts. Change management needs to be focussed and systematic, following a structured and transparent process which addresses, technical and behavioural issues. Most research has shown that the success of any change depends to a large extent on the employees of an organisation. An organisation or its management can initiate change, but its implementation is carried out by employees [19]. Several case studies provide evidence of the importance of "human factors" in organisational change [20] thus organizational change is often called "behavioural change" as employees have to change their behaviour to make the change happen. Understandably, employees' commitment to change has been identified as one of the decisive enabling factors [21].

Local authorities and especially their mobility departments have to adapt rapidly and continuously new skills and behaviours to cope with challenges relating to sustainable mobility and the drive for smart cities, informed by integrated, real time data. Organisational change is essential if they are to meet new climate change targets and harness the potential of smart city initiatives for the benefit of all their citizens.

Global climate change (and more recently the COVID pandemic) requires actions at the local level, with LAs assuming a central role on the critical examination and reorientation of mobility goals and strategies. LAs and their transport departments must become more effective and resilient to organisational change when developing and implementing new transport measures and strategies [3]. In this context, it is imperative to develop capacity to meet new challenges, making sure that full advantage can be taken of developments in areas such as innovative transport schemes, innovative financing and procurement, urban freight or safety and security. However, for transport measures to be successfully implemented, it is not enough to change the technology or technical aspects; it is the human side that needs even more attention, as argued by [3, 20]. Most organizational change management programmes end up failing when they focus solely on technological change, ignoring the importance of social and behavioural aspects.

By bringing technological/technical and social/behavioural change together, organizations, and in this case, local authorities, can achieve better operational performance. Socio-technical systems combine the human, social, organizational, as well as technical factors when designing organisational systems [22]. The leverage of the knowledge and capabilities of employees results in better operational performance as they become better placed to deal with technological uncertainty, variation and adaptation [23], and more resilient when to organisational changes. Ultimately, bringing together social and technical change will help to reduce the gap between social and technical capability.

6.2.2 Implementation of Organizational Change

When implementing organisational change, it is important that local authorities are open to input from their employees in the design of both social and technical systems, e.g., changing social behaviours or values that are anchored in their organisational culture and changing systems such as technologies or guidelines local authorities' employees have to work with in future. There is evidence that employees benefit from the challenge, variety, feedback and teamwork involved in the development of the organisational change process as change enhances, for example their skill set [23], and they may gain greater insights into the organisation and increase their stake and commitment to it. Therefore, it is important to get the employees on board at the start of the process.

In line with SUITS' overall objectives to increase capacity of mobility departments, employees and stakeholders to implement sustainable mobility measures and strategies, an organizational change process was initiated in seven local authorities taking into account technical and the social capacity. While the technical side of the organisational change comprises new technologies and services regarding innovative transport schemes or financing options, the social side of the organisational change focuses on the employees' openness and readiness for change.

Our first design of an organizational change process was built on the Eight-Step Process for Leading Change outlined by Kotter in 1996, [8] (see Fig. 6.1). This model is very transparent, clearly structured and practically oriented. Above all, it can be easily communicated to different change actors, facilitating the creation of shared meaning and ultimately contributing to the effectiveness of the change programme. The eight steps of the model are outlined below before discussing how this was used in the project.

Step One: Create urgency

First of all, a sense of urgency has to be created amongst employees. For SUITS, it was necessary to highlight why sustainable mobility planning was important, how change could be created and the benefits this could bring. This step creates the "need" for change, as opposed to a "want" for change.

Step Two: Form a powerful coalition (team)

A strong, proactive and influential coalition needs to be built from employees with a range of skills and experiences, across the different areas and departments involved in mobility planning to drive and support the change process. The team itself becomes a role model for the wider organisation, helping to spread messages throughout the organization, delegating tasks and ensuring support for the change organisation wide. Key to this is the appointment of an influential and widely respected change agent in the organisation, who has the backing of, and ability to influence, senior managers.

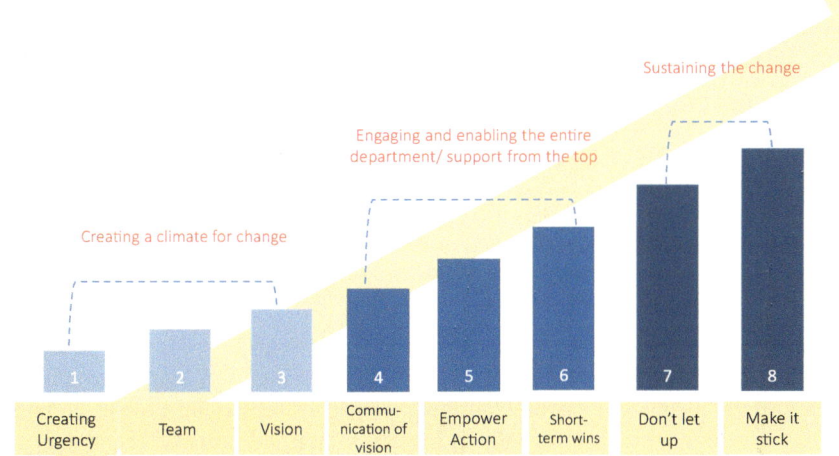

Fig. 6.1 Implementation process (based on Kotter [8])

Step Three: Create a vision for change

An inspirational, far reaching and comprehensible vision should be jointly developed by those immediately concerned with the change (in this case the mobility departments, change agent and local authority). It should be supportable by the whole organization.

Step Four: Communicate the vision

A vision that is developed but not communicated will not be known. Thus, the vision should be communicated throughout the whole local authority and understood by its employees. It should be continuously communicated in different ways in order to rise above competing messages.

Step Five: Empowering broad-based action

Empowering employees requires active listening, investing in them through training, and making them responsible for major accomplishments. Empowerment contributes to increased levels of employee engagement, and more specifically in their engagement in the change process, which is important in guaranteeing success. To increase employee empowerment, existing obstacles such as organizational structures, skills deficits, systems and supervisors must be addressed.

Step Six: Create short-term wins

Organisational change needs time, and few rewards may be visible at the start of the process. Therefore, it is most important to create opportunities for quick wins and the celebration of these to keep the employees' motivated to support ongoing processes.

Shorter-term targets are useful tools for motivation and direction. Using these wins to justify investment and effort can help to re-motivate staff to continue backing the change.

Step Seven: Build on the change

Many organisational change programmes fail because they lack an end point, continuation plans, or are not self-reflective. Organisational change is a continual process in which, in this case, the LA should keep setting goals and analysing what could be done better for continued improvement.

Step Eight: Anchor the changes in corporate culture

For a change to be sustainable and successful, it needs to be anchored in the corporate culture. The implemented sustainable transport measures, as well as change procedures and principles must be become an intrinsic part of the organisational culture of the LA, for example through guidelines that are known and utilised. The changes, and the need to evolve, must become part of the core way of working to have a lasting effect.

6.3   Implementing and Gathering Data on Organisational Changes in the Local Authorities

As part of the SUITS project, the mobility or transport departments of seven local authorities, i.e., Kalamaria (Greece), Valencia (Spain), Alba Iulia (Romania), Rome and Turin (Italy), Palanga (Lithuania) and West Midlands Combined Authority (UK) nervously embarked on a change journey to increase their capacity to engage in sustainable mobility planning. Details about each LA have been given in Chap. 3. The change process was conducted at their own pace, with each city being supported by a delegated project member from outside their organisation, under the leadership of Ann Marie Nienaber. Their journeys were recorded over the course of the project, enabling insights to be drawn from each case study which allows us to explore change management in this context and to draw similarities between organizations (e.g. in terms of barriers and enables) [24].

6.3.1   Data Collection

In order to continuously assess the effects of the change process, data was gathered through semi-structured interviews (using an interview protocol derived from Kotter's model), documentary analysis of minutes of meetings between the project team and city representatives, reporting sheets or work papers, and 11 workshops with city representatives.

6.3.2   Data Analysis

The aims of the analysis were to:

- Understand how the change process was conducted in each authority,
- Name, identify and discuss amongst the group barriers and enablers to change in action learning sets,
- Identify common patterns across local authorities which could be used to provide guidance to other local authorities,
- Demonstrate the generalizability and applicability of Kotter's model across very different transport departments,
- Create a replicable model for use after the project to help transport departments become more innovative, flexible and resilient, especially in relation to their development of sustainable transport plans.

Data analysis included familiarisation with the documents and transcripts followed by a coding process [25] and further analysis as suggested by the case study approach in particular pattern coding [24, 26]. The identified patterns are described in the following section.

6.4   Findings and Learnings

This section highlights the key results and learnings related to the implementation of organizational change to enhance organisational capacity during the SUITS project. As previously mentioned, the lessons learned were drawn from practical collaboration with the cities that carried out organizational change during the project. Most insights were gained from the 11 workshops held with representatives of the cities, in particular the transport or mobility departments.

The change procedure was critically reflected upon by SUITS' members and local authority employees during the whole implementation process and individually adapted to the particular needs and requirements of each local authority through use of "360° reflection and feedback circles".

In total 104 participants took part in these workshops. In addition, semi-structured interviews were conducted with 2–3 people from each local authority took part. (N = 17). The lessons learned are presented in line with the eight steps for organisational change as outlined in an earlier section of this chapter.

Increase urgency—Step 1

The very first step towards making a change, is raising awareness of the need for it. This required a '"benchmark" study of the weaknesses and strengths, opportunities and risks of each local authority (as outlined in Chap. 3). This analysis allowed the design of a systematic procedure for the implementation of organizational change.

From this, and in discussion with LA representatives, the key persons suitable for implementing change were identified. A pre-requisite being that such a person had to have knowledge of the areas/personnel in which the changes were required.

The identification of the "ideal" change agent was recognised as one of the most important contributory factors in creating and driving though successful changes. A change agent may be internal or external to the organisation, but in either case they must have strong relationships and trust amongst the key decision makers in the organisation. We found that the lack of direct relationships with the chief executive/mayor or departmental managers was a significant impediment to initiating change processes and the rate of change. Without senior management attention for the necessity and implementation of change and their continued support, existing bureaucratic organisational structures can stifle innovation and opportunities for change. Therefore, the change agent should have the power to directly contact senior management and gain support when needed, e.g., to change organisational structures or implement new working practises.

> the change agent's role is very challenging but so decisive to implement change. In my eyes, a very important decision to be made in every city [Deputy Mayor, City of Kalamaria].

In terms of personal attributes, the change agent needed to be flexible, proactive, results oriented, supportive of the changes and possess the necessary social skills to work with employees from all areas likely to be effected by change—so that they buy into, contribute to, and trust innovation.

As people respect courage and accountability, a change agent must take responsibility for his or her LA. A chief executive/mayor has to make decisions that go against dissenting opinions and this can cause conflicts, but doing so with conviction and being ready to handle the consequences will ultimately demonstrate that the local authority's intentions are motivated by the best interests of the city, thus gaining the trust of their employees and wider society.

In creating change, it was seen as helpful to tie specific priorities to overall organisational goals. Being able to see, contribute to, and influence the direction of travel of the organisation, required the change agents to have a level of maturity and standing within the organization and local network. The change agent has to be able to explore perspectives, be open to ideas, and take them into account when looking for solutions. This will help in getting buy in to a change; people want to feel that others are listening to their ideas. Those who do so develop stronger relationships with their people by gaining trust [27].

Build the guiding team—step 2

As change agents are more powerful when they are supported by a team, the second step requires building a coalition to support change and the change agent's activities at an operational level. The team itself, should be a role model for the wider organization, helping to spread messages. Delegating tasks and ensuring support for organization-wide change needs strategic thinking.

Our findings indicate that the best results will be achieved from a diverse coalition, built from employees with a range of skills and (life) experiences and from different departments. In some cases, depending on the objectives pursued, it may also be useful to expand the coalition externally. Especially when it comes to topics for which there is little awareness internally, external experts can provide important knowledge and evidence.

> To achieve this, the Change Agent of the City of Torino has built the working team 'MaaS', Mobility as a Service. [...] At the beginning, the managerial level of the city perceived MaaS as a personal initiative and, for this reason, it was cold on this topic. The only way to proceed was by building a technical team mainly composed of external experts (technological companies, ICT development companies, University). [...]. Over the last years, during SUITS project, the team started to meet regularly. Now the importance of MaaS is evident, it is seen as the solution to many mobility problems and it is not perceived as a personal initiative anymore. [Representative from the Transport department of the City of Torino].

Get the vision right—step 3

Our findings highlight that the development of a vision is often underestimated or not well understood. A vision must be easy to understand to ensure support from the whole organization, "inspirational" to have maximum effect and easily communicable. In the following, we present the vision of the City of Palanga and the argument for its development. The vision of Palanga is: *"City in nature, created by its citizens"*. The change agent of Palanga argued as follows:

> This vision has been chosen because Palanga is a resort city and synergy with nature is very important. It is also very important to ensure 'city comfort' not only for the resort guests but for the locals as well. In order to ensure that the population remains satisfied with urban development and smart mobility solutions, it is necessary to involve the population in the selection of the most appropriate alternatives for these measures. The Palanga local authority is aiming to involve citizens and business partners more closely in the decision-making process. [Change Agent, Palanga].

Almost all LA representatives reported that it was initially very hard to convince senior management to communicate a vision that reflects their city 's engagement with sustainable transport measures. The following statement was made by the change agent from Alba Iulia and reflects the general challenge related to the vision:

> As with all internal processes, there was some reticence at first, however in time all the departments understood the need and the relevance of the change vision in order to foster a new way of working together. The experience overall was a good one as it brought an increase in cooperation between the different departments at municipality level [Change Agent, Palanga].

The challenge is to create a vision that it simple, understandable, far reaching (so that it does not have to be changed continually) and precise, so that it does not leave too much room for interpretation.

Communicate to buy in—step 4

A vision that is developed but not communicated will not be known, supported or implemented. General awareness and enthusiasm needs to be created for the goal behind the vision. The "'vision" should embody the need for change in such a way that everyone can understand that change is needed and the long-term targets (e.g., in this case, for example zero emissions or sharing travel data).

Definition of concrete activities to enable the overall goal to be achieved requires the support and contribution of all employees. Everyone should be invited to participate, consider and suggest concrete steps that could be implemented in their workspace or organization. If proposed changes completely ignore the needs and abilities of the employees, then it will not be sufficiently supported at the work level.

All seven LAs reported that this step was of great relevance, but one of the most challenging. A process-oriented approach was the most promising way of communicating the vision. In the first step, the vision is communicated to those departments and employees that will be most directly affected (e.g., transport and planning departments and procurement), then those departments that are indirectly involved are informed, before moving out to the wider local authority.

Two aspects were found to be helpful when communicating the vision. Firstly, the vision itself needs to be clear and transparent so that it can be easily understood by all employees. Other departments should not necessarily be expected to fully support the vision as they will have their own, perhaps conflicting priorities. This can become a major barrier to the change process, accordingly great attention must be given to it by the change team, looking for synergies and win–win scenarios. Secondly, an "evidence-based" approach should be used to back up the vision, e.g., using good practice examples, feasibility studies, data, cooperation with interest groups and citizens. In the following, the statement of the change agent of Alba Iulia summarises the key learnings of this step.

> The change vision represented a process which began with the starting of the SUITS project and which continues till today. The vision was communicated in various meetings/workshops/seminars organised within the different departments of the municipality by the "innovation team" [guiding team]. Most of the departments were involved in the process which meant that they were either involved directly or indirectly in the actions and processes foreseen. [Change Agent, Alba Iulia].

Empower Action—step 5

Working with SUITS project members through training, development and knowledge exchange helped to empower change agents and guiding teams to drive organizational change in their local authorities. Based on a list of pre-selected "challenges/themes" in sustainable mobility planning, the guiding teams of each city selected the key priorities they wanted to target in the coming years, e.g. understanding and applying innovative financing methods, effective project management and monitoring, or citizens' participation (for detailed information regarding the development of these "challenges" see Chap. 3).

This selection helped the guiding team to focus on areas of interest and importance for sustainable mobility planning. While we cannot recommend switching the challenges to be tackled very often, in some cases a change may be necessary due to evolving technologies, environmental changes or other factors. To empower action,

Table 6.1 Challenge, Target and Key Performance Indicator for the organizational change of the West Midlands Combined Authority

Challenge	Target	Key performance indicator
Understanding and applying innovative financing methods	Staff understanding innovative financing methods	No. of staff attending training on innovating financing methods and reading guidelines on innovative financing methods
	Projects using innovative financing methods	No. of projects using innovative financing methods

it was further necessary that each local authority agreed on clear targets and measures (key performance indicators) for each identified challenge to be able to create and show impact. This was decisive also to show progress and success to the decision makers in the local authority and also to their stakeholders and the wider society.

Using the example of the West Midlands Combined Authority, the following table shows a selected challenge, the target pursued and a defined key performance indicator to measure impact (Table 6.1).

Create short-term wins—step 6

The very word "change" creates uncertainty and anxiety, especially when it is introduced by an external party, as in SUITS. Here, it should be pointed out that although organisational change was a fundamental part of the project, organisational churn within some LAs meant that those who had authorised the project at the proposal stage, were not in place when the project started. This is similar to the situation of external management consultants coming in suggesting sweeping changes, with those not involved in implementing, or affected by change removed from it.

In the SUITS process, after the capacity benchmarking, (Chap. 3) and vision have been agreed, change can take place in small steps, based on suggestions from within the organisation. Changes are grounded in the experience and culture of the employees. The goal(s) should correspond to the capacities of the departments, they should be achievable and, measurable.

In order to keep motivation for change high, small steps that lead to short-term success are meaningful, important and celebrated. Thus, organisational change is initiated with concrete, target-oriented activities by the guiding team in close cooperation with senior management, leading to clear achievable results.

While working with the mobility departments, it became clear that one of the biggest challenges is to keep up momentum. Very often daily-based operations overlap with the long-term strategy. The implementation of sustainable transport measures requires a lot of creativity and stamina, pulling together several threads to provide integrated solutions.

The following example from the West Midlands Combined Authority (UK) shows activities that helped to keep momentum and allowed celebration of small steps during the long process of organisational change.

To foster knowledge exchange inside their organisation, West Midlands Combined Authority organised, for example in-house workshops, on the one hand with external trainers, but also internal workshops in which staff jointly tried to develop ideas for concrete activities. Furthermore, employees were encouraged to use a special team-work software to share and discuss issues they found on specific topics of sustainable mobility. Weekly newsletters were created and sent to employees. These contain, for example news about ongoing projects, information about concrete team activities, information related to questions such as what challenges need to be tackled, what knowledge is available and what is needed. This created and still creates a general awareness for specific topics as well as the entire change process. Furthermore, these meetings together with the newsletter allowed the team to announce and celebrate the achievement of little steps during the long-term process of making organisational change to implement sustainable transport measures.

Creating short-term wins was intensely discussed with the LAs. In various work-shops, regular briefings took place, where the project team and LA representatives discussed intermediate results, experiences and new ideas. This also involved defining concrete targets and performance indicators for the achievement of objectives as well as scaling and structuring the projects. The following example provides an overview of the goals set to achieve a capacity change on a selected challenge, as well as the appropriate key performance indicators to measure success. Ideally, all activities in a LA should be aligned to one of the selected challenges. The following example (see Table 6.2) highlights the participation of a local authority in the European Mobility Week and demonstrates how it helps to improve citizen's participation in line with the local authority's challenge, target and key performance indicators.

Don't let up—step 7

The credibility of an organisational change depends upon whether it, and interest in it, fades over time—as enthusiasm wains, other tasks emerge, and staff move on. The LA's felt this was challenging, partly because the change process was part of a funded project. This means that either new staff were employed to service the e project or were allocated to it on a temporary basis. Where the change team is composed of such staff, it does not have the level of permeance and authority needed to effect change after the lifetime of the project, unless impact has been created and senior management understand the need for their organisation to continually evolve and adapt. Thus, it is highly recommended to foster permanent changes, for example through new organisational structures or guidelines that are widely known and utilised.

During the lifetime of the project, the cities achieved notable results in increasing of their capacity, and thus, became learning organisations which are more flexible and resilient [28]. A learning organisation is ideally a system that is constantly in motion. Events are taken as suggestions and used for development processes in order to adapt the knowledge base and scope for action to the new requirements. This is based on an open and individualised organisation that allows and supports innovative problem solving. Each LA was able to demonstrate such mechanisms that support learning processes. Examples listed below within the categories: organisational structure shift,

Table 6.2 Example of structuring activities along identified challenges

Challenge	Target	Key Performance indicator	Impact example
Citizen participation	(a) Improve safety levels for citizens through increase of road's level service, and installation of posts near pavement to prevent illegal parking and improve safety levels	E.G. % of roads with no congestion; Fewer Injuries and/or fatalities per unit of transportation	**Activity:** Municipality 's participation in European Mobility Week activities **Responsible department:** Culture/Education department
	(b) Increase awareness of citizens about sustainable mobility measures, such as infrastructure measures, new footpaths, bicycle lanes etc	E.g. Rate of citizen's participation (low—medium- high) in European Mobility Week	**Target groups:** Citizens, children, bicycle riders **Level of replicability:** High (this activity can be repeated easily) **Potential obstacles:** Lack of Political interest, weather, absence of campaign and marketing
	(c) Awareness and acceptance by citizens for sustainable mobility measures, high usage and acceptance for innovative technologies using renewable energy (solar power)	New smart pedestrian crossings at problematic points, especially near schools; E.G. number of citizens who use innovative technologies for urban mobility (smart pedestrian crossings)	**Implementation requirements:** Adequate advertising and campaign, infrastructure, good preparation
	(d) Increase quality of life for citizens	New green corridors (footpaths, bicycle routes, usage of innovative technology—online platforms)—number of citizens using them and average travel time reduction	**Final outcome and impact:** Increased awareness for urban sustainability

design and implementation of new guidelines and new ways of working to foster knowledge sharing. Examples at this stage included:

Changes to organisational structures

- The city of Valencia changed their organisational structure from a silo-organisation to a cross-functional project-based organisation which allows for more flexibility and innovative thinking and which prepares the ground for further organisational change. A silo-organisation is once in teams work in relative isolation from the rest of the organization, whereas a cross-functional project organisation is an organisational structure that is built by teams. In such teams, individuals, sometimes from different departments, work together on a project basis towards a common goal. This new organisational structure can be seen as the basis for a learning organisation.
- The city of Rome founded a new unit that is responsible to ensure that citizen's requirements are constantly identified and reflected in their long-term planning of sustainable targets (economic, social and environment). This new unit applies tools of behavioural economics, a new branch of economics (effects on organisational process). This unit demonstrates the local authority's openness to reflect their mental models, to become more innovative and flexible and citizens' focussed. The permanent implementation of such a unit also demonstrates that the city of Rome understood the relevance of a permanent change to become a learning organisation.

Design and implementation of new guidelines

The City of Alba Iulia worked with the procurement department to trial innovative procurement practises for the purchase of new buses. Initially, the procurement department was reluctant to adopt new working practises and applied the lowest price criterion in tendering. The lack of willingness to engage intensively with the comprehensive guideline on innovative procurement produced by SUITS was a big hurdle to overcome. In a number of internal meetings, the change manager worked out the principles of the guideline, showed the benefits and proposed concrete recommendations for action, and thus was able to get support from procurement team. This change became embedded when it led to successful tenders.

New ways to share knowledge

Transport for West Midlands (part of WMCA[1]) attempted to improve institutional cooperation and sustainable thinking within the authority by organizing in-house workshops with external trainers, together with internal workshops in which staff are encouraged to jointly develop concrete strategies and activities focusing on sustainable mobility. In addition, periodic newsletters were created and sent to employees to increase transparency on projects, team activities, challenges and experiences and open for discussion. In this way, WMCA tried to integrate change as a fundamental

[1] West Midlands Combined Authority, UK.

principle in their organisation. WMCA showed that it wants to learn from within and cultivate the culture of "learning and pleading", discussions about mental models tend to sound like this: "This is my opinion, and that's how I got there. What do you think of it?" Instead of fighting for the right point of view, the path on which facts and data assumptions and opinions are based is made transparent and openly discussed.

All of these institutional-based changes make the shift in the organisational behaviour of a local authority sticky and thus pave the way for more flexible and innovative thinking in local authorities and to become a learning organisation.

Make change stick—step 8

While structural and procedural changes are mandatory to allow a LA to be turned into a learning organisation, for change to be successful and sustainable it is crucial that the need to change become anchored in the corporate culture. If the need for change becomes a fundamental principle in an organisation, this organisation can be characterised as a learning organisation with high levels of organisational capacity. A learning organisation is defined by five principles [28].

1. Personal mastery. The constant pursuit of personal growth (personal mastery), one's own visions and a better understanding of reality.
2. Mental models. The constant striving to become aware of one's own beliefs and ideas, to check them and to develop them further.
3. Shared vision. The constant further development of goals and visions that motivate and inspire at the same time.
4. Team learning. The development in the team begins with the dialogue and the acceptance of the others.
5. Systems thinking, as the fifth discipline which integrates the other four by recognizing the complex interrelationships of the whole.

Learning organisations are in a continuous learning process and have the ability to question themselves and their behavioural patterns and to develop further from the answers. Beside the structural and procedural requirements that are described in step 7, the LAs were also able to demonstrate first aspects that they really anchored change in their corporate culture. One indicator was the different cross-learning groups that were established between the local authorities to foster permanent knowledge exchange inside but also between the different LAs, academics and stakeholders. Each group consisted of two or three cities that with the support of the project team, shared experiences and knowledge in regular meetings and supported each other. This cooperation is expected to continue after the project. One of these cross-learning sets is described more closely in the following:

West Midlands Combined Authority, *Coventry City Council* and *Coventry University* started to exchange experiences with the *City of Valencia*. Key issues of interest are, e.g., the design of a "roadmap"[2] with challenges and resources concerning

---

[2] A roadmap is a flexible planning technique to support strategic and long-range planning, by matching short-term and long-term goals with specific technology solutions. It is also expected that roadmapping techniques may help organisations to survive in turbulent environments and help them

sustainable mobility for the city, development and implementation of processes around the introduction of new ideas and marketing and designing of innovative solutions for close relationships with stakeholders. Through the exchange of experiences and knowledge, both local authority departments learn from each other and save time and costs when planning future mobility, e.g., recommendations that help to select a suitable software for the design of a roadmap or to identify a provider concerning the integration of a car-sharing system in the city.

6.5 Practical Guidance "Success Tips"

Five clear guiding principles were identified from the case studies which are key to successful organizational change in local authorities.

1. Identify a key person—the change agent—who will be the driver for successful change in your local authority. A change agent should be knowledgeable, committed and well respected, with a clear route of communication through to senior management.
2. Embed diversity in the change team by bringing in people with different skills sets, ethnicities, genders and life experiences from across the organization.
3. Support the change agent from the top and bottom in your local authority.
4. Communicate your vision authority wide to make it effective and resilient to change.
5. Celebrate little successes to keep employees motivated to support the change over the long term.
6. Make change continual so that the organisation itself has the ability to be flexible and resilient, e.g. setting up a permanent team.

6.6 Conclusion

This chapter highlights the relevance of a transparent and structured way to implement organisational change for enhancing the capacity of a LA to plan and implement sustainable future mobility. Implementing organisational change is a challenging undertaking, as the investments made are relatively high and thus, many LAs are reluctant to make changes inside their organisation, especially as they have to meet the needs of the city at the same time.

However, the impact of global climate change (and more recently the COVID pandemic) requires actions at the local level, with local authorities assuming a central role in the critical examination and reorientation of mobility goals and strategies.

---

to plan in a more holistic way to include non-financial goals and drive towards a more sustainable development.

Implicit in the actions needed to develop and deliver SUMPs and Smart city initiatives is the need for LAs and transport departments to work in new ways (e.g., with new data, concepts, partners, information and technologies), in a cross departmental/multidisciplinary way, taking care to ensure full and meaningful engagement with all citizens in any planning processes. The rate of technological innovation, citizen awareness, global challenges and socio demographic changes means that transport (and other departments across the LA) must be knowledgeable and agile if they are to deliver the best services for their city.

However, little consideration has been given to how very traditional, small departments can increase their capacity and become more resilient. Based on comprehensive data from working with seven local authorities in Europe over four years, we can show a clear, structured process that will support local authorities in Europe managing organisational change and allow their organisational capacity to grow. We close this chapter with the words from one of the change agents:

> Through SUITS, Kalamaria, as an organization, with the new administration, is ready to schedule and implement other sustainable mobility measures in the near future. [Change agent, City of Kalamaria].

References

1. Akgün, E.Z., Monios, J., Rye, T., Fonzone, A.: Influences on urban freight transport policy choice by LAs. Transp. Policy **75**, 88–98 (2019)
2. Nikolaeva, A., Adey, P., Cresswell, T., Lee, J.Y., Nóvoa, A., Temenos, C.: Commoning mobility: towards a new politics of mobility transitions. Trans. Inst. Br. Geogr. **44**(2), 346–360 (2019)
3. Nienaber, A., Spundflasch, S., Soares, A.:Sustainable Urban Mobility in Europe: Implementation needs behavioural change. SUITS Policy brief X. SUITS funded from the European Union's Horizon 2020 research and innovation programme under grant agreement no 690650. Mobility and Transport Research Centre, Coventry University (2019)
4. Nienaber, A., Spundflasch, S., Soares, A., Woodcook, A.: Distrust as hazard for future sustainable mobility planning. Rethinking employees' vulnerability when introducing new technologies in local authorities. Human Computer Interaction (forthcoming)
5. Cherns, A.B.: The principles of sociotechnical design. Hum. Relat. **29**, 783–792 (1976)
6. Cherns, A.B.: Principles of sociotechnical design revisited. Human Relat. **40**, 153–162 (1987)
7. Clegg, C.W.: Sociotechnical principles for system design. Appl. Ergon. **31**, 463–477 (2000)
8. Kotter, J.: Leading Change. Harvard Business School Press, Boston (1996)
9. Pollack, J., Pollack, R.: Using Kotter's Eight Stage Process to manage an organisational change program: presentation and practice. Syst. Pract. Action Res. (2015)
10. Applebaum, S., Habashy, S., Malo, J., Shafiq, H.: Back to the future: revisiting Kotter's 1996 change model. J. Manag. Dev. **31**, 764–782 (2012)
11. Lamé, G., Jouini, Q., Stal-Le Cardinal, J.: Combining the viable system model and Kotter's 8 steps for multidepartment integration in hospitals. In: 12ème Conférence Internationale de Génie Industriel, May 2017. Compiègne, France (2017)
12. Chowthi-Williams, A., Curzio, J., Lerman, S.: Evaluation of how a curriculum change in nurse education was managed through the application of a business change management model: a qualitative case study. Nurse Educ. Today **36**, 133–138 (2016)
13. Springer, P., Clark, C., Strohfus, P., Belcheir, M.: Using transformational change to improve organizational culture and climate in a school of nursing. J. Nurs. Educ. **51**, 81–88 (2012)

14. Lintukangas, K., Peltola, S., Virolainen, V.: Some issues of supply management integration. J. Purch. Supply Manag. **15**, 240–248 (2009)
15. Day, M., Atkinson, D.: Large-scale Transitional procurement change in the aerospace industry. J. Purch. Supply Manag. **10**, 257–268 (2004)
16. Ansari, S., Bell, J.: Five easy pieces: a case study of cost management as organizational change. J. Account. Organ. Chang. **5**, 139–167 (2009)
17. Joffe, M., Glynn, S.: Facilitating change and empowering employees. J. Chang. Manag. **2**, 369–379 (2002)
18. Beer, M., Nohria, N.: Cracking the code of change. Harv. Bus. Rev. **78**(3), 133–141 (2000)
19. Shah, N., Irani, Z., Sharif, A.M.: Big data in an HR context: exploring organizational change readiness, employee attitudes and behaviors, J. Bus. Res. **70**, 366–378 (2016)
20. Hoover, E., Harder, M.K.: Review: what lies beneath the surface? The hidden complexities of organisational change for sustainability in higher education. J. Clean. Prod. **106**, 175–188 (2015)
21. Choi, M.: Employees' attitudes toward organizational change: a literature review. Hum. Resour. Manage. **50**(4), 479–500 (2011)
22. Baxter, G., Sommerville, I.: Socio-technical systems: from design methods to systems engineering. Interact. Comput. **23**(1), 4–17 (2011)
23. Pasmore, W., Winby, S., Mohrman, S.A., Vanasse, R.: Reflections: sociotechnical systems design and organization change. J. Chang. Manag. **19**(2), 67–85 (2019)
24. Yin, R.: Case Study Research: Design and Methods, 5th edn. Sage, Thousand Oaks, CA (2014)
25. Kotter, J.P., Cohen, D.S.: Creative ways to empower action to change the organization: cases in point. J. Organ. Excell. **22**(1), 73–82 (2002)
26. Patton, M.Q.: Qualitative Research and Evaluation Methods, 3rd edn. Sage, Thousand Oaks, CA (2002)
27. Nienaber, A.M., Romeike, P.D., Searle, R., Schewe, G.: A qualitative meta-analysis of trust in supervisor-subordinate relationships. J. Manag. Psychol. **30**, 507–534 (2015)
28. Senge, P.: The Fifth Discipline: The Art and Practice of the Learning Organization, 2nd edn. Random House Business (2006)

Chapter 7
Introduction to the Capacity Building Toolset and SUITS Capacity Outputs

Anastasia Founta and Olympia Papadopoulou

Abstract This brief chapter serves as an introduction to the subsequent chapters in this section. It provides an introduction to the rationale behind the development of the SUITS capacity building toolbox, and how it was developed. The toolbox provides relevant, up to date information for LAs, in the latest developments in sustainable transport measures, closing a gap in information already available through ELTIS and CIVITAS web sites (see Chap. 2).

7.1 Introduction

Rapid technological innovation, climate change, the economy, socio-economic and cultural factors and inclusivity agendas are shaping mobility solutions.[1] Terms such as driverless cars, connected vehicles, intelligent transport systems, mobility as a service, new urban freight modes and last mile delivery are not only widely used, but also marketed as solutions to many of the problems associated with urban living.[2] However, their contribution to the development of sustainable cities and their impact on local authorities' (LAs) operations remains unclear. Real world trials and pilot projects receive initial funding and may be enthusiastically received, yet their longevity and long-term benefits/use has not been fully evaluated.[3]

Cities do need regeneration. They are so polluted and congested by the ubiquity of automobiles that they have become inefficient and unpleasant. With vast amounts of investment available, and technological innovations promising step changes in the ability to meet national and international sustainability and zero fatality targets, LAs are responsible for developing master plans, purchasing and implementing transport

---

[1] Available at https://cbt.suits-project.sboing.net/.

[2] https://www.eltis.org/.

[3] https://civitas.eu/knowledge-base.

A. Founta · O. Papadopoulou (✉)
Lever Development Consultants, Thessaloniki, Greece
e-mail: olympia.pp@gmail.com

© Transport for West Midlands 2023

A. Woodcock et al. (eds.), *Capacity Building in Local Authorities for Sustainable Transport Planning*, Smart Innovation, Systems and Technologies 319, https://doi.org/10.1007/978-981-19-6962-1_7

solutions which will make their cities more attractive to citizens and investors. They face challenges regarding the procurement and implementation of suitable measures, as well as the assessment of their added value in comparison with traditional mobility measures (i.e., cycle paths, traffic calming areas, urban tolls, etc.) and understanding their new role in the transportation ecosystem.

The EU has recognised the need to support LAs in this, and as outlined in Chap. 3 has developed a series of resources to, for example, encourage innovation and spread good practices across the community.

In line with this, SUITS has developed a range of material to build the capacity of transport departments to implement effective, sustainable transport measures. Our focus has been on small to medium cities, who have different informational needs. For example, they may be less familiar with the latest transport innovations and not have the time or capacity to peruse information on existing websites. Our capacity building programme is based around tailored, face to face (and during the covid epidemic) and distance learning which is crucially delivered in tandem with a tailored organisational change programme, so learning can be directly applied in the working context. The organisational change programme was discussed in the previous chapter; in this chapter, we outline the areas in which LAs need more information, and the way this was delivered via the 'capacity building programme' available online, in multiple languages at https://cbt.suits-project.sboing.net/.

The learning tools are intended to increase the adoption rate of sustainable mobility measures, that may form part of sustainable urban mobility plans (SUMPs) or a package of measures. Strengthening small–medium-sized LAs' knowledge in specific areas enables them to make more informed choices, engage in more detailed discussions with transport providers (or consultants) and evaluate how a new transport measure may work in their city. The areas selected for our learning modules emerged out of the capacity building audit undertaken with LAs at the start of the project (described in Chap. 3), and a review of gaps in existing information.

The four specific modules are:

1. Emerging Transport Technologies (ETT)
2. Innovative Transport Schemes (InnoTS)
3. Safety and Security for all/vulnerable users (SS)
4. Urban Freight Transport (UFT).

7.2   Delivery Mechanism

Each module includes a detailed facilitator's guide to run a one-day classroom course/workshop, the conceptual background and necessary training material. This approach was selected based on the results of a multilevel and multicriteria process. It integrates the outcomes of cities' capacity assessment to implement SUMP measures,

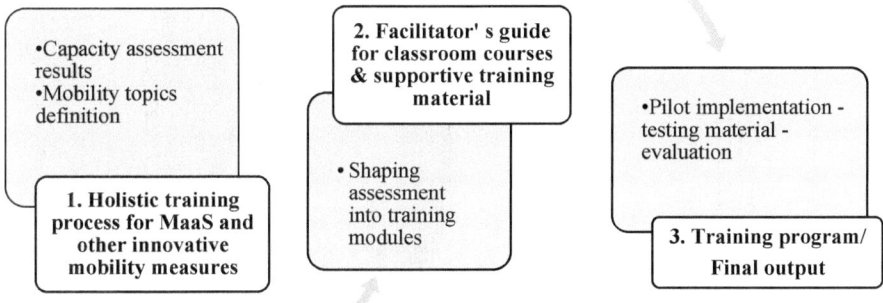

Fig. 7.1 Applied methodology

through an evaluation framework,[4] and the results of participatory activities which captured LAs' staff interest and experts' opinion.[5]

In particular, the theoretical background and the applied methodology are described by a three step process [1], shown in Fig. 7.1.

- Step 1 included activities related to topic definition and capacity assessment.

  - A multicriteria analysis was carried out to identify the more interesting mobility topics. MaaS, among other innovative mobility concepts, was revealed as one of the topics which should be included.
  - A baseline capacity assessment also defined the cities' specific capacity needs (financial, organisational, technical, etc.) related to SUMP measures' implementation. The capacity assessment results, both aggregated (all cities) and disaggregated (S-M cities), revealed the need of a holistic approach.

- Step 2 refers to transforming the findings from step one into a coherent training package. This resulted in the facilitator's guide along with the supportive material. The first approach and usefulness of the first module 'building capacity of S-M cities' local authorities to implement MaaS and other innovative transport schemes' were analysed.
- Step 3 validated the pilot training programme with all cities in the SUITS project. Pilots contributed to amelioration of outputs and prepared the ground for post capacity assessment.

The full version of the facilitator's guide to run a course/workshop inside the organisation and the supportive training material is available at [1]. A step-by step

---

[4] Chapter 16, this volume.

[5] Findings from this research resulted in the same approach for all topics described in the current Section. Online training approaches, using webinars for the topics are described in Chap. 6.

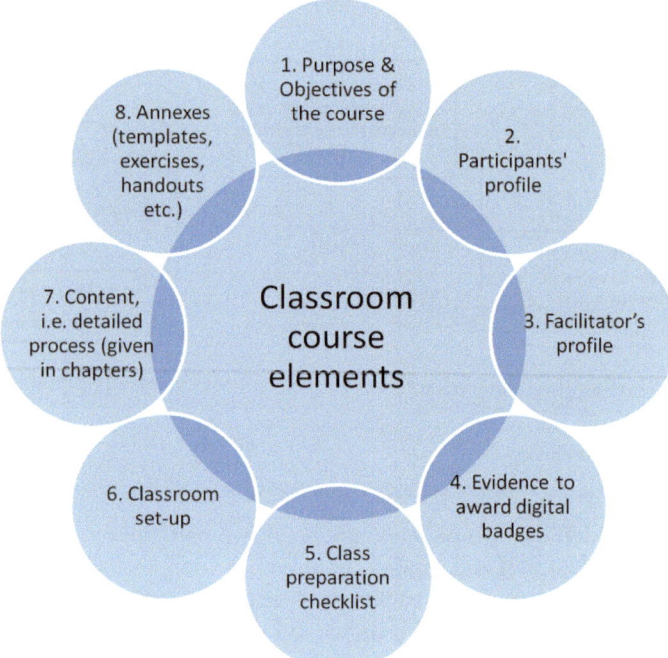

Fig. 7.2 Basic components of facilitator's guide for classroom course modules (*Source* [1])

guide is also provided in SUITS toolbox[6] to assist workshop implementation. The core comments are described below, with more details on specific modules provided in the following chapter.

It is assumed that the target audience will have some knowledge, interest or commitment to the development of sustainable transport. This might include someone new to the transport department, members of the LA who are going to be responsible for developing a SUMP or new transport measures, or wider stakeholders such as citizens or members of NGOs wishing to take a more active role in consultation /citizen engagement activities.

7.3   The Facilitators Guide

The facilitator's guide is an essential core element of the program. It aims to provide a 'novice' facilitator with all the necessary information to run a course/workshop. Its basic components are shown in Fig. 7.2.

---

[6] https://cbt.suits-project.sboing.net/.

Although we have provided a complete, standalone course, ideally the facilitator should be someone with knowledge of transport, and of the city in which training is being delivered, a transport consultant or learning provider,

The content is structured into chapters including teamwork activities to address training objectives shaped by capacity challenges and a final baseline capacity assessment [1].

Each chapter provides.

1. a condensed version of the content with reference to the respective workbook pages, where the content is further employed,
2. an estimation of its duration. The classroom course/workshop is designed to be conducted within a single day and with representatives from a single authority. However, the course has been designed so that it can be split into more days to better serve local needs and expectations and to allow participation of more than one LA.
3. instructions on how to run each training section including relevant information (teamwork exercises, a reference to the supporting power point slides, etc.).
4. references and additional sources are also provided to enhance knowledge on the topic.

7.4 Introduction to the Training Modules

Each training module follows a similar format and is based around core chapters, and the core content of which varies for each module. These are summarised below:

1. Introduction describing why sustainable mobility is important, and the workshop objectives
2. Description of the modules:

   2.1. Emerging Transport Technologies (ETT); including lean fuel and e-vehicles, cooperative systems, traffic information systems, pedestrian assistance systems, new systems and applications for parking management.
   2.2. Innovative Transport Schemes (InnoTS); including car sharing, ridesharing (carpooling, vanpooling), bike sharing, MaaS.
   2.3. Safety and Security for all/vulnerable users (SS); including awareness raising campaigns, advanced technologies for public transport, pedestrian and cycling infrastructure, road safety speed zones, security enforcement
   2.4. Urban Freight Transport (UFT); including urban consolidation centres, new technologies and telematics in last mile logistics, regulations regarding night deliveries and enforcement, eco-friendly vehicles, multi-use lanes, time-related (dynamic) loading space booking and/or multi-use of parking space, lockers as distribution points, limited traffic zones (LTZs).

3. Value for the cities, describing the benefits, stakeholders, participatory methods, relevant EU strategies, social impact assessment (see also Chap. 10).

4. Successful case studies or best practices drawn from the SUITS' cities include[7]:

   4.1. For Emerging Transport Technologies (ETT), case studies include Coventry's traffic and parking management regulations, Ljubljana's cleaner fuel vehicles (CNG buses), Rotterdam's electric mobility, electric vehicles and charging stations.

   4.2. For Innovative Transport Schemes (InnoTS), case studies include Helsinki's MaaS and Turin's free-floating bike sharing scheme. A supplementary chapter is included here on

       4.2.1. Business model canvases for MaaS and shared mobility which presents the business models methodology and exercise on MaaS business model canvas (see also Chap. 13)

   4.3. For Safety and Security for all/vulnerable users (SS), case studies include Rome's awareness campaign, pedestrian and cycling infrastructure, road safety speed zones; Coventry's speed safety zones and awareness campaign, Ghent's Awareness campaign, pedestrian and cycling infrastructure, advanced technologies application, Gdansk's anti-vandalism policy for safety and security in public transport.

   4.4. For Urban Freight Transport (UFT), case studies include Rome's Limited Traffic Zone (LTZ), Turin's multi-use lanes, parking regulation and LTZs., City of Utrecht's electric freight road vehicle, LTZs, UCC and lockers as distribution points.

5. A core chapter on innovative financing, procurement and partnerships include recommended mechanisms, methods and schemes depending on the module specific measures. Reference to relevant guidelines (more information on these topics can be found in Chaps. 11 and 12).

6. Process and implementation aspects include discussion of required data, available data collection methods, evaluation procedure, activities (both administrative and design/implementation), time plan, need for outsourcing, etc.

7. Available tools and guidelines provide references to available tools to support implementation steps specific for each module.

The complete capacity building programme is available online, in several languages at https://cbt.suits-project.sboing.net/. An overview of the module is provided in the following chapter.

Reference

1. Founta, A., Papadopoulou, O., Kalakou, S., Georgiadis, G.: Building capacity of small-medium cities' local authorities to implement MaaS and other innovative transport schemes (2020), Advances in Mobility-as-a-Service Systems. In: Proceedings of 5th Conference on Sustainable Urban Mobility, Virtual CSUM2020, June 17–19, 2020. Greece (2020)

---

[7] These are mentioned in several chapters in this volume.

Chapter 8
Implementation of Sustainable Mobility Measures for Passengers and Goods

Anastasia Founta and Olympia Papadopoulou

Abstract This chapter focuses on the key elements and steps to follow for the implementation of sustainable urban mobility measures. It is based primarily on material included in the capacity building framework. It provides an overview of latest developments in different categories of sustainable measures and is addressed to stakeholders responsible for, or who have an interest in the implementation of integrated and sustainable urban mobility.

8.1 Introduction

The widespread adoption of Internet-based technology and the increased use of big and real-time data is providing LAs and municipalities with the ability to better understand their urban transport challenges and develop and implement mobility measures in a more efficient and adaptable way. A variety of technologies are already embedded in transport systems either in vehicles or as part of the transport infrastructure such as advanced traveller information systems, driver assists, "smart" parking solutions through to cooperative systems, platooning and autonomous vehicles) (e.g. [1–6]). Combinations of these technologies address current transport problems in passenger and freight movement, while also providing planners with more accurate information to improve their transport systems.

The mitigation of congestion and climate change effects represents only one dimension of the problems which municipalities need to address. Past measures suffer from a lack of systematic evaluation and assessment of short- and long-term effects, while human factor was not taken into consideration upfront.

Safety and security are primary concerns for any transport system. Travellers expect transportation to be safe and inclusive. Citizens have a basic right to be able to travel without fear of being a victim of an accident or incident. While some studies have shown that fear of attack or harassment is a major factor in the rejection of public

A. Founta (✉) · O. Papadopoulou
Lever Development Consultants SA, Thessaloniki, Greece
e-mail: anasfou@gmail.com

© Transport for West Midlands 2023
A. Woodcock et al. (eds.), *Capacity Building in Local Authorities for Sustainable Transport Planning*, Smart Innovation, Systems and Technologies 319,
https://doi.org/10.1007/978-981-19-6962-1_8

transport [7, 8], others have shown that those who prioritise safety and comfort use public or active forms of transport [9].

Urban areas also represent particular challenges for freight transport, both in terms of logistical performance and environmental impact. Thus, there is a clear need for a comprehensive approach to urban freight solutions, particularly linking urban to interurban freight movements.

The purpose of this chapter is to increase the understanding of indicative innovative sustainable mobility measures, recognizing the opportunities and the operational environment for adopting them. It aims to provide guidance for overcoming current challenges and advance local priorities towards their implementation. The following sections provide an overview of the steps and considerations, which small-medium local authorities[1] need to make when selecting and implementing innovative urban mobility solutions for passengers and goods. More detail is provided in the workbooks found in the toolbox at https://cbt.suits-project.sboing.net/.

8.2 Making the Case for Sustainable Mobility

Prerequisites for designing a new transport measure, based on SUMP methodology, are

- Clarity and agreement about what needs to be achieved, framed as a vision and set of SMART objectives.
- Analysis of the current situation regarding the overall transport system.

A proposed starting step could be an analysis of the weaknesses[2] of each mobility mode (i.e. car-use, walking, cycling, and public transport system), both in general and in relation to the implementation of transport in the city, bearing in mind the differential effects each mode may have on transport users and non-users. This helps in developing analytical skills, a key asset for project management in all sectors and levels [11].

Weaknesses can relate to both the infrastructure and operation of the transport mode or system (see Table 8.1). They are also linked to human factor, as transport systems have an enormous effect on people's lives (e.g. on health, on social life, and on economy).[3] Based on the nature of the weakness, different solutions can be considered. Innovative transport technologies and mobility concepts address mainly operational drawbacks by optimising transport systems performance or by

---

[1] For whom a full SUMP may not be appropriate.

[2] Weaknesses, strengths, opportunities and risks synthesise the integrated approach of SWOT analysis, widely used in project management. Guidelines for developing and implementing Sustainable Urban Mobility Plans include analysis of current mobility situation in Phase 1 of plan development [10].

[3] The global view of the negative effect of the weaknesses is analysed in the next paragraph 3. Correlating local weaknesses of transport systems with wider problems and external transport cost Social Impact Assessment.

Table 8.1 Examples of weaknesses of transport modes on operational level, infrastructures and human factors

Weaknesses in "traditional" transport modes in relation to		
Infrastructure	Operational level or equipment	Human factors
Urban transport mode: Car/private vehicle		
Congestion in city centre main road network during peak hours due to lack of peripheral roads and increased traffic across city centre Congestion reduces safety, e.g. at intersections due to low visibility, poor road maintenance Limited parking facilities, space taken up by car parks; reductions in quality of the public realm	Congestion in main road network due to less effective traffic lights configuration Low safety levels due to lack of appropriate signalisation, etc. Illegal parking reduces quality of pedestrian and cycling infrastructure	High cognitive load and stress associated with driving in congested streets Lack of time for leisure and family activities. Effects of pollution on health Pedestrians and vulnerable road users at greater risk. As above, but with greater effects on those with limited mobility, e.g. elderly, visually impaired with pushchairs, wheel chair users
Urban transport mode: Public transport		
Lack of dedicated bus lanes Lack of tramway network Lack of safe and secure bus stops, etc	Low frequency of public transport journeys resulting in less passenger demand Lack of passenger information systems Old and not well-maintained bus fleet Lack of security during travelling	Transport poverty has intersectional effects on different groups Poor quality public transport reduces accessibility to resources (e.g. healthcare facilities, employment, education and retail outlets) and effects life chances/quality of life Disproportionate attention has been paid to servicing the journeys of principal wage earners Older, poorly maintained fleets increase pollution and may be uncomfortable Significant fear in travelling alone or at night, in certain areas by different groups
Urban transport mode: Bicycle		
Lack of cycle lanes in urban centre Lack of cycle network connecting city centre with neighbourhoods Lack of bicycle parking facilities, etc	Inefficient signals, awareness and priority of cycle paths resulting in severe accidents with cyclists Unsafe bicycle parking facilities	Poor cycling infrastructure leads to accidents, crime and reduces opportunities for more people to cycle Lack of opportunities to use active forms of transport has detrimental effects on health and wellbeing

(continued)

Table 8.1 (continued)

Weaknesses in "traditional" transport modes in relation to		
Infrastructure	Operational level or equipment	Human factors
Urban transport mode: Walking		
Narrow pedestrian pavements resulting in low accessibility for vulnerable users and low safety level Lack of pedestrian streets network in city centre, etc Lack of pedestrian crossings in most of urban road network	Pedestrian crossings are not safe for vulnerable users Increased pollution in the city centre due to congestion degrading walking conditions High speed roads in the city centre increase noise volume degrading walking conditions	Poor pedestrian infrastructure leads to accidents, crime and reduces opportunities for people to walk Lack of opportunities to use active forms of transport has detrimental effects on health and wellbeing
Urban transport mode: Logistics		
Lack of loading and unloading and parking bays Scheduling of deliveries	Delays in deliveries due to congestion Old and poorly maintained fleet resulting in pollution in the city centre	Effects due to pollution, impediments to right of way, noise associated with delivery vehicles

introducing new equipment. Nevertheless, depending on the mobility solution, its implementation may require different approaches to the design of networks and transport infrastructure. Hence, it is important to recognise weaknesses of all aspects of the transport system and enable integrated evaluation. Moreover, weaknesses of more than one transport mode result in the same wider problems (congestion, noise, poor air quality, safety issues, etc.). Consequently, a mobility solution might benefit more than one mobility mode, which is why an integrated analysis of the urban transport system is recommended.

8.3 Correlating Local Weaknesses of Transport Systems with Wider Problems and External Transport Cost

The identification of the wider problems and costs associated with current mobility patterns is important both to enhance sustainable thinking and affect decision-making towards sustainability.[4] Understanding transport external costs,[5] including human factors (such as mobility needs and rights, inclusivity, safety and security) next to infrastructure costs (i.e. enhancement costs, maintenance costs), and identifying whether these are internalised[6] or not, are all considered as key elements in the

---

[4] Two of the derived challenges, see Chap. 5.

[5] External costs arise when the social or economic activities of one (group of) person(s) have an impact on another (group of) person(s) and when that impact is not fully accounted, or compensated for, by the first (group of) person(s) [12].

[6] The internalisation of external and infrastructure costs means making such effects part of the decision-making process of transport users. This can be done by using market-based instruments

SUITS learning tools, for use in supporting arguments to advance sustainable mobility priorities and to move innovative mobility measures forward.

At European and national levels, key trends and issues on transport are reported by EU initiatives [13]. Based on the 2018 report "Transport in the European Union; Current Trends and Issues" [14], road accidents and congestion in combination with low levels of investment in transport infrastructure (maintenance and construction) require consideration at a European level.[7] Conventional city transport technologies and systems, in most cases, result in:

- Environmental degradation (noise, air quality, visual quality and nuisance).
- Delays due to high levels of congestion.
- High levels of fuel consumption due to congestion and use of cars by solo occupants.
- Higher levels of fatalities, casualties and injuries.

Important costs and impacts at a societal or individual level, have not been considered until recently, especially if they are difficult to quantify (e.g. issues related to community severance [15], the effects of transport poverty on life chances [16]) or have distributional [17] and intersectional effects [18, 19].

These costs are critical parameters to promote the change in current mobility habits and to help in raising awareness towards sustainable mobility. The estimation of the external costs of a transport system at a local level is important information for decision-making. This information about costs should be presented to contrast with the benefits of sustainable mobility measures, or other innovative solutions.

At a European and national level, the "Handbook on the external costs of transport" [20] provides an overview of the methodologies and input values that can be used to give insights to the main external costs of transport.[8] It also presents total, average and marginal external costs for all considered EU countries. Key findings include:

- An estimation of the overall external costs of transport, of € 1 000 billion (€ 981 billion) annually (almost 7% of the gross domestic product (GDP) of the 28 EU Member States) through air pollution, climate, habitat damage, well-to-tank emissions, noise, congestion, accidents.
- Almost 59% (€ 425 billion estimated) of total external costs of transport are due to the use of car, bus/coach and motorcycle (EU28 in 2016).
- Almost 27% of overall external costs in EU28 (2016) are due to road congestion (total delay costs € 270 billion estimated).
- Road traffic is responsible for the largest share of air pollution from the transport sector (71% of overall transport CO_2 emissions).

(e.g. taxes, charges and emission trading). Internalisation of external (and infrastructure) costs by using market-based instruments is one of the leading principles of the EU's transport policy [12].

[7] In "Transport in the European Union; Current Trends and Issues" report published in April 2018 [14], main current issues and key facts and figures on transport per EU country are available.

[8] The Handbook covers all main externalities of transport: accidents; air pollution; climate change; noise; congestion; well-to-tank emissions; habitat damage; other external cost categories (e.g. soil and water pollution).

In particular, the extended use of (conventional) cars in urban centres is responsible (along with other contributors) for health issues caused by pollution (such as heart attacks, asthma, anxiety, dizziness and fatigue), by noise (such as increased blood pressure and cardiovascular disease) [21] and lack of exercise (through an increase in obesogenic neighbourhoods [22]).

Car parking and whether the cost of public space occupation by parked cars is internalised (i.e. charges for the user) or not is one of the issues to be considered.[9] In combination with other factors, on-street parking hinders, the equal development of sustainable mobility means (public transport, bicycle, etc.) and reduces road capacity, especially when available space is limited. It has been suggested [23] that the high cost of free parking is an enormous public subsidy that makes driving less expensive than it should be, further skewing travel choices.

8.4 Gaining Familiarity with the Mobility Measure Concept to be Implemented[10]

Understanding the key characteristics of different mobility solutions and their application areas enables the most suitable technology to be selected and implemented for a given context. In the scope of SUITS learning tools, some examples of emerging transport technologies, innovative transport schemes, safety and security measures and urban freight measures that enhance sustainable mobility are described briefly regarding their main characteristics and areas of implementation.

Moreover, Social Impact Assessment (SIA) method is recommended for the systematic identification of measures' benefits and challenges (see Chap. 10). This should be used early in the development cycle as the results can be used to determine implementation parameters and support measures argumentation, highlighting the social and human aspect of mobility measures.

8.4.1 Emerging Transport Technologies: Main Characteristics

Technological evolution on different domains (e.g. telecommunications, data collection and processing, sensors, energy generation, etc.) offered the opportunity to develop and implement advanced transport systems. There is already a variety of technologies embedded to transport systems or transport means. Some are more widely implemented, while others are still being piloted, simulated or having limited implementation (e.g. cooperative systems [24], drones for last mile delivery [25]).

---

[9] "The right to have access to every building in the city by private motorcar in an age when everyone possesses such a vehicle, is actually the right to destroy the city" [23].

[10] The information presented at this subsection is part of "Chapter 2: Description of measures" of SUITS training modules.

Brief introduction of emerging transport technologies[11]

(a) *Clean fuel and e-vehicles*: The aim is to replace vehicles with conventional petrol or diesel engines with those that have a low-carbon or carbon-free operation. LAs may support the expansion of alternative fuel vehicle use by providing appropriate infrastructure, incentives (available) and upgraded to service vehicle fleets, buses, etc. [26, 27].

(b) *Cooperative systems*: Cooperative systems analyse data transmitted by road infrastructure elements to and by vehicles for several purposes (i.e. warn the driver against dangers, optimise traffic light phases, inform optimised speed and reductions in idling).

- **Intelligent Traffic Lights**: They allow the communication of traffic lights (and other infrastructure elements) with the vehicles (vehicle to infrastructure communication (V2I). The aim is to inform upcoming stream with traffic light phases duration and sequence. Like this, vehicles/drivers can adjust their driving avoiding idling and unnecessary acceleration [28, 29]. LAs can support the integration of these systems and allow the application of advanced driver-assistance systems (ADAS),[12] leading the transition towards autonomous vehicles.

- **Probe vehicle**: Data, such as traffic conditions, road surface and surroundings, can be generated by vehicles and communicated to the infrastructure. They can provide input for operational traffic management (e.g. to determine the traffic speed, manage traffic flows by informing drivers where the danger of accidents accumulates, etc.) [30, 31]. By using probe vehicles in their fleet, municipalities may improve data collection both for traffic management and infrastructure maintenance purposes.

(c) *Advanced traveller information systems* provide information about real-time traffic conditions and other mobility parameters. They may concern all mobility modes (car, public transport, walking, cycling and

---

[11] More information is available in the corresponding SUITS Workbook available in SUITS toolbox https://cbt.suits-project.sboing.net/.

[12] Advanced driver assistance systems are intelligent systems that reside inside the vehicle and assist the main driver in a variety of ways. These systems may be used to provide vital information about traffic, closure and blockage of roads ahead, congestion levels, suggested routes to avoid congestion, to judge the fatigue and distraction of the human driver and alert or make suggestions to the driver. They can even take over the control from the human on assessing any threat, perform easy tasks (like cruise control) or difficult manoeuvers (like overtaking and parking), and they enable communication between different vehicles, vehicle infrastructure systems and transportation management centres. This enables exchange of information for better vision, localization, planning and decision-making of the vehicles [27].

multimodal) and use different ways to disseminate this information (mobile phone, road signs, etc.) [27]. They can be based on communicative systems transmitted data or other types of sensors (on-road devices, optical sensors, cameras, GPS trackers, etc.).

(d) *Pedestrian assistance systems*: In this category, different systems might be considered such as (a) high-tech pedestrian crossing [32] (a new approach in the design of crossings where smartphones and in-car infotainment systems are used to alert pedestrians and drivers), (b) multi-media integration for pedestrians' varied needs [33] (a technology that caters for pedestrians having "non-standard" crossing needs, using smartphones and offering the potential of individualised solutions) and (c) cooperative traffic lights for vulnerable road users (VRU) [34], (a cooperative system that gives priority or additional crossing time—extending the green light phase or lessening the red phase—based on pedestrian characteristics or on special conditions, such as weather).

(e) *New systems and applications for parking management*: Parking sensors allow detecting the presence of vehicles parked in the parking space. The management software can analyse and manage all input data entering in the system in real time and helps drivers find parking. Pricing based on "consumption" or demand-responsive pricing policy may be complementary to these systems [35, 36].

From an understanding of available transport technologies (their status and uptake in similar cities (e.g. through use of the Eltis observatory[13]), their known strengths and weaknesses and the mobility problems the city faces, it is possible to make a selection of the most suitable technologies. Examples of emerging transport technology per mobility problems is shown in Table 8.2.

In particular, clean fuel and e-vehicles[14] address climate change and health issues caused by air pollution and noise in direct ways. However, they cannot respond to problems such as delays due to congestion or lack of parking space. On the contrary, intelligent transport technologies (ITS) better address issues related to road safety and demand management, both in parking and traffic.[15]

Nevertheless, it should be emphasised that at least some of the weaknesses identified in Table 8.2 not only have a behavioural component, but can be addressed partly, if not entirely through a combination of behavioural/technological and enforcement

---

[13] https://www.eltis.org/.

[14] More information regarding clean fuel and e-vehicle are available at CIVITAS training centre relevant field [37].

[15] More information regarding ITS technical parameters and areas of implementation are available at CAPITAL training platform [38].

Table 8.2 Examples of emerging transport technologies that address current transport system weaknesses

Weaknesses of current transport system	Transport technologies which may address weakness
Congestion in main road network due to ineffective traffic lights configuration	Intelligent traffic lights technology Advanced information systems for drivers
Illegal parking	Parking sensors and parking management software
Lack of public transport passengers' information systems	Advanced traveller information systems—telematics
Inefficient signalisation of cycle paths resulting in severe accidents with cyclists	Communicative systems—V2I communication to alert cyclist presence at intersections
Pedestrian crossings unsafe for vulnerable users	High-tech pedestrian crossing with multi-media integration for pedestrians' varied needs
High levels of pollution in city centre due to congestion degrading walking conditions High speed roads in the city centre increase noise volume degrading walking conditions	Clean fuel and e-vehicles

measures. Key to this is winning the interest and support of citizens for adopting more sustainable and thoughtful behaviours. Such an approach may be more appropriate for small-medium sized cities which can see an ethical and competitive advantage to becoming a liveable city.[16]

8.4.2 Innovative Transport Schemes: Main Characteristics

Based on US Department of Transportation [39], Mobility on demand (MoD) is an innovative transportation concept which uses mobility services such as shared mobility, courier services, unmanned aerial vehicles (UAVs) and public transportation solutions with the aim of offering consumers access to mobility, goods and services on demand. Trip planning and booking, real-time information and fare payment into a single user interface are advanced forms of MoD passenger services. Mobility as a service (MaaS) encompasses and emphasises these services, focusing on personal travel, by mobility aggregation, smartphone, app-based subscription access and multimodal integration. Nevertheless, shared mobility (car-sharing, bike-sharing, ride-sharing, scooter sharing) forms the backbone of much MOD for personal travel (along with public transportation and private transit such as shuttles, first and last mile micro-transit, etc.) and addresses both passenger travel and goods delivery. Forms of shared mobility have been applied in most EU cities,

---

[16] See for example, https://civitas-initiative.org/awards.html.

but continually evolve, harnessing new technological trends in transportation and telecommunications and use cases. The following forms a very brief overview.

Brief introduction of sharing schemes and MaaS

(a) Car-sharing: Car-sharing is a form of transport by which several persons in turn make use of one or more collective cars. This can be arranged both by the parties mutually and by a car-sharing provider [40].

(b) Ride-sharing (carpooling–vanpooling [41]): Ride-sharing is the concept of "offer a ride" on vehicle where seats are available. It covers various options, the most common is when the owner of a vehicle has a predetermined journey and offers a seat to passengers going in the same direction in exchange for sharing the costs of the journey [40]. In this way, the additional mileage is minimised. Carpooling generally uses participants' own automobiles [42].

(c) Bike-sharing: Bike-sharing schemes can be defined as "short-term" urban bicycle rental schemes that enable bicycles to be picked up at and returned to any self-service bicycle station, which makes bicycle-sharing ideal for point-to-point trips. The basic premise of the bike-sharing concept is sustainable transportation, and this differs from traditional, mostly leisure-oriented bicycle rental services in many ways. Bike-sharing schemes could be with station-based bike-sharing (SBBS) or without docking stations (free-floating bike-sharing FFBS) [43].

(d) Mobility as a service (MaaS): MaaS is defined as the integration of various forms of transport services into a single mobility service accessible on demand. The key concept behind MaaS is to put the users, both travellers, and goods, at the core of transport services, offering them tailor-made mobility solutions based on their individual needs. This means that, for the first time, easy access to the most appropriate transport mode or service will be included in a bundle of flexible travel service options for end users [44]. Research undertaken by SUITS showed that the implementation of MaaS may be problematic for some cities [45].

Again, the success of these measures depends on the appropriateness of the selection and the level of uptake among citizens. The shift away from automobility requires a paradigm shift by those planning future transport systems and the system users [46–48]. The capacity building programme provided by SUITS, along with our other outputs and this book, have been designed to help create this shift (Table 8.3).

Table 8.3 Examples of innovative transport schemes that address some of the weaknesses of current transport systems

Weaknesses and problems of current transport system	Innovative transport schemes
Lack of bicycle parking facilities	Station-based bike-sharing system
Underuse of existing bike network/cycle paths	Bike-sharing system and scooter sharing systems[17]
Under use of public transport lines	MaaS
Lack of car parking spaces	Car-sharing/ride-sharing
High fuel consumption per capita	Ride-sharing

8.4.3 Safety and Security Measures for All/vulnerable Users: Main Characteristics

Safety and security are of primary concern in any transport system.[18] Urban transport measures aim to increase safety and security for all users, thereby increasing inclusivity of transport services and meeting the needs of vulnerable transport users[19] while at the same time promoting sustainability. Using advanced technology such as video cameras along with a more optimal usage of existing resources are some of the suggested measures.

> **Brief introduction of safety and security measures**
>
> *Awareness raising campaigns*: Awareness campaigns are a form of official motivation towards the public to encourage a certain type of behaviour. Such campaigns can have the form of advertising material or in some cases provide training and are widely used as part of an integrated approach. They are valid for both safety and security issues [51].

---

[17] See for example, UK guidance on e-scooters: https://www.gov.uk/government/publications/e-scooter-trials-guidance-for-local-areas-and-rental-operators/e-scooter-trials-guidance-for-local-areas-and-rental-operators.

[18] The problems of transport safety are defined as vulnerability to accidental injury (usually involving at least one vehicle as the instrument causing the injury). The problems of transport security are defined as vulnerability to intentional criminal or antisocial acts suffered by those engaged in trip making [49].

[19] Vulnerability related to transport policies can be split into two types: firstly (and more commonly), a definition based on health and safety aspects of transport activities, and secondly, one based on social aspects, which is closely related to the idea of accessibility. Vulnerability as applied in the EMPOWER Project is considered to include the following groups: (a) low-income groups, (b) children, youths and the people caring for them, (c) women, (d) the elderly, (e) people with disabilities, (f) lower education people and (g) non-locally born people [50].

Advanced technologies for public transport (PT): Advanced technologies for PT consist of measures that use technological equipment to enhance and improve the level of service provided. Such measures can be implemented on different points of the infrastructure while accounting for both traffic safety as well as individual security [52, 53].

Pedestrian and cycling infrastructure: Such measures include segregation of cyclists and pedestrians, development and maintenance of infrastructure such as cycle paths, parking stations and pedestrian crossings. Moreover, such measures can be supplemented with legislative and educational actions to improve safety and security traffic conditions [54].

Road safety speed zones: Traffic restriction measures include actions such as road design elements or use of technological tools that are able to either provide or deny access to an area or control and adjust traffic conditions in order to optimise safety [55].

Security and road safety enforcement: Security enforcement contains usage of technology such as surveillance cameras or physical attendance of trained staff that aim to maintain security in vehicles and on stations [56].

Safety and security measures may focus either on infrastructure or on operations. Awareness raising campaigns and enforcement measures can be applied in all cases, even though some of the measures may address specific transport means or specific users (Table 8.4).

It is also important that security is not so intrusive as to make travel an unpleasant experience. However, certain groups, such as women and older users prefer more overt security measures [57]. Transport security refers to everything from terrorist

Table 8.4 Examples of safety and security measures that enable facing transport system weaknesses

Weaknesses and problems of current transport system	Safety and security measures
Pedestrian crossings are not safe for vulnerable users	Upgrade pedestrian infrastructure to be appropriate for vulnerable users Increase amount of time allowed to cross road, provide audible and tactile signs Increase amount of pedestrian only access areas
Decrease of public transport use due to security incidents	Advanced technologies for public transport such as technological equipment to enhance and improve the level of service provided
High speed roads in the city centre increase accidents with pedestrians and cyclists involved	Road safety speed zones Maintain/improve pedestrians and cyclists network infrastructure

attacks to prevention, harassment, vandalism and graffiti. Perceived lack of maintenance and poor lighting increases insecurity [57, 58] and offer opportunities for increased criminality. Above all, urban transport measures aim to increase safety and security for all users, thereby increasing inclusivity of transport services and meeting the needs of vulnerable transport users while at the same time promoting sustainability. Suggested measures include using CCTV, better lighting, more patrols and behavioural nudges along with a more optimal usage of existing resources.

8.4.4 Urban Freight Measures: Main Characteristics and Areas of Implementation

Urban freight transport (UFT) is considered as the movement of freight vehicles whose primary purpose is to carry goods into, out of and within urban areas [59] UFT is a vital part of the economy of cities, resulting, however, in significant environmental impacts. Urban freight transport constitutes approximately 10% of total transportation activities but due to the usage of heavy-duty vehicles it contributes to approximately 40% of total transport emissions.

Brief introduction of urban freight measures [59, 60]

Urban consolidation centres (UCC): An urban consolidation centre is a new logistics platform, either private or public that is designed to serve the urban centre or other large magnitude sites. The principle and main objective is to receive large scale freight operations, break them down and deliver them to its target area while ensuring sustainability through the usage of light goods vehicles (LGVs) and other smaller vehicles. The key purpose of UCCs is the reduction in total distance travelled and the avoidance of poorly loaded goods vehicles making deliveries in urban areas.

New technologies and telematics in last mile logistics: This measure includes applications that provide fleet management along with route optimisation. Drivers are receiving real-time information regarding their route choice, in order to minimise accumulative costs (financial, environmental, time, etc.). It provides the optimal management of the company's fleet.

Regulations regarding night deliveries and enforcement: Such actions aim at enforcing night-time deliveries by taking advantage of lower congestion at these time slots. It contributes to avoiding congestion due to freight operations.

Eco-friendly vehicles: Promotion and adoption of alternative and more environmentally friendly vehicles such as electric vehicles, LNG vehicles, cargo bikes both traditional and electric, tricycles, scooters and drones. The future deployment of autonomous vehicles will result in even more logistics delivery solutions replacing traditional modes of freight transport [61].

Multi-use lanes: This measure is based on the idea that the capacity of the network (lanes in this case) can be dedicated to different specific transportation modes based on the time of the day, the traffic conditions, etc. Such a solution is very promising for both urban freight systems and public transport. The allocation of the affected lane can be designed using different time windows among different users, and restrictions can be applied by vehicle type, scope of transportation, etc. In this aspect, it supports the operation of buses in urban centres.

Real-time (dynamic) loading space booking and/or multi-use of parking space: Through digital solutions the logistics service provider can schedule and book available parking spaces for a limited amount of time in order to load/unload cargo. The allocation of the parking spaces could be designed using different time windows among different users and restrictions can be applied by vehicle type, scope of transportation, etc.

Lockers as distribution points: A network of automated delivery points located in convenient sites such as transportation stations or large grocery stores. The system works similarly to ATM machines, where with a temporary personal password the person is granted access and receives the shipped item [62, 63].

Limited traffic zones (LTZs): Access to urban areas is limited to freight vehicles that meet certain emissions standards. LTZs are becoming increasingly common in major European cities as mean for city authorities to meet European air quality standards. There is a positive impact by reducing emissions from freight vehicles both by renewing the fleet and reducing trips.

The European Commission is consistently putting efforts into planning and integrating urban freight transport into the European Urban Mobility Strategy and transforming its future into a more sustainable and equitable system (Table 8.5).

8.5   Identifying Benefits, Beneficiaries and Challenges of Mobility Measures Implementation[20]

The ability to identify the benefits of mobility measures implementation at the local level is one more factor for successful decision-making. Depending on the weaknesses and taking into consideration the wider context of the problem (i.e. external

[20] The information presented at this subsection is part of "Chapter 3: Value for the Cities", of SUITS training module in question.

Table 8.5 Example of urban freight measures that enable facing transport system weaknesses

Weaknesses and problems of current transport system	Urban freight measures
Air pollution, noise and optical annoyance increased fuel consumption in big urban areas due to the circulation of big trucks (>3.5 ton)	Limited traffic zones Eco-friendly vehicles
Increased fuel consumption in big urban areas, due to large travelled distances and extra travels caused by poorly loaded cargo vehicles	Urban consolidation centres (UCC)
Load and unload takes place in the middle of the street in urban centres, resulting either in congestion or in limiting pedestrian and cyclist active space	Real-time (dynamic) loading space booking and/or multi-use of parking space Eco-friendly vehicles that require less space (i.e. cargo bike, drone, scooters, etc.)
Congestion during peak hours also caused by the cargo vehicles circulation. Congestion results both in deliveries and commuting delays	Regulations regarding night deliveries and enforcement New technologies and telematics in last mile logistics Multi-use lanes

and infrastructure costs of not acting), the identification of the benefits and beneficiaries of each measures' category lead to a preliminary set of mobility measures options which address the current weaknesses of the transport system. Then, more levels of assessment are required to select the mobility measures that they are complementary to each other.[21] For example, clean fuel and e-vehicles address in a direct way climate change and health issues caused by air pollution and noise. However, they cannot respond to problems such as delays due to congestion or lack of parking space. On the contrary, intelligent transport technologies (ITS) better address issues related to road safety and demand management, both in parking and traffic.

Employing Social Impact Assessment (SIA) at appropriate stages of the design and implementation of sustainable measures is recommended for a systematic and holistic identification of measures' benefits and challenges. A detail description of SIA and associated tools and methodologies can be found in Chap. 10.

The benefits vary among the measures' categories, but in a wider context they tackle two main mobility problems: lack of safety and air pollution. It is also important. An overview of the contribution per transport technology is shown in Table 8.6. Challenges such as the mitigation of congestion, the decrease of conventional fuel consumption, the increase of user's satisfaction, the optimisation of available resources, the participation and consequently the reduction of external and infrastructure costs are also confronted.

The implementation of emerging transport technologies replies, among others, to the participation challenge. It requires strong collaboration between many different stakeholders (private companies, research institutes, national and international

---

[21] The importance of packages of measures is highlighted in SUMP methodology [10].

investors, central government, etc.) and it is a prerequisite for informed decision-making. This collaboration sets the ground for stronger stakeholder[22] engagement, useful for upcoming actions or initiatives. Additionally, innovative concepts (transport technologies and schemes) provide a potential for the local economy to expand through private investment and/or tourism based on "smart" city concept. Safety and security measures address, in principle, the safety challenge, while also supporting the participation challenge, especially from the perspective of citizens' involvement.

Stakeholders mapping is necessary at this stage. It is a fundamental step at any project employment [69] in order to draw up an efficient strategy to manage them. There are stakeholders that might be positively affected by the implementation of an emerging technology in a direct way, while there are others without direct benefit but with high involvement at the implementation process or high influence at the local community level. Many stakeholders' mapping techniques are available. The most popular one is the power/interest matrix shown in Fig. 8.1 proposed by Mendelow [70].

In particular, the SUITS learning tool proposes the identification of stakeholders by also noting the probable reasons why objections may arise and providing arguments or solutions to ensure implementation support, taking also into consideration emerging technology benefits.

Table 8.7 gives an example of stakeholders, their probable reactions and possible responses to be provided, with regard to the introduction of clean fuel and e-vehicles at public transport system bus lines. A similar approach can be followed for each type of mobility measure. The benefits of the measures are also considered as responses to challenges. Based on the measures implementation area, stakeholder analysis should be broken down, with the identification of specific actors, challenges and responses at a very local level.

In most cases, people express their objections for sustainable mobility measures which lead to the internalisation of external cost. In general, citizens are more negatively predisposed to the introduction of payment policies than to other types of sustainable measures (introduction of cycle paths, pedestrianisation of streets, etc.). The negative reaction is more intense when the average income of the households is low and the cost of living is already high. A typical example is parking management measures, where parking cost is introduced (or increased) and available, free of charge parking seats, are diminished. In these cases, LAs and relevant actors need to design measures which allow direct financial and social positive impact for the citizens and local entrepreneurs, taking into account both their needs and stated preferences. For example, it is crucial to estimate and highlight the positive impact in terms of time and money saving, when "smart" parking sensors with navigation systems are introduced and allow drivers to reach directly the available parking seat and avoid congested road sections. Pricing based on "consumption" or demand-responsive pricing policies seems more "fair" and might inspire more confidence to the users. Another approach for increasing positive reactions by stakeholders is to

---

[22] Any actor having a vested interest in the decision process, either directly affecting or being affected by its resolution, including experts and the public, is named stakeholder [68].

Table 8.6 Contribution of mobility measures to overcome environmental, social and other city challenges[23] (synthesis from [10, 64–67])

Weak/indirect correlation	Moderate correlation	Strong correlation	

Challenges / Measures	Health	Congestion	Safety & security	Participation	Strategic planning	Global climate change
Clean fuel & e-vehicles						
Traffic lights optimisation						
Traffic information systems						
Pedestrian assistance systems						
New parking management systems & applications						
Car-sharing						
Car-pooling						
Walking & cycling						
Urban Consolidation centres						
Parking regulations for freight vehicles						
Limited traffic zones						

---

[23] Health challenge : How to create a healthy environment for citizens; Congestion challenge: How to create an economically viable and accessible city; Safety and security challenge: How to ensure a safe and secure urban environment and mobility; Participation challenge: How to involve citizens and other stakeholders; Strategic planning challenge: How to achieve policy goals while ensuring that mobility needs of society and its citizens are met; Global climate change challenge: How

Fig. 8.1 Example of power/interest matrix for stakeholders mapping. "Who needs what" is indicated (*Source* Smartsheet)

adopt reciprocal measures which can be directly implemented and make obvious that income of the payment policies is reinvested in the society.

8.6 Compliance with EU Legislation

The planning and implementation of sustainable mobility measures is necessary to comply with European legislation (avoiding environmental fee, claiming EU funding with higher probabilities, etc.). Table 8.8 provides a list of the most recent of these. Developing integrated master plans following SUMP guidelines may also increase success in applying for investment or winning tenders (also see Chap. 11). Benefits of mobility solutions and their correlation with local, national and European goals can further support argumentation.

Documentation with specific focus on emerging technologies and innovative transport technologies is also available (* MERGEFORMAT Table 8.9).

In particular, road safety actions and policies are also described in the strategic action plan on road safety and in the Valetta declaration (Table 8.10).

8.7 Using Case Studies to Strengthen Argumentation and Preparedness of Implementation

One of the most useful practices to strengthen skills regarding mobility measures implementation is to learn from other cities' experience. Case studies provide insights regarding possible obstacles, challenges and outcomes of measures implementation helping both the performance of an ex ante impact assessment (SIA) and the anticipation of issues that may arise. In most cases, positive impact is highlighted, and general

---

to reduce climate change related emissions from urban transport to contribute to achieving local, national and global climate change goals.

Table 8.7 Identification of stakeholders and correlation with challenges and responses to ensure their support. The example of the introduction of clean fuel and e-vehicles at public transport systems

Stakeholders	Possible challenges	Responses
Public transport operators	High purchase cost Unavailable recharging or refuelling infrastructure Vehicles do not present the same engine performance as conventional vehicles, risking to not provide air conditioning as necessary, not serve high vehicles occupancy and not be able to climb up hill areas	Lower lifetime cost for vehicles operations Joint procurement procedures (see next paragraph) LAs to provide the recharging or refuelling network Select appropriately bus lines to introduce new fleet based on urban areas characteristics, local climate and demand
Vehicle manufacturers	Improve technical parameters to address engine performance and recharging, refuelling needs	LAs to collaborate with vehicle manufacturers to support R & D departments (ask for pilot implementation, etc.)
LAs	Support public transport network and provide relevant infrastructure to motivate more citizens to use public transport. Provide conditions to raise public transport operator income	High social positive impact by the use of clean fuel vehicles in public transport: less pollution, less noise, citizens become familiar with new vehicles fleet (paradigm shift)
Central government	Legal and regulatory framework does not facilitate the procurement process and operations	Work closely with LAs, public transport operators and vehicles manufacturers to adjust legal and regulatory framework accordingly
Citizens (passengers or not)	Citizens may react negatively if the introduction of these vehicles results in higher pricing Low performance of these vehicles may result in negative predisposition against these technologies	More comfortable experience with less noise and air pollution during travelling Central government or LAs need to provide financial incentives to assimilate any extra charging to passengers Examine or check thoroughly the bus lines to be supported by the new fleet with the help of vehicles manufacturers/providers-better design

information is provided while drawbacks and/or quantitative data are not easily available. However, it is important to gather information that allows better understanding of measures requirements, implementation conditions and potential impact. Therefore, it is recommended to contact directly the implementation authority of the concerned city, promoting at the same time the collaboration between municipalities and the exchange of experience.

Table 8.8 Most recent EU strategy or policy documents supporting sustainable mobility measures implementation

Document name/year of publication	Short description
A 2030 framework for climate and energy policies. European Commission. /2013 (Green Paper)	This framework integrates different policy objectives such as reducing greenhouse gas (GHG) emissions, securing energy supply and supporting growth, competitiveness and jobs through a high technology, cost effective and resource efficient approach. These policy objectives are delivered by three headline targets for GHG emission reductions, renewable energy and energy savings. There are additional targets for energy used by the transport sector
A European Strategy for Low-Emission Mobility./ 2016 (Communication from the Commission to the European Parliament)	A forward looking and long-term policy approach with the aim of ensuring a regulatory and business environment that is conducive to meeting the competitiveness challenges that the transition to low-emission mobility implies, is a vital precondition. The analysis carried out in this paper provides insights on the necessary tools to do this
Strategic plan 2016–2020 – DG for Mobility and Transport. / EU Strategic Plan/ 2016	This sets out the department's vision for a five-year period, up until 2020
Directive on the promotion of clean and energy-efficient road transport vehicles EU Directive of the European parliament and of the Council/2019	The directive on the promotion of clean and energy efficient road transport vehicles aims at a broad market introduction of environmentally friendly vehicles. It requires that energy and environmental impacts linked to the operation of vehicles over their whole lifetime are taken into account in all purchases of road transport vehicles, as covered by the public procurement directives and the public service regulation
Proposal for post-2020 CO2 targets for cars and vans Climate Action—European Commission/2019	The new regulation maintains the targets for 2020, which were set out in the former regulations. It adds new targets that apply from 2025 and 2030. The regulation also includes a mechanism to incentivise the uptake of zero- and low-emission vehicles, in a technology-neutral way

A list of case studies[24] is given in Table 8.11.

SUITS CBP training modules propose the following key elements to focus on when examining a case study so to be able to identify whether a measure is useful and applicable to a city:

---

[24] For detailed information on each case study, please refer to the Factsheets included in each workbook and as separate documents in SUITS toolbox https://cbt.suits-project.sboing.net/.

Table 8.9 Additional EU strategy or policy documents for emerging transport technologies and innovative schemes

Document name/year of publication	Short description
COM 283: An agenda for a socially fair transition towards clean, competitive and connected mobility for all (2017)	The communication document identifies the challenges and the key strategy directions towards sustainable mobility in Europe in 2025. Among other aspects, it promotes harnessing of the benefits of digitisation, automation and intelligent mobility services. Moreover, it refers to strategies such as "smart" road charging, "smarter" mobility in cities, better compliance and enforcement through "smart" digital technologies, acceleration of alternative fuel infrastructure deployment and other fields
Strategic Transport Research and Innovation Agenda (STRIA)/2017	STRIA aims to set out common priorities to support and speed-up the research, innovation and deployment process leading to radical technology changes in transport. The agenda outline future transport research and innovation priorities to decarbonise the European transport sector. The roadmap factsheets aim at providing a summary of the main features and targets identified in each of the seven STRIA Roadmaps: • Cooperative, connected and automated transport; • Transport electrification; • Vehicle design and manufacturing; • Low-emission alternative energy for transport; • Network and traffic management systems; • Smart mobility and services and • Infrastructure
Transport Research and Innovation Monitoring and Information System (TRIMIS)/2019	The Transport Research and Innovation Monitoring and Information System (TRIMIS) supports the implementation and monitoring of STRIA and its seven road maps. Through TRIMIS reports on progress made in the different R&I areas, it is possible to identify the gaps and consequently the actions to be taken by LAs to advance emerging transport technologies

a. **City's basic characteristics**, such as the city size, the geography of the city (seaside areas, riverside areas, mountainous areas, etc.), the architecture of the city (narrow streets, historical centre, modern houses, etc.) and the land use (mostly residential areas or economic centres, etc.). This information helps in understanding the level of relevance with your own case.

b. **Initial problem and target goal**: This information also helps in understanding the level of relevance with your own case.

c. **Scalability/replicability**: This defines whether this measure can be applicable to other cities and under which circumstances.

Table 8.10 Additional EU strategy or policy documents for safety and security inclusive measures

Document name/year of publication	Short description
Europe on the move. Sustainable Mobility for Europe: safe, connected and clean Annex 1, Strategic Action Plan on Road Safety /2018	This Annex outlines specific actions envisaged under the current commission mandate, i.e. until end Q3 2019. These actions are set out with a specific target time frame. It cannot of course prejudge action for the next commission but instead sets out additional envisaged actions for the post 2019 period indicatively and without a specific target time frame
Valetta Declaration on Road Safety, Malta/2017	Transport ministers of the Member States of the European Union, meeting in Valletta on 29 March 2017 under the Maltese Presidency of the Council of the European Union, reconfirm their commitment to improving road safety

Table 8.11 Case studies from cities involved in SUITS project and other EU cities

Category of measure	Location
Module: Emerging transport technologies	
Electric mobility, electric vehicles and charging stations	Rotterdam, The Netherlands
Parking management	Coventry, United Kingdom
Cleaner fuel vehicles—CNG buses	Ljubljana, Slovenia
Module: Innovative transport schemes	
Bike-sharing system	Turin, Italy
Mobility as a service	Helsinki, Finland
Module: Safety and security inclusive measures	
Speed safety zones, awareness campaign	Coventry, United Kingdom
Advanced technologies for public transport, awareness campaigns	Gdansk, Poland
Awareness campaign, pedestrian and cycling infrastructure and advanced technologies application	Ghent, Belgium
Awareness campaign, pedestrian and cycling infrastructure and road safety speed zones	Rome, Italy
Module: Urban freight measures	
Limited traffic zone (LTZ)—freight distribution system	Rome, Italy
Multi-use lanes, parking regulation and LTZs	Turin, Italy
Electric freight vehicles (The Cargo–Hopper and the Beer-boat)	Utrecht, The Netherlands

d. **Implementation requirements**, such as total budget, time framework and number of implementation phases.

e. **Indicators to measure success** and final outcome/impact, as necessary tools to monitor and evaluate measures implementation. Quantitative data help argumentation and provide guidance on social impact assessment of the measure.

f. **Social groups affected**, as important information to better understand the needs of stakeholders and adjust measure parameters if necessary.
g. **Barriers and drivers** during implementation, from different perspectives such as cooperation/coordination, financial resources, process, technical skills, data resources, political and legal conditions, societal conditions and staff.

8.8 Identifying Financing Mechanisms, Partnerships and Procurement Methods for Mobility Measures Implementation

The identification of financing mechanisms, procurement procedures and partnership schemes are key steps in the implementation of mobility measures and need to address capacity gaps related to financial resources and technical skills. Selecting the most appropriate approach often determines the success of the measure. New approaches may be introduced which can be adapted to current financial and societal conditions, which makes all these three topics, more or less, interconnected. For example, based on the chosen financing mechanism, new partnerships may be formed, and procurement procedures may need to adjust to this innovation. On the other hand, based on measures requirements, specialised procurement criteria may need to be added and finally determine the partnerships and financing mechanism to be applied. Table 8.12 provides an example of innovative financing mechanisms, partnerships and procurement procedures for intelligent traffic lights technology (cooperative on-road systems). Financing mechanisms, procurement processes and partnerships are discussed in more detail in Chaps. 11, 12 and 13.

8.9 Identifying Implementation Aspects and Activities to be Considered

Integrating measures in a wider strategic plan such as the Sustainable Urban Mobility Plan (SUMP) is indispensable. In "Guidelines for Developing and Implementing a Sustainable Urban Mobility Plan [10]", planning and implementation principles are presented. The plan is a useful tool supporting local public decision-makers and stakeholders in "governing" urban mobility systems with medium/long-term vision. With this integrated approach, efficiency of measures is more likely guaranteed.

Besides the strategic action plan, an implementation plan for each mobility measure is required, where activities within a measure, implementation time framework of each activity, human resources, cost and stakeholders involved per activity must be defined [70].[25]

---

[25] In standards for developing a SUMP Action Plan, main principles and case studies are available [70].

Table 8.12 Example of innovative financing mechanisms, partnerships and procurement procedures for cooperative on-road systems (*Source* [SUITS Capacity Building Toolbox])

Category	Title	Key elements
Financing mechanism	Collaborating with other cities, research consortia and private companies	Cities provide specific data while at the same time they offer demo and pilot sites, while they also provide support to other partners. These projects offer to cities benefits from investments into its infrastructure and capacity building programmes along with the benefits derived from pilot projects while at the same time additional funding may be available
Partnership	Innovative public private and research and development partnership	Formulation of a partnership between LAs, universities, companies which makes use of each partner's expertise. Efforts require political will in order to eliminate constraints and willingness to participate and create a learning network which will eventually enhance innovation and applied research throughout the city
Procurement approach	Functional performance specifications	This substantiates the item to be procured in relation to the problem to be solved. This is at the very heart of the procurement procedure. It is also the point of reference and central basis of the bidder's tender calculation

Example of implementation activities that could be considered for mobility measures implementation.

- Audit of mobility needs
- Data identification and collection.
- Risk analysis and (barriers and solutions).
- Construction study/installation study.
- Study approval
- Construction/installation
- Promotion of the measure/marketing
- Monitoring implementation activities
- Evaluating measure impact in relation with measures goals objectives.
- Contribution to SUMP goals and objectives and those of the region

One key implementation activity, independently of the type of the measure, is the identification and the collection of the required data sets. Data are needed for the designing of measures' technical parameters (and data gathering can be part of the construction/installation study), for monitoring the implementation procedure and impact evaluation. The early identification of indicators for monitoring and evaluation helps in collecting the "right" data and performing baseline assessment while optimising data collection and data management. The most suitable data collection and data selection methods[26] can then be defined, by taking also into consideration the available resources (human, financial, hardware/equipment, etc.).

The initial screening of weaknesses and wider problems of the current mobility system provides orientation of what quantitative data need to be collected to justify the weaknesses and clarify the extent of the problem. Data gathering and baseline assessment take place at the very early stage of implementation in order to specify with accuracy implementation aspects and quantify implementation objectives. Then, it is possible to measure implementation.

Indicators selection determines whether the evaluation and monitoring procedure will be fruitful or not. The key performance indicators (KPIs) which are available in the literature, enable the reporting of both the degree and impact of measures implementation, depending on related goals and objectives.

Selection criteria for transport key performance indicators (proposed by Litman and Victoria Transport Policy Institute in 2007 [73]):

- **Comprehensiveness.** Indicators to reflect various economic, social and environmental impacts and various transport activities (i.e. personal and freight transport).
- **Data quality.** High standards for collection methods are needed to ensure that information is accurate and consistent.
- **Comparability.** Standardisation of data collection can make results suitable for comparison between different jurisdictions, times and groups.
- **Easiness to understand.** Decision makers and the general public have to be able to understand them.
- **Accessibility and transparency.** Primary data, analysis details and indicators results should be available to all stakeholders.
- **Cost effectiveness** to collect and estimate the indicators. The decision-making value of the indicators must outweigh the cost of collecting them.

---

[26] Examples of data requirements for each type of measure are included in chapter "Processes and Implementation aspects" of the relative workbook found in SUITS toolbox https://cbt.suits-project.sboing.net/.

- **Net effects**. Differentiation between total impacts and shifts of impacts to different locations and times is needed.
- **Performance targets**. Indicators should be suitable for establishing usable performance targets.

Identifying and checking potential solutions for potential difficulties/barriers per measures category could be considered as part of risk analysis and planning. During the identification of beneficiaries, challenges arising from stakeholders' reactions and possible responses have been discussed. However, challenges might also be linked with the particularities of the site of implementation, the legal framework, the technical parameters of the measures, the financing mechanism that has been selected, etc. To be prepared, it is recommended to examine other cities' case studies (preferably from the same country), and contact the responsible authority, if needed, exchanging knowledge and experiences. Furthermore, it is required to analyse the local legal, economic and technical framework and check for solutions, before a problem occurs.

Advancing implementation is the main objective of risk analysis. But the identification of potential barriers or difficulties may result in quitting the idea of the implementation, especially from public administration staff and policy makers. In contrast, LAs staff and in general implementation authorities, need to focus on solutions and take advantage of this proactive approach to adapt procurement criteria accordingly and coordinate their actions with other authorities to overcome legal, financial and societal barriers that may delay implementation or risk its success. Effective marketing, in combination with alliances creation, qualified personnel and sound financing, is a basic condition for success [4] and a useful tool to overcome different kinds of barriers.

8.10   Using Available Tools and Guidelines

Guidelines and tools to support the design and implementation of such measures are numerous (see Table 15 for a selection of the most relevant for the specific modules). Besides SUMP guidelines that include generic suggestions (available in all EU languages by ELTIS),[27] there are guidelines and tools depending on the

_____

[27] https://www.eltis.org/search/site/sump%20guidelines.

measure type developed relating to several EU projects. The EU platform which integrates most of EU project results (tools, guidelines, case studies, reports, etc.) is the CIVITAS learning centre [37]. National guidelines or tools tailored to your specific country, if available, should be prioritised (Table 8.13).

Table 8.13 Selection of the most relevant available online tools supporting the implementation of measures (*Source* [SUITS Capacity Building Toolbox])

Mobility measures categories	Tool name
Emerging transport technologies	ELAN https://civitas.eu/content/elan
	ELIPTIC—electrification of public transport in cities http://www.eliptic-project.eu/results https://www.mobility-academy.eu/course/view.php?id=74
	SMARTSET—efficient urban freight transport http://smartset-project.eu/downloads
	CODECS—cities requirements for C-ITS https://civitas.eu/tool-inventory/codecs-cities-requirements-c-its
Innovative transport schemes	CIVITAS ECCENTRIC tool: MaaS readiness level indicators for local authorities http://civitas.eu/news/maas-readiness-level-indicators-local-authorities-launched
	MOMO car-sharing https://ec.europa.eu/energy/intelligent/projects/en/projects/momo-car-sharing
	CHUMS carpooling http://chums-carpooling.eu
Safety and security inclusive measures	MIMOSA (i) road safety for pedestrians and cyclists and (ii) safe and friendly transport infrastructure http://civitas.eu/content/mimosa
	SafetyCube—road safety decision support system (DSS) https://www.roadsafety-dss.eu/#/
	NODES—security and safety management at interchanges https://nodes-toolbox.eu
Urban freight measures	NOVELOG—freight distribution and service trips http://evalog.civ.uth.gr/Default.aspx
	STRAIGHTSOL—strategies and measures for smarter urban freight solutions http://www.straightsol.eu/results.htm
	Freight TAILS—logistics http://urbact.eu/freight-tails

Oversupply of tools, guidelines and literature is not easily manageable by organisations' staff. On the contrary, it might discourage them from using the available material, since it requires a minimum review to identify the most useful and practical tool for the circumstance. In this respect, for EU cities' staff, it is suggested to start the review from EU integrated inventory platforms such as [37], where the use of filters facilitates the search. This embeds material mainly addressed to EU cities that share similar living conditions, lifestyles and mobility issues. After all, EU projects take into consideration EU goals and objectives which are reflected in the tools and guidelines provided.

8.11   Conclusions

The successful implementation of sustainable mobility measure is highly dependent on factors such as the ability to involve and convince stakeholders at an early stage and to overcome managerial and technical barriers. The integrated analysis of mobility problems and their consequences on the three pillars of sustainability (society, economy and environment) are the foundation for any decision-making. Then, the familiarity with the technical parameters of the available mobility solutions and the understanding of their value and drawbacks is very critical. In this procedure of selecting, planning and implementing sustainable mobility measures, small-medium local authorities (and not only) have at their disposal tools and mechanisms to ensure success, such as the EU regulations and documentation, the available case studies, the partnerships and the rich capacity building material which is available by EU initiatives and projects.

In this context, this chapter focused on innovative sustainable solutions under the spectrum of understanding their benefits, the actors/stakeholders need to cooperate with, the required processes and resources, the financing and procurement opportunities, the relevance of local strategy with the European one.

References

1. Chorus, C.G., Molin, E.J.E., Van Wee, B.: Use and effects of advanced traveller information services (ATIS): a review of the literature. Transp. Rev. **26**(2), 127–149 (2006). https://doi.org/10.1080/01441640500333677
2. Bengler, K., Dietmayer, K., Farber, B., Maurer, M., Stiller, C., Winner, H.: Three decades of driver assistance systems: review and future perspectives. IEEE Intell. Transp. Syst. Mag. **6**(4), 6–22 (2014). https://doi.org/10.1109/MITS.2014.2336271
3. Yan, G., Yang, W., Rawat, D.B., Olariu, S.: SmartParking: a secure and intelligent parking system. IEEE Intell. Transp. Syst. Mag. **3**(1), 18–30 (2011). https://doi.org/10.1109/MITS.2011.940473

4. Festag, A.: Cooperative intelligent transport systems standards in Europe. IEEE Commun. Mag. **52**(12), 166–172 (2014). https://doi.org/10.1109/MCOM.2014.6979970
5. Bergenheim, C., Shladover, S., Coelingh, E.: Overview of platooning systems. Proceedings of the 19th ITS World Congress, Oct 22–26, Vienna, Austria (2012)
6. Faisal, A., Yigitcanlar, T., Kamruzzaman, Md., Graham Currie, G.: Understanding autonomous vehicles: a systematic literature review on capability, impact, planning and policy. J. Transp. Land Use **12**(1), 45–72 (2019)
7. Cozens, P.M., Neale, R.H., Hillier, D., Whitaker, J.: Tackling crime and fear of crime whilst waiting at Britain's railway stations. J. Public Transp. **7**(3), 23–41 (2004)
8. Smith, M.J., Clarke, R.V.: Crime and public transport. In Tonry, M. (ed.) Crime and Justice: A Review of Research, 27. https://doi.org/10.1086/652200 (2000)
9. Nordfjærn, T., Şimşekoğlu, O., Lind, H.B., Jørgensen, S.H., Rundmo, T.: Transport priorities, risk perception and worry associated with mode use and preferences among Norwegian commuters. Accid. Anal. Prev. **72**, 391–400 (2014)
10. Rupprecht Consult (editor).: Guidelines for Developing and Implementing a Sustainable Urban Mobility Plan. 2nd edn. SUMPs-Up Horizon 2020 Project (2019)
11. Heerkens, G.R.: Project Management. Mcgraw-Hill Education (2015)
12. European Commission.: Mobility and transport: state of play of internalisation in the european transport sector, Report (2019). https://ec.europa.eu/transport/sites/transport/files/studies/internalisation-state-of-play-isbn-978-92-76-01413-3.pdf (2019). Accessed 01 June 2020
13. EU transport policy.: https://europa.eu/european-union/topics/transport_en. Accessed 01 June 2020
14. European Commission.: Mobility and transport: transport in the European Union; Current trends and issues, Report. https://ec.europa.eu/transport/sites/transport/files/2018-transport-in-the-eu-current-trends-and-issues.pdf. Accessed 01 June 2020
15. Mindell, J.S., Paulo, R., Anciaes, P.R.: Transport and community severance. In: Advances in Transportation and Health, 175–196 (2020)
16. Pojani, E., Boussauw, K., Pojani, D.: Reexamining transport poverty, job access, and gender issues in Central and Eastern Europe. Gend. Place Cult. **24**(9), 1323–1345 (2017). https://doi.org/10.1080/0966369X.2017.1372382
17. Jones, P., Karen Lucas, K.: The social consequences of transport decision-making: clarifying concepts, synthesising knowledge and assessing implications. J. Transp. Geogr. **21**, 4–16 (2012)
18. Ryan, J., Wretstrand, A., Schmidt, S.M.: Disparities in mobility among older people: findings from a capability-based travel survey. Transp. Policy **79**, 177–192 (2019)
19. Moodley, J., Graham, L.: The importance of intersectionality in disability and gender studies. Agenda **29**(2), 24–33 (2015). https://doi.org/10.1080/10130950.2015.1041802
20. European Commission.: Mobility and transport: handbook on the external costs of transport, Report (2019). https://ec.europa.eu/transport/sites/transport/files/studies/internalisation-handbook-isbn-978-92-79-96917-1.pdf (2019). Accessed 01 June 2020
21. European Environmental Agency, Transport and public health, Online article.: https://www.eea.europa.eu/signals/signals-2016/articles/transport-and-public-health (2016). Accessed 01 June 2020
22. Bethlehem, J.R., Mackenbach, J.D., Ben-Rebah, M., et al.: The SPOTLIGHT virtual audit tool: a valid and reliable tool to assess obesogenic characteristics of the built environment. Int J Health Geogr **13**, 52 (2014). https://doi.org/10.1186/1476-072X-13-52
23. Shoup, D.C.: The high cost of free parking. Planners Press, American Planning Association, Chicago, U.S.A (2011)
24. Lu, M., Türetken, O., Mitsakis, E., Blokpoel, R., Gilsing, R.A.M., Grefen, P.W.P.J., and Kotsi, A.: Cooperative and connected intelligent transport systems for sustainable European road transport. Transport Research Arena 2018, Vienna, Austria (2018)
25. Aurambout, J.P., Gkoumas, K., Ciuffo, B.: Last mile delivery by drones: an estimation of viable market potential and access to citizens across European cities. Eur. Transp. Res. Rev. **11**, 30 (2019). https://doi.org/10.1186/s12544-019-0368-2

26. Hauw, N.: E-mobility: from strategy to legislation. Civitas site (2016), https://civitas. eu/sites/default/files/civitas_insight_13_e-mobility_from_strategy_to_legislation.pdf (2016). Accessed 26 March 2019

27. European Commission.: Mobility and transport: clean vehicles directive =-mobility and transport. https://ec.europa.eu/transport/themes/urban/vehicles/directive_en (2019). Accessed 26 Mar 2019

28. Eltis.org. (2019) Traffic lights 'talk' with cars in North Holland pilot | Eltis. http://www.eltis. org/discover/news/traffic-lights-talk-cars-north-holland-pilot (2019). Accessed 26 Mar 2019

29. Jonkers, E., Gorris, T.: Intelligent transport systems and traffic management in urban areas. CIVITAS: http://www.eltis.org/sites/default/files/trainingmaterials/civ_pol-not6_its_web.pdf, (2015). Accessed 26 Mar 2019

30. Kala R.: On-road intelligent vehicles. Motion planning for intelligent transportation systems, pp 59–82 (2016)

31. Martin, J., Shchuryk, O.: Course syllabus topic study 2: ITS and C-ITS user services, CAPITAL Consortium. https://www.its-elearning.eu/assets/courseware/v1/2c2e947faa78211 4878b86c0231c5145/asset-v1:Capital+ITS1+test+type@asset+block/CAPITAL_WP3_ITS1. pdf. Accessed 26 Mar 2019

32. Partnership Talking Traffic. https://www.talking-traffic.com/en/. Accessed 01 July 2020

33. Figg, H.: New high-tech pedestrian crossing design to improve safety. Walking and cycling. South Korea, http://www.eltis.org/discover/news/new-high-tech-pedestrian-crossing-design-improve-safety (2019). Accessed 26 Mar 2019

34. Hayes, S.: Accessible pedestrian crossing solution (Innovative pedestrian traffic signal crossing device). Barcelona City Council (2015), http://www.transform-europe.eu/wp-content/uploads/ 2015/09/D3.2-Barcelona-City-Council-Case-Study-draft-final.pdf (2015). Accessed 26 Mar 2019

35. Perallos, A., Hernandez-Jayo, U., Onieva, E., Garcia-Zuazola, I.: Intelligent transport systems: technologies and applications. Wiley, UK (2016)

36. Polis Smart Parking Systems-A Division of Intercomp S.p.A. Management software.: http:// smartparkingsystems.com/en/management-software-polis/ (2019). Accessed 26 Mar 2019

37. CIVITAS learning centre.: https://civitas.eu/learning-centre. Accessed 18 Mar 2020

38. CAPITAL Online Training Platform.: https://www.its-elearning.eu/courses. Accessed 01 June 2020

39. Shaheen, S., Cohen, A., Yelchuru, B., Sarkhili, S.: Mobility on demand operational report. U.S. Department of Transportation, https://rosap.ntl.bts.gov/view/dot/34258 (2017). Accessed 05 July 2020

40. European Commission.: Study on passenger transport by taxi, hire car with driver and ridesharing in the EU. https://ec.europa.eu/transport/sites/transport/files/2016-09-26-pax-tra nsport-taxi-hirecar-w-driver-ridesharing-final-report.pdf (2016). Accessed 11 July 2020

41. Nechita, E., Crişan, G.C., Obreja, S.M., Damian, C.S.: Intelligent carpooling system. In: Nakamatsu K., Kountchev R. (eds) New Approaches in Intelligent Control. Intelligent Systems Reference Library, vol 107. Springer, Cham. https://doi.org/10.1007/978-3-319-32168-4_2 (2016)

42. Ridesharing Carpooling and Vanpooling. TDM encyclopedia. Victoria Transport Policy Institute (2018)

43. CAPITAL CIVITAS project.: Bike-sharing as a link to desired destinations. https://civitas. eu/sites/default/files/civitas_insight_10_bike-sharing_as_a_link_to_desired_destinations.pdf (2016). Accessed 11 July 2020

44. Mobility as a Service Alliance Homepage. https://maas-alliance.eu. Accessed 11 July 2020

45. Woodcock, A., Spundflasch, S., Fadden-Hopper, K., Miller-Crolla, K., Rudolph, F., Glensor, K.: MaaS implementation: local Authorities' perspectives, CIVITAS, Brussels, Downloadable from https://civitas.eu/sites/default/files/civitas_forum19_session_7_andree_woodcock. pdf.pdf (2017)

46. Urry, J.: The 'System' of automobility. Theory Cult. Soc. **21**(4–5), 25–39 (2004). https://doi. org/10.1177/0263276404046059

47. Kent, J.L., Dowling, R.: Puncturing automobility? Carsharing practices. J. Transp. Geogr. **32**, 86–92 (2013)
48. Hoffmann, S., Weyer, J., Longen, J.: Discontinuation of the automobility regime? An integrated approach to multi-level governance. Transp. Res. Part A Policy Pract. 103, 391–408 (2017)
49. World Bank.: Cities on the move: a world bank urban transport strategy review. International Bank for Reconstruction and Development. http://documents1.worldbank.org/curated/en/928 301468762905413/pdf/Cities-on-the-Move-A-World-Bank-Urban-Transport-Strategy-Rev iew.pdf (2002). Accessed 19 July 2020
50. Glensor, K.: Development of an index of transport-user vulnerability, and its application in Enschede, The Netherlands. Sustainab Open Access J **10**(7), 2388 (2018)
51. CIVITAS EU. Awareness raising campaign for sustainable mobility. https://civitas.eu/measure/ awareness-raising-campaigns-sustainable-mobility (2013). Accessed 27 Mar 2019
52. CIVITAS EU.: Safety and security for seniors and public transport passengers. https://civitas. eu/measure/safety-and-security-seniors-and-public-transport-passengers (2019). Accessed 27 Mar 2019
53. How Public Transport is using Technology.: Digital J. http://www.digitaljournal.com/tech-and-science/technology/how-public-transport-is-using-digital-technology/article/503467 (2017). Accessed 28 Mar 2019
54. Dinh-Zarr, T.B.: Pedestrian and bicycle safety overview. *10th University Transportation Centers Spotlight Conference Proceedings*, pp. 2–5, Transportation Research Board, Washington. https://onlinepubs.trb.org/onlinepubs/conf/cpw21.pdf (2016)
55. European Commission.: Speed and speed management. https://ec.europa.eu/transport/road_s afety/sites/roadsafety/files/ersosynthesis2015-speedspeedmanagement25_en.pdf (2015). Accessed 28 Mar 2019
56. Security magazine.: Security on mass transit to improve service and reduce crime. https:// www.securitymagazine.com/articles/86338-security-on-mass-transit-to-improve-service-and-reduce-crime (2015)
57. Osmond, J., Woodcock, A.: Traveller groups and public transport. In: Tovey, M. Woodcock, A., Osmond, J. (eds.) Designing Mobility and Transport Services Developing Traveller Experience Tools, Routledge (2020)
58. Cozens, P., Neale, R., Whitaker, J., Hillier, D.: Managing crime and the fear of crime at railway stations—a case study in South Wales (UK). Int. J. Transp. Manag. **1**(3), 121–132 (2003)
59. DG MOVE European Commission.: Study on Urban Freight Transport, MDS Transmodal (2012)
60. CIVITAS Policy Note.: Smart choices for cities: making urban freight logistics more sustainable. https://civitas.eu/sites/default/files/civ_pol-an5_urban_web.pdf (2015). Accessed 11 July 2020
61. UPS.: The road to sustainable urban logistics (2017)
62. Landmark Global.: The development of parcel lockers in Europe. https://landmarkglobal. com/en_CA/trends-insights/the-development-of-parcel-lockers-in-europe/ (2014). Accessed 11 July 2020
63. Intelligent Lockers for Parcel Delivery: The last mile. American Locker.: https://americ anlocker.com/electronic-parcel-lockers-for-parcel-delivery-the-last-mile/ (2017). Accessed 28 Mar 2019
64. CIVITAS.: Guide for the urban transport professional; results and lessons of long term evaluation of The CIVITAS initiative. https://civitas.eu/sites/default/files/civitas_guide_for_the_ urban_transport_professional.pdf (2019). Accessed 17 Mar 2019
65. CIVITAS.: Policy Note. Intelligent transport systems and traffic management in urban areas, CIVITAS WIKI consortium. https://www.eltis.org/sites/default/files/trainingmaterials/ civ_pol-not6_its_web.pdf (2015). Accessed 17 Mar 2019
66. European Commission.: COM (2016) 501, A European strategy for low-emission mobility. https://eur-lex.europa.eu/resource.html?uri=cellar:e44d3c21-531e-11e6-89bd-01aa75ed71a1. 0002.02/DOC_1&format=PDF (2016). Accessed 07 Mar 2020

67. Rupprecht Consult (editor).: Guidelines for developing and implementing a sustainable urban mobility plan. 1st edn. ELTIS (2016)
68. Ferretti, V.: From stakeholders analysis to cognitive mapping and multi attribute value theory: an integrated approach for policy support. Eur. J. Oper. Res. **253**(2), 524–541 (2016)
69. SUITS CBP.: Guidelines to new business models, bankable projects and innovative partnerships, EUROKLEIS, Italy (Summarised in Chapter 13, this volume) (2019)
70. Mattsson, H.: Standards for developing a SUMP Action Plan. CIVITAS SUMPs-UP Project (2018)
71. SUITS CBP.: Guidelines to innovative financing, ARCADIS, U.K. (Summarised in Chapter 12, this volume) (2019)
72. SUITS CBP.: Guidelines to innovative procurement. Integral Consulting R&D (INTECO), Romania (Summarised in Chapter 11, this volume) (2019)
73. Litman, T.: Well measured: developing indicators for comprehensive and sustainable transport planning. Victoria Transport Policy Institute, Canada (2007)

Chapter 9
Integrating Transport Programmes for Sustainable Reduction in Urban Road Congestion—Best Practise Examples from Local Authorities Working with SUITS

Janet Saunders, Sunil Budhdeo, Keelan Fadden-Hopper, Ann-Marie Nienaber, Sebastian Spundflasch, Marco Surace, Fabio Nussio, Miriam Pirra, Angel Navarro, Ieva Girdvainienė, Tudor Drambarean, and Krinos Ioannis

Abstract This chapter highlights selected sustainable transport measures that were implemented in nine partner cities during the SUITS project and discusses outcomes and learnings. Mutual exchanges, either in workshops or focussed meetings between the cities, provided inspiration for planning and implementing sustainable transport

J. Saunders (✉) · A.-M. Nienaber
Coventry University, Coventry, UK
e-mail: janetsaunders90@gmail.com

S. Budhdeo
Coventry City Council, Coventry, UK

K. Fadden-Hopper
West Midlands Combined Authority, Birmingham, UK

S. Spundflasch
TUIL, Ilmenau, Germany

M. Surace · F. Nussio
Roma Servizi Per La Mobilità S.R.L, Rome, Italy

M. Pirra
DIATI - Department of Environment, Land and Infrastructure Engineering, Politecnico di Torino, Torino, Italy

A. Navarro
Las Naves, Valencia, Spain

I. Girdvainienė
Smart Continent LT UAB, Palanga, Lithuania

T. Drambarean
Alba Iulia Municipality, Alba Iulia, Romania

K. Ioannis
Municipality of Kalamaria, Kalamaria, Greece

© Transport for West Midlands 2023
A. Woodcock et al. (eds.), *Capacity Building in Local Authorities for Sustainable Transport Planning*, Smart Innovation, Systems and Technologies 319,
https://doi.org/10.1007/978-981-19-6962-1_9

measures. For the project team, such exchanges provided insights into challenges the cities faced on a day-to-day basis which were used to develop training material and further understand the organisational process. During the project, cities worked on transport measures in five broad topic areas: clean fuels and low emission vehicles, collective passenger transport, vulnerable road users, safety and security and intelligent transport systems and services. Here, we present a selection to illustrate the scope of the undertaking and, how the project was able to enhance the implementation of each initiative.

9.1   Introduction

During the SUITS project, the partner cities Alba Iulia, Kalamaria, Rome, Turin, Valencia, Coventry and the West Midlands Combined Authority and the 'follower' cities Palanga, Stuttgart and Dachau worked on a variety of different transport measures. Within the project, we used the term 'transport measures' to refer to any initiatives undertaken by the city partners affecting all forms of transport and urban mobility, from public transport to private transport and active travel such as walking and cycling, along with programmes to reduce congestion and traffic emissions.

In following the cities, we have been able to understand the challenges they face and have offered concrete support to city teams. Transport and mobility planning is a complex task and of course the cities have a very good understanding of the necessary procedures. However, it is a major concern of the EU that cities, and especially small-to medium-sized cities, should place an even greater focus on aspects of sustainability when developing future transport programmes. This chapter outlines best practise examples that show how the SUITS partner cities implemented plans and strategies towards the goal of more sustainable mobility. Some of these were initiated during SUITS, while others had already started. Case studies of best practise were brought as 'shared experience', with local authority delegates collaborating, with SUITS guidance, to help their cities develop into 'learning organisations'. The strategic transport measures which the cities worked on can be grouped under the following themes, each reflecting important aspects of sustainable mobility:

(1)   *Clean fuels and low emission vehicles.* Clean air is a basic requirement for human health and wellbeing. Studies by the World Health Organisation 2016 [1], have shown that ambient air pollution contributes to 7.6% of all deaths. Road traffic affects citizens' quality of life as well as their health. According to the EEA Environmental Indicator Report 2018, key EU air quality standards for the protection of human health were not being met in large parts of the EU, mainly attributed to emissions from road traffic and residential combustion in urban areas [2]. Therefore, local authorities are faced with challenging agendas in their regions to achieve these targets and to demonstrate their engagement to support citizens' health and well-being. The best practise examples we want to outline in this chapter are the implementation of urban goods freight distribution

with clean vehicles as shown in Turin, the clean air programme in Stuttgart and the provision of public electric charging points and electric vehicle sharing in Palanga.

(2) *Collective passenger transport.* This area focusses on the continuous development of a modern and energy-efficient public transport system as a driver for reducing car traffic and thus decreasing traffic congestion and air pollution. SUITS worked with the cities on innovative ways to maximise the potential for local public transport through an accessible service that is a fast and convenient alternative to the use of a private car. Best practise examples include Stuttgart's ambitious goals, with a suite of measures started before SUITS in 2015, shared with SUITS partners during workshops and continued during the SUITS project, the implementation of traffic light priority by Rome and the new plans developed by Turin on intermodality around new rail interchanges, including car-sharing services, shared taxi apps, bike-sharing schemes and e-scooter sharing. In addition, we show how Alba Iulia utilised innovative procurement and financing methods to purchase electric buses with numerous features to encourage wider use and to facilitate other sustainability goals for the city.

(3) *Vulnerable Road Users.* In the global assessment of road safety, the WHO 2018 demonstrated that just over half of the estimated 1.35 million fatalities occurring each year on the world's roads concern vulnerable road users, (pedestrians, cyclists and motor cyclists). Road traffic injuries are the leading cause of death for children and young adults [3]. Vulnerable road users are defined in the European Union Intelligent Transport Systems Directive as 'non-motorised road users, such as pedestrians and cyclists as well as motorcyclists and persons with disabilities or reduced mobility and orientation' [4].

The SUMP Guidelines quote statistics for 2017, stating that 70% of those killed on urban roads in the EU were vulnerable road users—39% pedestrians, 12% cyclists and 19% motor cyclists [5, p. 14]. Pedestrians, pedal cyclists, and motorcyclists are all considered as vulnerable since they benefit from little or no external protective devices that would absorb energy in a collision. In SUMPs, vulnerable road users intersect with 'vulnerable groups', such as children, disabled people and older people whose needs must be considered when planning and developing safe and sustainable transport systems [4, 5]. Best practise examples to enhance the safety of vulnerable road users are provided by the city of Rome in establishing road safety-speed zones and a safety campaign aimed at school students (DesiRe training initiative) and from Coventry (West Midlands, UK), implementing a neighbourhood speed safety zones awareness scheme.

(4) *Safety and Security.* Enhancing safety and security in all transport modes is a key objective of the European Commission. In the 2011 "Roadmap to a Single European Transport Area—Towards a competitive and resource-efficient transport system", the European Commission stated that by 2050, Europe should become a world leader in safety and security of transport in all transport modes [6]. In thinking about safety and security, our cities focussed on the safety of

pedestrians and cyclists. We highlight this key area by providing best practise examples from Rome, becoming a more walkable and cycle-able city; from Valencia, promoting a bike culture and thus developing a cycling infrastructure; and Dachau, using innovative citizen participation methods to develop its mobility vision which includes more cycling and improved safety and security.

(5) *Intelligent Transport Systems and Services.* Intelligent transport systems and Services (ITS), also called smart mobility, include the integration of smarter information and communications technology with transport infrastructure, vehicles and users. By enabling the sharing of vital information on roads, supply chains and public transport services, ITS allow for a more efficient use of the transport network. In addition, safety can be increased, and the environmental impact can be reduced. We will highlight the benefits of intelligent transport systems and services with examples from Kalamaria, implementing an intelligent waste collection system and using a real-time traffic information system to optimise vehicle travel time, and from the West Midlands, developing a testbed for connected and autonomous vehicles on public roads. Additionally, we also highlight an example from Coventry featuring implementation of an intelligent variable message system (iVMS) with the trial of an interactive App to improve traffic flow.

9.2   The Cities: Background Information

While we focus on transport measures that have the potential to be adopted by other small- and medium sized cities, it is important to always be aware of the specific needs of each city to allow measures to be implemented successfully (e.g., macro-environment such as number of cars, inhabitants, region; and micro-environment, such as the size of the local authority and its traditions, structures and processes). This has a major influence on which measures are implemented to ensure a sustainable reduction of congestion and how this is done. There is no magic formula for solving mobility problems; measures must be tailored to local circumstances to be effective. In the following, we revisit the earlier description of the SUITS cities (see also Chap. 4) focussing this time on the challenges they wished to overcome.

9.2.1   Alba Iulia, Romania

Alba Iulia is a city of 74,000 inhabitants and capital of Alba County in the Central Region of Romania. Public transport in Alba Iulia is considered one of most efficient in the country, but at the beginning of the SUITS project, transport in the city was mostly car-centred, with accompanying high pollution, and problems with parking and traffic management. This presented major challenges to Alba Iulia Municipality regarding the achievement of sustainable mobility. Moreover, the different departments at municipal level found it challenging to cooperate efficiently in the area of mobility and lacked a clear vision on how to improve the situation.

However, the municipality has been highly active in attracting EU funds. In the period 2007–2019 over 200 million Euros from EU funding were used for local development and overall, the municipality represents a good practise at national level in terms of projects implemented through non-reimbursable funds.

We outline the city's use of innovative procurement and finance to support enhancements to its public transport fleet and other sustainability measures.

9.2.2 Dachau, Germany

Dachau is a small German city, with 50,000 inhabitants located in the metropolitan region of Munich. The car is the number one form of transport in Munich's surrounding region. During peak traffic hours, Dachau struggles with traffic jams, especially because a busy main road runs right through the middle of the city. Dachau also wants to strengthen cycling and introduced a new cycling concept in 2019, within which 10% of Dachau's citizens are expected to switch to the bicycle in the foreseeable future. In recent years, Dachau has placed great emphasis on citizen participation. Improving the quality of life is at the top of the agenda. Dachau was a 'follower' city in the SUITS project. We describe how a vision and guidelines for the future development of Dachau were drawn up in cooperation with the citizens.

9.2.3 Kalamaria, Greece

Kalamaria, with a population of 92,000 (2019), is one of the largest municipalities in Greece's second largest city, Thessaloniki. As a residential and recreational area for Thessaloniki, Kalamaria is facing severe environmental pressure due to rapid urbanisation [7]. The municipality receives transit traffic from Thessaloniki's centre, airport and the ring road, and some main roads have heavy traffic throughout the day. Internal targets are aimed at improving safety levels for citizens, decreasing accidents, increasing citizens' awareness and acceptance of sustainable mobility measures such as new pedestrian footpaths and bike lanes and also a focus on use of innovative technologies such as renewable energy. We describe Kalamaria's introduction of innovative pedestrian crossings with solar power and the city's trial and adoption of an innovative technology traffic management system.

9.2.4 Palanga, Lithuania

Palanga is a seaside resort town in western Lithuania, on the shore of the Baltic Sea with a population of just over 15,000 people (but during the summer the number of inhabitants including tourists exceeds 120,000). The most popular mode of travel is by private car. Because Palanga is a resort city, synergy with nature is considered particularly important and the city aims to ensure 'city comfort' not only for the resort visitors but for the locals as well. The city's SUMP was approved in 2017.

Palanga is a 'follower' city of the SUITS project. We will describe how it has focussed its activity on deploying car sharing points, and developing an electric vehicle charging network, while increasing the local population's awareness about sustainable mobility.

9.2.5 Rome, Italy

Rome has 2,873,000 residents. In 2015 the city approved the Traffic Masterplan for Rome which formulated a new understanding of mobility. In common with many Italian cities, Rome has implemented restricted driving zones where only freight and passenger vehicles with permits can enter at certain times, known as 'Zona a Traffico Limitato' or limited traffic zones (LTZ).[1]

Rome was facing challenges regarding road safety and a need to increase sustainable mobility in general, including an increase in active travel modes. Building trust with citizens and including them in the planning, played a key role in the city's approach to preparing its SUMP. In late 2017, a public consultation campaign via a web portal was launched and enjoyed lively participation. This was part of a concerted effort to bring about a cultural change in the city to raise awareness of mobility, discourage a car-centred attitude and become a cycle-able and walkable city. In this chapter, we will discuss Rome's transport measures that focussed on improving public transport to increase its usage, extending the bicycle lane network, promoting bike sharing and improving bicycle parking.

9.2.6 Stuttgart, Germany

Stuttgart is the state capital of the federal state of Baden-Wurttemberg. While the city has only 620,000 inhabitants, it is the centre of a metropolitan region with more than 5.3 million inhabitants. As part of an important economic zone within the European Union and located in one of the densest conurbations in Germany, Stuttgart has to cope with a very high volume of traffic. Every day, around 800,000 cars enter and leave the city and are major contributors of traffic congestion and air pollution.

---

[1] These are variously referred to as ZTL, or LTZ, in this chapter we will use LTZ.

Due to its location in a basin, Stuttgart suffers from particularly severe urban climatic problems, with low rate of exchange of air in the valley, relatively high average annual temperatures, and low precipitation. This can lead to strong inversion weather conditions in which the air pollution generated by industry, households and traffic is concentrated in the city for a particularly long time.

Stuttgart was a 'follower' city in the SUITS project, and we describe how it is implementing a programme of integrated transport measures to bring about change in travel behaviour.

9.2.7 Turin, Italy

Turin is a city of about 887,000 inhabitants located in the north-western part of Italy and surrounded by a metropolitan area of about 2,000,000 inhabitants. Turin adopted its SUMP in 2010 and through its participation in several European and National projects, it has become a leading city in Italy for sustainable urban logistics. The PUMAS project (started 2013) consisted of a pilot for the delivery of goods in the limited traffic zone (LTZ) in the central area of the city. The results provided an impetus for the collection of traffic data, enabling a holistic picture of goods delivery in urban areas for the first time in Italy.

It also led to the signature of a Memorandum of Understanding between the city, the Chamber of Commerce, Transport and Commerce Associations, and the creation of a Freight Quality Partnership, which included the definition of incentives or so-called 'pull' measures in an innovative approach.

Turin continued those activities through the project H2020 NOVELOG, which started in 2015 and focussed on gaining knowledge about freight distribution and service trips for implementing effective and sustainable policies and measures. The involvement of the city of Turin on these themes continues through active participation in the ongoing projects SOLEZ, SUMPS-UP and SUITS. We outline how Turin has implemented more sustainable freight delivery with incentives for acquisition of clean vehicles, and how it plans for intermodality and sharing schemes around its new rail interchanges.

9.2.8 Valencia, Spain

The city of Valencia is the third largest city in Spain, with a population of around 787,000 inhabitants in the municipality. Valencia has experienced similar issues to many cities in its wider urban area: private car dependency, poor public space usage, low use of cycling and lack of bike lane connectivity, high vehicle speeds, air pollution and traffic accident casualties. This was despite a state-of-the-art traffic control centre and a good intermodality between walking and public transport. Valencia adopted its SUMP in 2013, but it was not until 2015 that changes in local government provided the political endorsement needed to promote a sustainable mobility approach. We feature Valencia's work with the SUITS project to promote a bike culture, improve cycling infrastructure and implement a 30 km/h speed zone within the city centre.

9.2.9   Coventry and the West Midlands Region, UK

Coventry, with a growing population of approximately 363,000 is a small-to-medium sized city and a regional and industrial centre in the UK's West Midlands region, with a very strong link to the automotive sector. Across the region, Coventry is considered as the leader in transport innovation, with extensive connections with the motor industry and a history of innovative project collaborations. Nonetheless, it is similar to many other cities of its size, with a heavy dependence on commuting by car. This may be reflected in its modal share of commuter transport, with 77% of morning peak trips being by car and 22.7% by public transport (2015 figures, [8]).

Coventry has been involved in the SUITS project since its inception, and here, we share two examples of best practise from the city in speed safety zones, and an intelligent variable message system (iVMS).

The West Midlands region encompasses three cities (Birmingham, Coventry and Wolverhampton) together with post-industrial towns and suburban areas, and 2.8 million residents. The area is characterised by the diversity of both its population and urban landscape. In line with many urban regions across the world, the region faces a number of transport challenges: poor air quality (the health effects of which are increasingly well known), congestion (which causes delays to driving and bus journeys) and safety (where the need to reduce harms resulting from use of the transport system are recognised).

The West Midlands Combined Authority (WMCA) was created in 2016 as a partnership between the seven local authorities in the West Midlands region, including Coventry City Council, and has worked to bring powers from central government to regional level. Transport for West Midlands (TfWM) is part of WMCA and took over the responsibilities of the transport authority covering the West Midlands region. As TfWM developed, it was well positioned to commence new projects, building on the experience of its member Coventry City Council in the field of connected and autonomous vehicles (CAV), and to bid for projects to stay at the forefront of development in CAV technology. The Midlands Future Mobility CAV Testbed project is described as a best practise example in the next section.

9.3 Best Practise Examples

9.3.1 Clean Fuels and Low Emission Vehicles (Turin, Stuttgart and Palanga)

Turin—Strategic Programme: Urban Goods Freight Distribution with Clean Vehicles

Turin proposed a set of transport measures for the SUITS project connected with improving freight distribution with clean vehicles. Since April 2014, the city has had extensive involvement in urban logistic freight monitoring and delivery. It has developed a set of transport measures dealing with restrictions (with penalties or 'push' programmes) and incentives (rewards known as 'pull' programmes) for logistics operators delivering their operations under the Freight Quality Partnership (FQP) Agreement. Despite this, increasing commercial traffic had continued to affect traffic flow and burden the environment. Turin's goal for its new transport measures was to reorganise the loading and unloading of goods within the limited traffic zone (LTZ), while enhancing the loading/unloading parking areas, in the city centre. This would be achieved by granting LTZ privileges to operators using only methane or electric engine vans, with tracking through GPS by the city traffic management centre, to provide data that will help to improve loading/parking efficiency.

One of the main points of interest for local authorities while dealing with urban freight delivery lies in the location and the use of loading/unloading bays. The central area of Turin is a LTZ, which in this case consists of a 3 km^2 area that contains a large proportion of mobility attractors, including shops, business districts, and public offices. This area is characterised by a significant number of traffic flows, mainly during peak hours.

In terms of delivery vehicles, the municipality is encouraging progressive substitution of the most polluting vehicles commonly owned by operators. Moreover, the city is moving towards the use of logistic platforms and vehicles that meet a set of minimum requirements for the distribution of goods in the urban area. These actions are supported by special incentives for the movement of 'accredited vehicles'.

In May 2018, a 'Memorandum of Understanding' was made between the Turin municipality and the leading logistics operators. The latter would benefit from special permissions to enter the central area LTZ, conditional on using vans with only methane or electric engines and being tracked through GPS by the city traffic management centre run by 5 T (5 T manages the traffic control room of the Turin metropolitan area and is the owner of data collected from traffic sensors). The memorandum applies to most sectors of goods distribution excluding a few specific categories (e.g. drugs, newspapers and fuels), and those requiring special vehicles.

As a result, specific conditions and benefits have been granted to a selected additional group of vehicles (classified 'Euro 5', up to a specified maximum weight, and equipped with an 'on-board unit' linked to the Turin traffic operation centre). The operators taking part in the agreement and accepting the requests have free access to the LTZ and can use the bus lanes outside the LTZ zones providing faster access to the city centre. At the same time, these vehicles have exclusive use of loading/parking areas.

The 'on-board units' provide data for analysis. Studying trends in this data can help with planning corridors for the delivery of goods and learning which places are used most by delivery vehicles. This information can be exploited to arrange the creation of loading/unloading parking areas that can be used effectively.

Thanks to this experimentation, the city has tested 'V2I—Vehicles to Infrastructure' connection systems and has collected a significant amount of data. The analysis of those datasets helps in demonstrating an increase in the operators' commercial production and speed, together with a decreased emission of pollutants per delivery.

Challenges and Lessons Learned

One of the main problems the municipality faced in implementing this transport measure was the lack of specialised staff for data analysis, which is a common issue for smaller cities. For Turin this absence was covered through support from the SUITS project partners, mostly through assistance provide by staff from Politecnico di Torino. Recognising this issue, the material developed within the SUITS framework was focussed on developing the knowledge of local staff to raise their awareness of tools and data, such as what data might be needed and what skills to be developed or sourced elsewhere. Resources from the SUITS CBT will remain available to support Turin and other cities in future activities (see the discussion of the MyPolis tool in Chap. 12). Transport departments may be limited in their day-to-day activities, but they have learned there is scope to have additional analysis from experts and what these skills can provide. This is an example of how a city administration can grow into a 'learning organisation'—gaining new awareness of what is needed, how it can be done and what kind of internal skills or external help they may need in future.

One of the key results of the SUITS project was to create specific cost effective, scalable, and easy to use tools that could help the cities in facing such challenges. (See Chap. 12 for more information and associated publications, [9, 10]).

The experience gained by Turin's team through participating in the project revealed that sustainable urban mobility solutions are increasingly the result of a mix of technological and policy innovation. The value of participating in European projects was also demonstrated, through interacting with other entities that can give valuable support in activities that otherwise would not be possible.

Outcome and Impact

The measures put in place by Turin to provide faster access to the city for less polluting delivery vehicles, have shown positive results. Faster access to the city centre and exclusive access to loading/unloading bays enabled more goods to be delivered in a faster time, resulting in an overall increase of 20% in the commercial productivity rate for all logistics operators taking part.

At the same time, there was an important reduction of PM10 emissions due to the use of the new low emission vehicles. The proven increase in delivery efficiency and fall in emissions, have made it easier to demonstrate to operators that it pays to invest in technologies and vehicles with lower emissions, because they can take advantage of these city access measures with their associated benefits.

The main achievement of these transport measures was the stimulation provided to the logistics operators to renew their fleet of delivery vehicles by adopting less polluting ones and phasing out older diesel engines, which required considerable cost and commitment from operators. The user-specific transport measures that Turin introduced demonstrated to operators that this cost was worthwhile, in return for increased productivity and to reduce the environmental impact of urban freight. The feedback from operators is good and the municipality plans to renew the agreement in 2020, a good result thanks in part to the input from SUITS' team, and an important legacy of the results produced in the project.

In addition to the greater use of clean vehicles, the city-wide pilot in urban freight logistics allowed Turin to understand critical aspects of the new management system focussed on the use of ITS. On the data side, GPS traces were analysed by SUITS partners from Politecnico di Torino, to define both a disaggregated congestion Key Performance Indicator (KPI) [9] and a method to identify the most critical freight loading/unloading points in the city [10]. The congestion KPI can be used to evaluate the most critical routes for given travel purposes, such as parcel services, so that it is possible to assess the effective use of traffic lanes. At the same time, information can be extracted about the most congested areas influencing the travel time of vehicles: so that here, it is possible to propose the creation of new reserved lanes to ensure an efficient travelling time for those vehicles.

In an additional exercise, a specific survey was conducted in the Turin LTZ to collect information on the dynamics of freight deliveries (and pickups) at these locations. This included which operators usually deliver in the selected areas and where the vehicles are parked, to check the exploitation of the available loading/unloading areas. Moreover, the survey also checked the duration of stops, the number and dimension of packs delivered and the final destination of such deliveries, together with courier collection and deliveries. The analysis of this data helped in understanding the effective use of loading bays and access to restricted areas in the Turin LTZ [10].

Turin's experience shows that a range of city-level policy actions related to freight delivery can be informed through understanding the effectiveness and impact of delivery operations in critical areas of the city. Moreover, the additional activities have shown that by focussing at the micro level, for example on deliveries in a particular street and on retailers' and shops' exploitation of express courier services, new actions can be identified that local authorities could make to improve urban freight policies at specific locations.

Stuttgart—Strategic Programme: Working Together for a Human-Centred Mobility, Less Pollution, and Less Noise

Stuttgart was a 'follower' city in the SUITS project and is implementing a programme of integrated transport measures to bring about changes in travel behaviour.

With the Mobility Plan 2030, Stuttgart defines primary objectives to reduce emissions, noise and congestion and to improve quality of life in the city. This plan is regarded as the overall SUMP for the municipality and includes an action plan on sustainable mobility in Stuttgart. This action plan has a short-to medium-term life, as it is revised and updated by the political authorities of Stuttgart every two years. The action plan includes goals to improve conditions of public transport, pedestrians, and cyclists as well as a reduction of the modal share of motorised individual transport. In addition, the topics of reachability and accessibility are given high priority.

Before the SUITS project, Stuttgart had actively participated in projects at a European level, coordinating two EU projects 2MOVE2 and CARAVEL and participating in several other European projects (Go Pedelec, Active Access, SUMPA MED). In 2MOVE2[2] [11], the main objective was to improve urban mobility by advancing or creating sustainable, energy-efficient integrated urban transport systems in participating cities. CARAVEL[3] [12], aimed to improve quality of life in the participating cities by tackling urban mobility issues through public–private partnerships, stakeholder consultations, awareness-raising activities and research.

To reduce air pollution, Stuttgart has introduced numerous transport measures and launched different initiatives in the transport and mobility sector. An overview of Stuttgart's traffic management measures is provided here. In a later section we will outline the enhancement of the public transport network and its attractiveness.

Since 2008, Stuttgart has implemented a low emission zone in the entire city (with a few exceptions), i.e. driving and parking is prohibited for vehicles that do not have a green environmental sticker (vehicles with emission classes lower than Euro 2 and vehicles without a regulated catalytic converter). In 2019, the traffic bans were extended to vehicles with emission classes below Euro 5, and in 2020, the diesel traffic ban was further extended to vehicles with diesel engines of emission class Euro 5 and lower on certain routes through the city centre.

In addition, bans on the passage of trucks on certain routes have been in force since 2010. To support this, a truck route recommendation network was developed, which recommends optimal routes through Stuttgart and at the same time displays routes that are closed to truck traffic.

Environmentally sensitive dynamic speed adaptation has been implemented on some main urban routes. This system regulates traffic depending on the weather and traffic situation, thus improving traffic flow, and reducing stop-and-go situations. Furthermore, continuous improvements are being made to the parking guidance system and parking management. This on the one hand leads to a reduction in traffic from vehicles searching for parking spaces, and on the other hand creates incentives for the use of Park and Ride car parks and public transport because of the comparably high parking fees in the city centre.

From 2016 to 2020, the 'particulate matter alert' was introduced and used as an instrument for air pollution control. If air quality falls below the designated limit, citizens are called on not to use their cars in the city if possible and to switch to

---

[2] https://civitas.eu/project/2move2.

[3] https://civitas.eu/content/caravel.

environmentally friendly alternatives or to form carpools. In the domestic setting, the operation of comfort chimneys, i.e. chimneys which do not serve basic needs, are also prohibited on these days according to a regulation of the state government.

Since 2015 Stuttgart has actively promoted a campaign called 'Stuttgart steigt um' (Stuttgart is changing), to encourage citizens to avoid using the car and to use other modes of transport.

In addition, there are a number of other transport programmes aimed at environmental protection and pollution reduction. These include: comprehensive funding programmes for e-mobility in the commercial sector, e.g. through support for the purchase of new vehicles with high emission standards or support for the purchase of e-cargo bikes; a programme to increase the modal share of cycling, which includes the expansion and improvement of the cycle path network (e.g. creation of special cycle routes through the city centre), programmes for environmental protection in companies; intensification of the greening of roads and light train rails; a ban on combustion (solid fuels and green waste) in the city, restrictions on dust-intensive operations and construction sites on days with high air pollution and subsidies for further expansion of car and bike sharing services in the city.

Challenges and Lessons Learned

The introduction of Stuttgart's' transport measures posed several challenges. Cities need to consider that during the planning and implementation of transport measures, changes are likely to happen in different areas, all of which need to be managed. These can be, for example changes in political leadership or changing regulatory framework conditions, but also unforeseen problems such as delays in the implementation of projects as well as social developments that lead to changes in mobility needs. The municipal bodies/technical departments need to have both patience and determination; they must be empowered and motivated at the political level to be courageous when it comes to the design and implementation of transport measures. Cities have to increase the visibility of their transport programmes, enhancing transparency and enabling citizen participation. Programmes must be consistent with the longer-term strategic objectives of the municipality (Sustainable Urban Mobility Plans). To avoid an abrupt end of transport measures it is essential to build up strong political support (Mayor and City Council) and create a broad network of local actors that support the deployment of sustainable transport measures in the city.

Outcomes and Impact

The example of Stuttgart clearly shows the complexity of urban pollution problems and that a bundle of coordinated transport measures is needed to improve it in a long-term and sustainable manner. In addition, a well-developed public transport system is required, which is further discussed in the next section.

Stuttgart has developed exploitable results in 2MOVE2 and other national and European projects which can become integral parts of the overall mobility strategy of the municipality. By developing a dynamic 'Action Plan for Sustainable Mobility', Stuttgart has gained a living SUMP instrument that focuses not only on planning but also on the implementation of concrete actions. The various transport measures have contributed to better air quality, with no limits being exceeded since 2018. The air in the city has become cleaner and subsequently the particulate matter alarm was ended in April 2020. It has led to people in Stuttgart and the region actively developing awareness on the issue of air pollution control.

Palanga—Strategic Programme: Developing an Electric Vehicle Charging Network and Car Sharing Points

Palanga is a small coastal town in Lithuania and the issue of mobility poses challenges particularly in the summer months due to the huge influx of tourists. The city's SUMP was approved early in 2017, and its full implementation is expected by 2030. The SUMP is focussed on the implementation of universal design principles in the transport system (including mobility on its sandy beach) and infrastructure modifications to improve and promote bicycle and pedestrian mobility.

In the framework of SUITS, Palanga focussed on the challenges of citizen participation as well as on developing interaction and cooperation with business partners. The Palanga team wanted to broaden their understanding of the principles of sustainable mobility, as well as increase their capacity and agility in these areas. Since both the city administration and the citizens had limited experience with citizen participation, some quite elementary questions arose in the beginning. This is regarded as typical for smaller cities, where there is often simply a lack of capacity, both in terms of personnel and experience, to initiate and moderate processes for citizen participation. The initial question for Palanga was how to get its rather conservative population to participate at all. An important prerequisite for active participation is information. Only if citizens are informed about strategies and planned projects can they engage with issues, develop their own opinions, and participate.

Hence, it was necessary to educate and explain the benefits of sustainable transport and the essence of the innovations to the local population, to gain acceptance and support. Before that, the local administration had to build up its own knowledge in this area, which, as Palanga has impressively shown, is possible if the will and the appropriate support is there, such as that provided by the SUITS project.

As a resort city with a population heavily dependent on car use, Palanga focussed its attention on increasing the use of clean vehicles as a first step towards more sustainable mobility in the city. Palanga has attracted the biggest car sharing service provider in Lithuania (CityBee) and implemented several car-sharing points at regional level. In addition, an electric charging network was implemented consisting of seven charging posts, one of them state-owned, the others privately funded.

Key success factors include the communication and cooperation with business partners (setting the location of the parking and pick-up spots in the city), and involvement of local politicians to gain the necessary support and consultations with citizens in order to take their requirements into account during planning and ensure acceptance. A bike sharing system is in the planning stage, which will also be implemented in cooperation with the car-sharing provider. However, this project poses great challenges, as there are numerous private bicycle rental companies in the city already.

Challenges and Lessons Learned

Sustainable mobility was a very new concept for Palanga and its citizens. It was a concern that Palanga's generally conservative population would be reluctant to accept any kind of innovative transport measure. Hence, it was necessary to educate and explain the essence of the innovations to local citizens, to gain their acceptance and support.

This represented a big challenge for the Palanga local authority, as the municipality is very small, and few people are involved in mobility planning. National policies and political agendas provide the underlying context within which city initiatives must be constructed and introduced. With SUITS support, the city has developed a vision for future developments in the mobility sector and prepared relevant strategic documents. These are being used as the foundation for communicating its strategy for sustainable mobility within the local authority, to high-level management and to the wider city population.

Outcome and Impact

With SUITS project help, Palanga has moved forward considerably from its starting position, of being a city and population with very little knowledge about sustainable transport, or experience of introducing it and with relatively low resources in terms of the size of its team. The implementation of the charging points and developing partnerships with relevant businesses has been a successful first step, with accompanying political support for car sharing services. Charging points are a highly visible indicator, to both citizens and visitors, of the city's commitment to sustainable mobility. It is planned to renew Palanga's SUMP and add more mutually integrated transport measures focussing on sustainability. The city also plans to evaluate the social impact for all the transport measures.

In line with the approach advocated by SUITS, the Palanga local authority is exploring how to involve citizens and business partners more closely in the decision-making process about appropriate transport measures, to ensure that all stakeholders remain satisfied, that they make the best choice from the alternatives available and that local concerns about maintaining Palanga's status as a 'resort city' are fully considered.

9.3.2 Collective Passenger Transport (Alba Iulia, Rome, Turin and Stuttgart)

Providing fast and convenient public transport is vital in enabling cities to reduce car traffic and provide a clean, safe and appealing urban environment. First, we will show how Alba Iulia used innovative procurement methods to finance new buses to improve public transport. Then we discuss Rome's focus on bus lanes, traffic light priority and measuring passenger experience, followed by Turin's development of inter-modality around new rail interchanges, such as car-sharing services, shared taxi apps, bike sharing and e-scooter sharing. Finally, we revisit Stuttgart, where many integrated transport measures were developed as part of the Public Transport Pact 2015 and continued through the SUITS period.

Alba Iulia—Strategic Programme: Using Innovative Procurement and Financing Methods to Enhance Public Transport and Work Towards 'Smart City' Goals

During the SUITS project, the municipality of Alba Iulia was involved in an innovative pilot project testing the three Guidelines developed within the project: Guidelines on Innovative Procurement, Innovative Financing and New Business Models.[4] These were used to enhance the development of a sustainable mobility policy for the city, with an emphasis on strengthening public transport.

The transport measures tested by Alba Iulia during the SUITS pilot project included:

- Organising an innovative public tender with the Ministry of Development for the acquisition of electric buses for Alba Iulia together with several other municipalities—the first public tender of its type both at national level and in the mobility field.
- Novel criteria for public procurement contracts in transportation: these were highly innovative and a 'first' at national level. Although some of the criteria introduced were considered somewhat restrictive, the majority were approved by the national authorities for public procurement, and thus can be seen as innovative. For example, buses would have free Wi-Fi, GPS navigation with real-time monitoring in stations, video surveillance in buses, account-based ticketing, an emergency situation management system and part of the fleet would use biofuels.
- Developing the new public parking policy with support from EU technical assistance facility JASPERS [13] which provided the necessary expertise to develop an efficient and novel parking policy.
- Implementation of the pilot project, 'Alba Iulia Smart City'—with several smart and innovative transport measures in the mobility area such as smart parking using intelligent parking sensors and mobile apps, monitoring air quality in the city through air quality sensors installed on buses, smart cameras installed at the busiest intersections in the city and installing free Wi-Fi and other enhancements in buses.

Challenges and Lessons Learned

The municipality was the first of its kind in the country to attempt to issue municipal green bonds—a financing instrument usually used solely by governments; however, due to some legislative drawbacks the opportunity for issuing green bonds has been temporarily postponed but will be tried as soon as an opportunity arises.

---

[4] https://cbt.suits-project.sboing.net/guidelines.

Outcome and Impact

There was a strong emphasis in these transport measures on enhancing public transport provision, by procuring new buses with superior facilities that would encourage people to use them. The public transport company for the metropolitan area covers the municipality of Alba Iulia and the surrounding eight villages. Now, with one simple SMS, any citizen can pay for a bus ticket and go anywhere in the city. In addition, the buses were adapted for people with disabilities, and Alba Iulia's public transport was one of the first in the country to provide free Wi-Fi and air conditioning in its buses. The 15 newly purchased buses have air quality monitors, which are used to measure the level of pollutants at the city level.

The municipality went through a process of systemic change during the SUITS project. The departments that work on mobility issues were involved in a series of workshops and training sessions. This resulted in them starting to actively cooperate in a more integrated way not only in transportation and mobility, but also in other fields of activity.

Last but not least, during the SUITS project Alba Iulia Municipality developed and submitted three large projects on transport/mobility, funded under ERDF (Regional Operational Programme), of which two were successful. After implementation, these will drastically change the face of mobility for the city [14].

Rome—Strategic Programme: Public Transport Improvements

Rome was facing challenges regarding a need to increase sustainable mobility, through better public transport, active travel and road safety. In this section, we outline enhancements in public transport in particular.

The Municipal Assembly of Rome adopted its SUMP-Sustainable Urban Mobility Plan ('PUMS' in Italian) on August 2, 2019, under the coordination of the Rome Mobility Agency (RSM), (during the period of participation in the SUITS project). Rome's SUMP is based on EC Guidelines for SUMP, adopted by the Italian Ministry of Transport (MIT) in 2017 in line with the mission of the National Guidelines for SUMP in Italian Cities.

It is a strategic plan that defines and develops infrastructures for mobility services on a medium-term basis (10 years). It promotes safety, accessibility for all and implements smart technologies towards a connected vehicle-infrastructure-pedestrian environment and intelligent transportation system. The urban transport system has been designed to:

- guarantee all citizens transport options to access key destinations and services.
- improve safety conditions.
- reduce air and noise pollution, greenhouse gas emissions and energy consumption.
- increase the efficiency of transport flows of people and goods.

SUITS partner Rome Mobility Agency (RSM) supported the municipality of Rome in preparing its SUMP and advocated a participatory approach according to the SUITS development process and discussions at early SUITS workshops in 2016.

Following the SUITS model, RSM acted as 'Change Agents' to accompany and support the city in cultural change, in terms of working approach and the development of tools to improve understanding and integrate aspects concerning mobility.

From 2017 RSM worked with the city to develop an effective 'communication strategy' to carry out research and engage broader public participation. This process involved public meetings, research interviews and online media. The communication strategy also included the launch of a SUMP website in September 2017, the 'pumsroma.it' Fig. 9.1. This website portal[5] hosted a large-scale online participatory process to collect proposals from citizens to inform preparation of the SUMP, and this gained impressive results in terms of participation, [15]. At the end of a 4-month period, over 1,600 citizens had registered to participate in the SUMP consultation; over 2,600 citizen proposals were published; 50,000 visits had been made to the portal and over 20,000 votes made.[6]

A follow-on 'listening phase' took place in July 2018, consisting of two separate investigations: a survey of 2,000 telephone interviews with a representative sample of citizens, and an online questionnaire through the PUMS web portal (with 4,800 usable responses). This provided clear opinions for input to the preparation of Rome's SUMP. Residents identified Rome's main mobility issues as: reduce the accident rate (improve safety), reduce congestion, improve air quality, promote urban cycling, and strengthen the public transport infrastructure, [16].[7]

In respect of public transport, during the period of engagement with the SUITS project, Rome implemented a range of measures focussed on increasing the use of public transport in order to decrease private car use. In addition to renewal of the bus fleet to reduce its environmental impact, improvements to the service were implemented by the creation of dedicated bus lanes; the implementation of traffic light priority for public transport to increase performance and the analysis of user satisfaction in public transport, outlined below.

Renewal of the Bus Fleet

Rome's plan is to renew the surface fleet by replacing older vehicles with low and zero-emission vehicles by 2030—there have already been 682 new vehicles in 2020. In addition, 60 electric buses will be renovated within the next 3 years.

E-mobility is supported by 118 electric vehicle charging stations, a number that is predicted to increase to 300 in the near future.[8]

---

[5] https://www.pumsroma.it/ (accessed 27/10/2020).

[6] More can be found about Rome's participatory approach at: https://www.eltis.org/resources/case-studies/giving-people-what-they-want-romes-sump-and-its-participatory-co-creation (accessed 26/01/2021).

[7] Information provided by RSM and also from presentation at POLIS 2019 https://www.polisnetwork.eu/wp-content/uploads/2019/11/3G-Fabio-Nussio.pdf [16].

[8] As above.

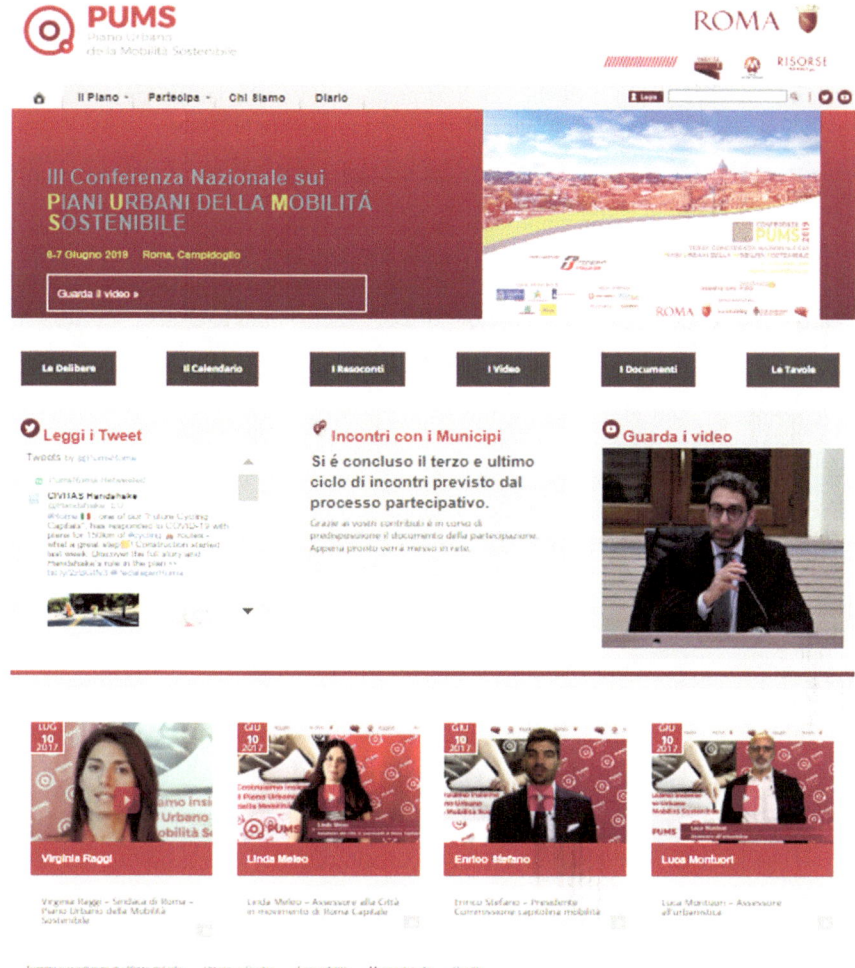

Fig. 9.1 The PUMSRoma website. *Source* www.pumsroma.it (accessed 27/10/20)

Dedicated Bus Lanes

Dedicated (protected) bus lanes increase the speed of public transport. By extending the protected bus lane network, Rome shortened journey times, making public transport more attractive. Two routes were completed by 2020, one in a densely populated commercial road (2 km, both directions) the other along a safeguarded archaeological site (0.5 km), and further routes are being implemented.

Traffic Light Priority

Implementing traffic light priority for various main public transport routes has increased the performance of the public transport network. Traffic light priority is planned on all tramway lines (about 50 km) and along six bus lanes.

Measurement of Quality of Public Transport Experience

In line with its participatory approach, in 2019 the Rome Mobility Agency carried out a customer satisfaction survey into public and private transport, to evaluate public transport experience and the safety and security of vulnerable road users. Evaluation of satisfaction related to surface public transport, metro lines and other services such as interchange parking, traffic lights systems and road parking fees. In total, over 18,600 interviews took place in 2019. The results indicated a need for improvement to the reliability and efficiency of the Rome surface public transport bus service, which was losing appeal with the public. In 2020, public transport was also subject to 'pandemic effect' with reduction of capacity due to the need for social distancing. Consequently, further improvements to public transport are planned for the coming years, even if the 'pandemic measures' in the public transport system are generating less revenues and increased costs for public transport operators.

Challenges and Lessons Learned

Rome adopted a 'guiding coalition' approach towards leading the SUMP activities, in line with SUITS recommendations on organisational approach. (See Chap. 6 for more information about this approach). Several different groups, each with specific skills, came together early on, to form a 'guiding coalition' to progress Rome's SUMP. These consisted of the Interdepartmental Working Group, and Technical Secretariat, with coordination by RSM's Expert Steering Committee. These groups worked together to build integrated plans and act until the conclusion of the SUMP process.

Rome's participatory approach towards preparation of its SUMP consisting of multiple interactions with stakeholders and citizens and planned 'listening' phases, played a key role, firstly in preparing the SUMP and then post-implementation in the surveys that followed—for example to monitor the effects of public transport improvements.

This participatory process was an important step in seeking to bring about a culture change in the city to raise awareness of sustainable mobility, encourage sustainable travel behaviour, discourage a car-centred approach and become a cycle-able and walkable city. Rome shared this example with other SUITS cities as an example of 'best practice'.

Outcome and Impact

The 2019 surveys showed a general increase in satisfaction with all the services investigated, with substantial peaks compared to 2018 for: 'Traffic light systems' (+10 points), 'Mobility Information Systems' (+8 points), 'Individual Transport for the Disabled' (+6 points), 'Metro lines' (+5 points) and Variable Message System 'iVMS' (+4 points). Only 'Public Transport Surface Lines' showed decreased satisfaction, losing 4.5 points. This latter decrease in satisfaction was despite shortened bus journey times, so will need further investigation.

As further improvements are made, the culture of engaging with citizens will allow Rome to continue to monitor satisfaction with its transport measures—and react accordingly. Rome's engagement of citizens on a large scale from the outset may well have encouraged long-term engagement with sustainability [17].

Rome's web portal continues with a high level of citizen engagement, communicating ideas and concepts regarding sustainable transport measures. Sustainable mobility has become the responsible norm as a strategic planning instrument for the wellbeing and improvement of citizens' daily lives.

Turin—Strategic Programme: Developing Intermodality Around Public Transport Infrastructure

Through the period of engagement with the SUITS project, Turin has been seeking to achieve intermodality around new interchanges for its underground and regional railway system (in addition to the extensive transport measures with clean vehicles, mentioned earlier in this chapter). The transport measures outlined here consist of infrastructural development projects aimed at completing the regional railway system within the city, designed to include new car-sharing services, shared taxi apps, free-floating shared bikes and e-scooters/mopeds.

Challenges and Lessons Learned

During the initial phases, it was realised that the main obstacle to achieving intermodality was the lack of an integrated ticketing system. A solution is being investigated from the Piedmont region aimed at transforming the current public transport smartcards into a MaaS platform, that will enable the purchase and use of integrated tickets for all regional transport services through smartphone apps on a 'pay per use' basis [18]. Turin municipality has been working with the operators of the different mobility services to make them aware of the MaaS concept and gain their support.

Outcome and Impact

The initial feedback from the mobility service operators is positive and the city plans to introduce the system in 2021. This is a clear example of a city following the approach recommended by SUITS: recognising barriers to change, searching for innovative solutions and working with multiple partners to achieve goals.

Stuttgart—Strategic Programme: Integrated Measures for Public Transport

As mentioned earlier, another goal of Stuttgart's Action Plan, which fits very well with similar goals of other SUITS partner cities, focusses on the improvement of conditions for public transport, pedestrians and cyclists as well as a reduction of the modal share of motorized personal transport. Since 2015, the city has been implementing a programme of several transport measures to achieve this goal.

Public Transport

Stuttgart has implemented a set of integrated transport measures to enhance public transport in the city. In 2015, the Public Transport Pact was adopted, which aims to increase the efficiency of public transport, especially bus and rail, enabling at least 20 percent more people to use it by 2025. The public transport measures include the following:

- Expansion and enhancement of the light rail and suburban railway network
- Expansion of Park and Ride and Bike and Ride spaces along the urban railway lines
- Use of express buses to close gaps in local transport, for example to connect one of the most populous city districts with the city centre. A separate bus lane can be used on a congested route, thus by-passing traffic jams. Two other lines connect city districts with the airport and the exhibition centre, by-passing the city centre.
- Increasing transport capacity through more frequent services to cope with increasing numbers of passengers and the purchase of new cleaner buses (hybrid and diesel with high emission standards) and buses with higher passenger capacity
- In the long term, it is planned to only purchase hybrid buses that recover energy during braking. This reduces diesel consumption by almost 7 L per 100 kms, resulting in a CO_2 saving of 30 percent
- Intermodal connection of different transport modes

Mobility Management

These improvements were accompanied by a number of transport measures to improve 'mobility management' to achieve simpler, more efficient, and thus more attractive use of public transport and to reduce road traffic:

- More intensive use of new information and communication technologies to improve user-friendliness (e.g. regarding passenger information or ticketing)
- 'Jobticket' Initiative: For employees of the city administration and for employees in participating companies—a subsidy of public transport fares provides incentives for commuting by public transport
- POLYGO Card: a smartcard for the use of public transport, which also enables simple access to car and bike sharing services and the use of e-charging stations
- Public transport interchange car park—the parking ticket can be used as a public transport ticket at the same time
- Streamlining of the fare zone system. More than 50 fare zones (so far) will be turned into five 'ring zones', making the fare system simpler, clearer, and also cheaper for many passengers, especially for the numerous commuters.
- Implementation of a special mobility guidance programme for new citizens.

In addition, Stuttgart implemented numerous mobility awareness campaigns (posters, radio advertising, videos, Facebook posts, interviews and newsletters) aimed at increasing the population's awareness of poor air quality and triggering changes in citizens' mobility behaviour towards the use of the well-equipped public transport system. These initiatives are for example: 'Stuttgart is changing'; 'Together for clean air' (increasing the use of public transport) and 'Together towards the goal' (promotion of carpooling and car sharing).

Challenges and Lessons Learned

The example of Stuttgart, working with SUITS as a 'follower' city, and sharing its goals and experiences, clearly shows that a well-developed public transport network and services are prerequisites for encouraging people to switch to public transport. It also shows how closely the different areas are interlinked and that impacts on air quality should be considered at every point.

An important learning point from Stuttgart's experience is that it takes patience and perseverance to achieve the desired change in transport. Many integrated transport measures are needed, which must have an overall objective and be applied in a wide range of areas.

Outcome and Impact

In recent years, the local public transport system in Stuttgart has developed positively. Stuttgart offers its citizens one of the densest local transport networks in Germany. On average, around half a million people travel by public transport in the metropolitan region every day. The city's goal is to ensure that even more people use public transport in future, which will further reduce the volume of traffic and improve air quality and thus improve the quality of life. The corresponding transport measures to achieve this are described in the Transport Development Concept 2030, [19].

9.3.3 Vulnerable Road Users (Rome, Kalamaria and Coventry)

All the SUITS partner cities introduced transport measures for increasing safety and security for road users and promoting active travel through cycling and walking. Vulnerable road users include pedestrians, cyclists, children and young people, people with disabilities, older people and others with reduced mobility or cognitive ability. Here, we describe transport measures from Rome, Kalamaria and Coventry, which addressed issues affecting these groups in particular.

Rome—Strategic Programme: Vulnerable Road Users in Road Safety-Speed Zones, with a Safety And Security Campaign

Rome has established a number of transport measures to improve the situation for vulnerable road users. These include 'environmental islands' where cycling and walking is safer, together with a safety and security campaign aimed at young people, to raise awareness and improve future road use.

Environmental Islands

'Environmental Islands' are portions of land bounded by the mesh of the trunk road network, and supported by the chamber of commerce, freight transport associations and the local trade operators. The scheme identified protected pedestrian routes and pedestrian spaces; other interventions concern temporary and permanent extensions to existing pedestrian areas, and the creation of new ones. These zones guarantee a low vehicle speed where cycling and walking is safer and vehicle emissions reduced thus encouraging more active travel at the same time as providing greater safety for more vulnerable users.

In addition, within the SUMP framework, an initiative 'Vialibera' was established to incentivize car-free mobility—once a month, cars are banned from 15 km of streets around the city centre so that citizens can use the areas for different activities such as walking, cycling, parties and other events. This initiative arose from a proposal made during the public consultation phase.

Safety and Security Campaign

While working with the SUITS project, a new safety awareness campaign was initiated to underpin cultural change efforts. Safety awareness campaigns are mainly aimed at young people. They raise awareness about road safety issues and are widely advocated as an effective way to guarantee good behaviour from future road users [20]. A new and extensive training initiative was launched during the period of participation in SUITS, in 2017/2018, 'De.Si.Re. – La città che vorrei' (the city I dream of), with a focus on road safety and sustainable mobility, aimed at the primary school age group. The initiative involved 14 primary schools and 4,300 pupils in the 2017/2018 school year.

Challenges and Lessons Learned

It is not always easy for organisations from different sectors, with sometimes conflicting interests, (for example cyclists and motorists) to work together, but this was achieved successfully for the safety campaign 'De.Si.Re'. The campaign was promoted by Rome Municipality, with the Rome Mobility Agency (RSM) acting as coordinator, with additional collaboration from the Italian Automobile Club of Rome, the Italian Cyclist Federation, and the local police.

Outcome and Impact

Before these measures there was a lack of citizen engagement in Rome regarding sustainable mobility. The road safety awareness campaign and the customer satisfaction exercise followed-on from Rome's initial engagement with citizens through the 'PUMSRoma' web portal, when preparing the SUMP. A perceived outcome has been a growing engagement by citizens and greater public participation regarding sustainable transport measures. With measures like these, Rome is following the *Vision Zero* of the European Union [21] and is initiating a series of steps aimed at reducing the number of accidents, on the one hand through infrastructural transport measures, and on the other, by creating awareness of the issue amongst citizens, and thus improving safety on Rome's roads in the long term.

Fig. 9.2 Example of a 3-D crossing.[9] *Credit* iStock.com/olaser

Kalamaria—Strategic Programme: Smart Solar-Powered Pedestrian Crossings Near Schools

Working with the SUITS project, Kalamaria developed internal targets aimed at improving safety levels for citizens and decreasing accidents. In addition, the city aimed to increase citizens' awareness and acceptance of sustainable transport measures such as new pedestrian footpaths and bike lanes and particularly of innovative technologies such as renewable energy.

Smart pedestrian crossings using sustainable solar-powered technology were identified by Kalamaria as combining an increase in safety, with high visibility to the public. Kalamaria implemented these near different schools resulting in an increased level of safety and a reduction of injuries and fatalities at the new crossings.

The next step is to implement additional crossings which will be new '3D' smart pedestrian crossings, also sited near schools to achieve higher road safety; in addition, one is planned on a commercial road near the commercial centre.

These 3D crossings are painted on the road surface in a way that they appear to 'float' in 3D form and are much more effective than 2D crossings in visually compelling motorists to stop. They will also be highly visible to the public as an innovation (Fig. 9.2).

Challenges and Lessons Learned

---

[9] See also an article about use of these in Valencia: https://www.themayor.eu/en/a/view/a-3d-zebra-crossing-to-be-painted-in-almussafes-in-the-valencia-region-988.

Implementation requirements included securing the budget and establishing the need for and acceptance of the technologies. Participation in European Mobility Week, encouraged by SUITS involvement, helped to raise the profile of safety and sustainability with citizens.

The municipality is also aiming to promote renewable solar power energy resources as part of this transport measure. Although the installation of the crossings was delayed due to the coronavirus pandemic crisis, the scheme gained the support of the new Mayor and it was expected that two smart pedestrian crossings (one ordinary and one 3D), would be completed in the first months of 2021.

Outcome and Impact

Kalamaria was able to combine a focus on safety for vulnerable groups with its goal of raising public visibility of road safety and sustainability. There was an increased level of safety and a reduction of injuries/fatalities at the new crossings. The innovative solar powered beacons are very visible, as they are sited near the crossings, so will promote innovation in sustainability and urban mobility to the public. The pedestrian crossings initiative shows a valuable combination of citizen participation, public visibility of transport measures and promotion of innovation in sustainability. It is hoped their public visibility alongside support from the new Mayor will in turn lead to a higher probability for widening this transport measure.

Coventry—Strategic Programme: Community Speedwatch

High vehicle speeds in urban areas pose a safety hazard to vulnerable road users, but speed limits need to be actively enforced and drivers' awareness raised to ensure compliance. Community Speedwatch (CSW) is a national initiative in the UK where active members of local communities join together, supported by the police, to monitor speeds of vehicles using speed detection devices such as speed guns, speed cameras and tablets. Vehicles exceeding the speed limit are referred to the police with the aim of educating drivers to reduce their speeds. In cases where education is blatantly ignored and evidence of repeat or excessive offences is collated, enforcement and prosecution follow.

In Coventry, the scheme has been implemented alongside safety and security awareness campaigns and speed zones strategies and is further enhanced using state-of-the-art technology for better accuracy. The CSW online platform[10] provides CSW groups and local police with training and a range of tools for making community involvement more efficient. It also produces a range of feedback notifications and collates statistics useful to both the police and neighbourhood groups. The statistics cover case status, live offence rate and volume reports of repeat and excessive speeding. Amongst many other functions, it also keeps track of operator accuracy and interactively engages practitioners in online training to increase their efficiency.

Challenges and Lessons Learned

---

[10] https://www.communityspeedwatch.org/,(accessed 07/09/2020).

Some input from Coventry City Council staff was initially required to disseminate and promote the scheme, but no extra staff were needed. There was some initial opposition from vehicle drivers, but despite this, public acceptance overall was high, and citizens were extremely positive about the scheme, making continuous requests to extend the scheme to more areas within the city [22].

Outcome and Impact

The benefits in fewer accidents, reduced fatalities, improved quality of life and public awareness, balance well against the relatively small funding cost and fast implementation. This is a community-led project that is scalable to larger or smaller cities and was shared within the SUITS project as a good practise. Funding came from local funds and only three months is needed from traffic zone planning to equipment purchase, training and implementation. It has created a framework of cooperation between the local authority, police and the public, which can lead to future cooperation.

9.3.4 Safety and Security (Rome, Valencia and Dachau)

Promoting safety and security for all was a constant theme in the SUMPs of our partner cities and many of the transport measures implemented during the SUITS project. In the previous section we outlined some examples showing an increased focus on providing safety for 'vulnerable road users' in particular pedestrians, including older people and children, through traffic calming and pedestrian areas. In this section we explore examples which focussed on a broader theme of road safety, especially for cyclists, and on providing an infrastructure to promote the safety and security of cycling and make it more attractive. There is some overlap with the themes in the previous section. Here, we focus on three examples from Rome, Valencia and Dachau. With the COVID-19 pandemic, 'safety' gained a new meaning, in terms of providing less crowded city space, discussed here in the Rome example.

Rome—Strategic Programme: Improving the Bicycle Network for Safer Cycling

In all corners of Europe, cities face alarming levels of congestion and air pollution and a scarcity of public space, whilst urban environments remain dangerous for vulnerable road users. Cycling is a powerful way to address these challenges and steer cities away from dominance of private cars and towards being more sustainable, equitable and economically prosperous places for citizens. Rome's SUMP includes actions to increase active mobility as well as to better integrate these transport modes with public transport.

As a consequence, Rome's SUMP includes key initiatives to promote cycling, bike sharing and bike parking, partly developed through the lifetime of the SUITS project.

Rome extended the city bicycle lane network, two bike lanes were completed along major routes (3.6 km opened June 2019, and 2.2 km planned during first half of 2020) with another of 5.8 km still under construction. The SUMP reference scenario (interventions already financed) expects 91 km of new cycling routes in the next three years, while the SUMP scenario plan (a 10-year plan not yet financed) includes a further 304 km of new cycling routes.

Education campaigns and activities took place to raise awareness about road safety and cycling, particularly aimed at young people. Standards and incentives have been developed for parking bicycles in common spaces and the creation of local cycling networks, together with continuous repair and development of the main cycle network. In addition, 2,600 parking stalls for bikes have been installed.

Intermodality will be promoted by combining cycling with public transport in multi-modal hubs, providing integrated facilities enabling users to organise modal exchange between private cars, public transport and bikes or electric vehicles, with bicycle parking in exchange nodes; bike sharing and transporting bikes on public transport. The hubs are planned (for example a 69-hub bike park) but not yet implemented at time of writing, dependent on budget availability.

An Uber service for bike sharing with 2,700 electronically assisted bikes (e-bikes) began at the end of 2019. Due to COVID-19, the operation of this has not yet been assessed. A five percent modal share for cycling is estimated for 2022 [17].

Challenges and Lessons Learned

With the COVID-19 emergency, the Rome Administration is focussing even more on 'intermodality', as travelling by private car may be perceived by citizens to be 'safer' than public transport during the pandemic. The challenge is to minimise the use of cars with a set of innovative measures, like the extension of sharing systems, launch of new micro-mobility services, large use of smart working, different regulation of city hours, opening times and related duration of activities to avoid the typical 'rush hours', avoid crowding and enable implementation of social distancing. These measures are complemented by another special set of implementations to encourage cycling and walking:

- Implementation of emergency cycle lanes of 150 km based on SUMP bike plan.
- Local measure to promote active modes, especially for movements below 5 km.
- Incentives for the purchase of electric bikes.

The establishment and construction of these new routes on the main streets of the city and on other 'strategic' itineraries will encourage safe, active and sustainable mobility in the later phases of the health emergency, and beyond.

Outcome and Impact

The new cycling infrastructure developed during the period of Rome's engagement with the SUITS project encourages more sustainable active mobility along with numerous health benefits associated with less pollution and increased physical activity in walking and cycling. Rome was already aware of these benefits through its earlier involvement in specific projects such as PASTA [23]. The new cycle routes offer a viable alternative to the use of private cars, especially for short journeys less than 5 km, and they will be integrated with public transport, by creating a network of cycle lanes mainly on the right-hand side of the roadway that is both inter-connected—and safe.

Valencia—Strategic Programmes: promoting Safety Through a 'Bike Culture' and 'City 30'

Since 2015, when changes in local government provided political endorsement to begin implementing its 2013 SUMP, Valencia has developed many activities directed towards creating a city with more active travel and sustainable transport. This included many activities to raise public awareness, creating participatory instruments such as the 'Mobility Round-table', new cycling lanes, creation of a better cycling network or 'cycling ring', reclaiming public space through pedestrianisation in specific areas, calming down transit, communication campaigns, workshops and collaboration with stakeholders. These measures were particularly focussed on improving safety in the city for pedestrians and cyclists. The measures described here were carried out with support from the SUITS project and feature initiatives to create a 'bike culture' and the implementation of 'City 30' speed limits.

A specific communication strategy was developed, communicating the SUMP sustainability plans and new regulations to citizens through two people-centric and user-friendly documents: a short booklet with key points about the SUMP, 'Towards a sustainable mobility in València. Policies and objectives in the area of mobility and public space of the City Council of Valencia', [24]; and also, a booklet providing the new Mobility Regulations in a friendly format "Keys to the sustainable mobility ordinance for living and co-living in the city of València", [25]. Encouraging cycling is an important part of the new transport measures and the city has created a special department called the 'Bike Agency' (*Agencia de la Bicicleta*) which plays a key role in communicating and promoting a 'bike culture' and providing information to citizens about the cycling network, [26].

An important milestone for the city and the Sustainable Mobility Service was the 'Valencia City 30' transport measure. This introduced a 30 km/h speed limit on streets with single carriageways in both directions, with taxi/bus lanes as additional lanes. In this way, practically all the streets in the inner-city neighbourhoods, or 64% of the city, are now safer and calmer with drivers respecting a maximum traffic speed of 30 km/h. Higher speeds are only permitted in the larger avenues, with two or more lanes in each direction, where up to 50 km/h is allowed but drivers are encouraged to give priority to pedestrians and drive with necessary caution to avoid accidents. This is one of the main innovations introduced by the new regulatory standard with a view to reinforcing road safety and quality of life in the city.

Challenges and Lessons Learned

In implementing these measures, and working within the SUITS approach, the Valencia team learned that planning and coordination are key when implementing sustainable transport measures, which often have inter-dependencies.

Many groups were involved in the road safety consultations, involving meetings with different entities such as the Local Police, the Municipal Transport Company (EMT), Ferrocarriles de la Generalitat Valenciana (FGV), as well as neighbourhood associations and other areas of the City Council de València. The outcome was a new Road Safety Master Plan 2018–2023, which outlines eight strategic directions to alleviate deficiencies and reach common mobility objectives.

The eight strategic directions seek to improve safety conditions in pedestrian mobility, cyclist mobility, improve the urban road network and calm traffic, improve control and monitoring of road discipline, as well as management mechanisms, education and training, promoting participation and dissemination in road safety and promoting the use of sustainable transport.

A change of mind-set from both government and citizens was required in order to implement City 30. Communication and citizen engagement were essential, to avoid potential opposition and misunderstanding, while there were additional challenges such as lack of budget, human resources and dealing with bureaucracy.

Outcome and Impact

The development of an improved cycling infrastructure has helped to calm traffic, improving road safety for all, while the cycling ring has had a multiplicative effect in promoting the adoption of cycling as a feasible alternative. City 30 is expected to reduce accidents and limit severe casualties with an ultimate aim of reducing fatalities to zero during the coming years.

The Road Safety Master Plan for the city of Valencia 2018–2023 has played an important role in this strategy and consolidates the efforts to make Valencia a safer and more sustainable city for its citizens. This plan was drawn up in line with the EU objective to halve the total number of fatalities on the roads of the European Union by 2020, with the aim of zero road deaths in line with 'Vision Zero' [21].

Valencia continues to work on many projects including improving the cycling network, pedestrianisation of Valencia City Council Square and redesign of the main commercial ring road, [26].

Dachau—Strategic Programme: 'Dachau Thinks Ahead'—Citizen-Oriented Urban Planning/Using Participation Techniques To Gain Citizen Engagement

Dachau was a 'follower' city in the SUITS project, and therefore not obliged to participate in any of the project work, although taking part in the benefits of attending at workshops and receiving information produced by the project. Like many small-to-medium-sized cities, Dachau struggles with traffic congestion, especially because a busy main road runs right through the middle of the city. In the transport/mobility area, Dachau's goals are to achieve improved safety and security for road users, and to improve the cycling network.

In this case study, we show how Dachau used citizen participation techniques to engage citizens in these goals, as part of a wider 'urban planning' initiative. While this example has a wide focus of which 'safety' was only a part, we have included it here, because it highlights how a small city can use participation techniques successfully to harness citizen buy-in for sustainable mobility measures.

The city has been focussing on citizen participation in its urban planning for several years and the 'Dachau thinks ahead' programme was launched in 2018 with concepts very much in line with SUITS' approach to citizen participation. The aim of the programme is to involve citizens at an early stage in the considerations of future urban development and to jointly develop a vision for the city. Citizens are called upon to actively participate in the development of a model and the planning of concrete transport measures through suggestions and ideas for improvement.

In the first participation module, citizens were asked to mark ideas, suggestions, criticisms, and possible solutions on various fields of action by using an interactive online map on the city's webpage. Within three months, almost 2100 entries were made, some of which were of very high quality. The results were assigned to the corresponding thematic fields: mobility, nature, urban design and public space, housing, economy and jobs, sport and leisure, urban technology, education and social infrastructure, culture and environment. Most of the contributions related to the broad topic area of transport/mobility (56%). Here, participants' suggestions focussed on improving public transport, reducing the high level of 'transit' traffic and improving conditions for cycling—in which safety plays a major role. (In second place with 13% was the thematic field 'nature'—which can be perceived as having broad links to sustainability and quality of life).

In addition, working groups were set up involving members of the city council, the city administration and various interest groups to develop the guiding principles of the city's vision. The presentation and discussion of the results with the citizens was accompanied by a large social event (workshops, music and childcare). At this citizens' workshop attended by almost 300 people, another 350 contributions were made by the participants on the various topic areas.

Challenges and Lessons Learned

The 'Dachau thinks ahead' public participation campaign was very different from the 'classic' participation approach, where schemes are presented, and citizens are asked to provide feedback and suggestions. Rather, the aim was to involve the citizens at an early stage, even before the concrete planning, and to take their problems and wishes into account in the development of overarching guiding principles for urban planning.

The organisation of such a participation process and communication with the public at different phases involves a huge amount of work for the local authority. The unlimited support of the city administration staff was an important factor, particularly as the public events took place mainly in the evenings or at weekends. Good moderation is important for the success of such events, as a big challenge is to guide participants and motivate them to contribute. The events were well received, and the citizens participated very actively in the different phases and events.

Besides the citizens, the involvement of different stakeholders (representatives of interest groups, associations, politicians, council members, planning experts, etc.), and their activities in different working groups was another key success factor. The early and active participation of interest groups in the development of guiding principles and concepts helped to enable a very broad consensus to be reached, which led to a perception of less intense resistance to the implementation of concrete measures at a later stage. In the council decision-making meetings, in different situations, it became clear that many aspects had already been clarified and agreed during the campaign. Thus, excessive discussions on decisions and possible adjournments of council meetings could be avoided in many cases, enabling progress to be made.

One project that was inspired by the contributions of the citizens is the so-called 'Parklet', which was implemented as a fast way to collect feedback. The Parklet (see Fig. 9.3) is a mini-garden, replacing two parking spaces in the city centre and offering seating and additional parking facilities for bicycles. It has a feel-good factor that offers space for creative ideas, for example it contains an 'idea tree' where people can leave behind messages, about urban mobility issues. Citizens are encouraged to not only make suggestions but also to change their own mobility behaviour and to communicate this to other citizens, leaving messages on the tree [27]. At the same time, it should raise awareness of how much space a parking space takes up and how it can be used alternatively.

The Parklet was received very positively and resulted in a total of about 150 contributions—interestingly much of this feedback was received from children and young people—a group often left out of consultation but important as vulnerable road users, pedestrians and cyclists. Their large participation is mainly due to the location of the Parklet very centrally in the city and next to an ice cream shop. The Parklet is mobile and will be set up at other locations in future.

Outcome and Impact

The major outcome of the 'Dachau Thinks Ahead' initiative was the development of a vision paper, serving as the basis for the new land use plan, and which was unanimously approved by the city council at the end of 2019. In addition, a strategy

Fig. 9.3 Dachau's 'Parklet', *Photo* J Hoffmann

to promote more cycling was introduced through a new 'Cycling Concept' in 2019, within which 10% of Dachau's citizens are expected to switch to the bicycle in the foreseeable future.

Delegates from SUITS partner Technische Universität Ilmenau (TUIL) worked with the city, assisting with the initiatives, and contributing information on several mobility topics, including input to the planning of a car-reduced district with enhanced safety through fewer cars. The new residential quarter will feature reduced car traffic, residential garages (so less on-street parking), efficient urban freight delivery and greater pedestrianisation—moving towards a car independent lifestyle.

Particularly in this last example, the participation of citizens has enabled greater understanding of needs and issues, and an agreed solution, which will provide a concrete example of a pedestrian-safe environment, hopefully paving the way for further traffic reduction measures.

The results of the intensive participation were set up as an exhibition in the foyer of the town hall, while the ongoing process of engagement and feedback continues through such initiatives as 'The Parklet'. This participative approach can be targeted at further initiatives over time. Dachau's experience (although not directly connected to SUITS activities) aligns very clearly with SUITS' approach that even small cities with comparatively low capacities can initiate extensive participation procedures, which are ultimately supported to a high degree by the citizens – and that this brings benefits in helping to define and gain agreement of a vision for sustainable transport programmes.

9.3.5   Intelligent Transport Systems and Services (Kalamaria, WMCA and Coventry)

During the lifetime of the project, most SUITS partner cities were implementing transport measures to integrate smart information and communications technology with transport infrastructure, vehicles and users aiming to improve traffic flow and reduce 'churn' through smart parking. Common challenges include coordinating input from multiple partners with different skills, achieving 'buy-in' from both citizens and municipalities and long-term planning. These are seen in examples from Kalamaria, in smart parking management and steps to optimise vehicle travel time through a real-time traffic information system, and from the UK West Midlands region, developing a testbed for autonomous CAV on public roads. An additional example is included from Coventry with the trial of an iVMS 'smart tech' traffic flow system and a mobile App to connect with drivers.

Kalamaria—Strategic Programme: Smart Parking Management and a Real-Time Traffic Information System

One of Kalamaria's SUMP goals was to optimise vehicle travel time to reduce congestion. In line with this the city investigated innovative technology, in smart parking management and implementing a real-time traffic information system, which began through being the test city for the MyPolis Live platform. We present the key points below.

Smart Parking Management

In line with its goal of investigating innovative technologies, Kalamaria is developing a sensor-controlled parking management system, to optimize usage of urban parking spaces and to reduce traffic churn generated by 'park search' driving.

Requiring coordination between several different groups of municipal staff and with European funding, the sensor-controlled parking system went through the planning stage and was put out to tender, extended to the end of 2020. The system is planned to consist of 150 smart parking spaces, planned for three roads in the commercial centre of the city, equipped with on-street sensors which measure the occupancy of the parking lots, and send this information to a system that can be used by drivers to find a vacant parking slot.

Using a Real-Time Traffic Management System for Intelligent Waste Collection

Collecting and using historical and real-time traffic data enables municipal staff and drivers to have access to intelligent features such as estimated arrival time and accurate routing allowing them to make informed transport assessment and planning decisions. However, the collection of this data can be a costly exercise.

Kalamaria participated in the pilot of the SUITS MyPolisLive platform,[11] (developed by SUITS partner SBOING), investigating the use of crowd sourcing as a low-cost solution for small-medium sized Local Authorities to capture, visualise and use traffic data and produce guidelines for implementation of such programmes, [28], (also see Chap. 12).

MyPolisLive was trialled in Kalamaria for four months from October 2018 until the end of January 2019, used by 50 assorted vehicles including city vehicles, taxis and company vehicles. The main target was the decrease of freight and passenger traffic with the optimization of the traffic flow. In addition, the platform can improve accessibility and economic development of the area by optimising the speed of distribution of goods, [29].

Challenges and Lessons Learned

Kalamaria's experience with innovative technologies has been one of incremental change. Initially, colleagues within the team were resistant to change, but this gradually declined as the programme progressed. A new municipality administration in September 2019 was a catalyst for getting new projects underway.

The parking scheme project required coordination between several different groups of municipal staff and also the use of innovative financing mechanisms. An early SUITS workshop held in Kalamaria indicated that participants from the trade transport sector had little awareness of the social and environmental impacts

---

[11] https://www.mypolislive.net/.

associated with the externalisation of freight transport. It was found that raising trade associations and citizens' awareness was important to overcome this [30].

In piloting the MyPolisLive crowd-sourced traffic management system, some issues were encountered with driver concerns and reassurance about data privacy, but once these were overcome, the pilot was very successful, providing data which enabled a reduction of traffic and in the number of trucks at peak times.

Outcome and Impact

The intelligent parking scheme showed a combination of visible benefits to drivers, through reduced driving time and a better driver experience, together with a better quality of life for all citizens through reduced traffic congestion, reduced emissions and better safety.

Following the successful MyPolisLive traffic management trial, Kalamaria is planning to use the system for its fleet of waste collection vehicles, developing a real-time information system for waste collection within the municipality. This will be used to control waste collection vehicle routes and to optimize travel times for the municipality's fleet of 127 vehicles (including waste collection trucks, construction vehicles, buses and municipal passenger cars).

Through using the platform, Kalamaria aims to reach the following goals:

- Reduction in operational costs
- Decrease in air pollution during peak traffic freight hours
- Optimization of waste collection in the municipality's boundaries
- Reduction in traffic nuisance
- Reduction of costs in vehicle maintenance through reduced mileage

Success will be monitored through measurement of travel time and optimization of travel routes. Discussions with the new municipality's administration are in progress regarding how to use and benefit further from the platform. Kalamaria ran an open international competition for the implementation of the real time information system during 2020 and announced a contractor in January 2021. The project is expected to complete by the end of 2021.

The trial of the MyPolisLive traffic platform helped all stakeholders to visualise, understand and manage traffic data. It is a cheap and cost-effective system for smaller municipalities to improve their traffic conditions, thereby reducing vehicle pollution and energy consumption and finally to improve their citizens' living standards.

Kalamaria's combined transport measures indicate how embracing new technologies can lead to realisable benefits on several levels, both for the municipality's own duties (such as in waste collection and fleet management) and in terms of raising public visibility of new technology and promoting the profile of sustainability to citizens, which we outlined in the earlier example of highly visible smart and solar powered crossings.

West Midlands—Strategic Programme: Testbed for Connected and Autonomous Vehicles (CAV) on Public Roads

There is a long history of mobility innovation in the UK West Midlands region, and the motor vehicle industry still forms an important part of the region's economy. The exciting and developing technology of Connected and Autonomous Vehicles (CAV) offers the opportunity to address the familiar challenges of poor air quality, congestion, and safety concerns in the West Midlands, and elsewhere, through more efficient and safer use of the road network. In the West Midlands region, it also helps to nurture a local industry to develop new technologies which can be exported around the world, driving economic growth.

Coventry City Council has experience in various innovation projects trialling connected and autonomous vehicles. The UK Autodrive[12] project, aimed to integrate connected and autonomous vehicles in a live real-world environment and involved trialling autonomous vehicles and electric autonomous 'pods' on live streets of Coventry. The trial included challenging areas of shared space environments. UK CITE[13] created a living lab environment for companies in the CAV industry to test how vehicles interact with V2V, V2X and V2I,[14] various modes of communications infrastructure on the roads of Coventry, Highways England roads and a Jaguar Land Rover test track.

These projects enabled Coventry City Council to develop deep collaborations with a range of different parties in the industry, including local universities, vehicle manufacturers, technology suppliers and provided prior experience for the WMCA CAV Testbed project.

The Midlands Future Mobility CAV Testbed aims to provide a real-world environment for the testing of Connected and Autonomous Vehicles on public roads in the West Midlands, connecting Birmingham, Solihull and Coventry. The project has an overall budget of £25.3 m, made up of a mix of public sector funding from Innovate UK and private sector contributions from the project partners. Once the initial development is complete, the testbed will continue into an operational phase for 8 years.

The testbed covers 80 km of public road, representing the largest public road test facility for CAVs available in the UK and all phases of work were scheduled for completion in 2020. The physical infrastructure includes deployment of specific CCTV cameras to enable the monitoring of CAVs during on-road tests; roadside units to enable communication between test vehicles and road infrastructure; weather stations and global navigation satellite system (GNSS) correction equipment to provide accurate positioning information to users of the test facility.

Through its involvement in the project, Transport for West Midlands (TfWM) can give the project access to the required public assets, such as Urban Traffic Control systems. This also enables TfWM to be at the forefront of the development

---

[12] http://www.ukautodrive.com.

[13] https://ukcite.co.uk/.

[14] vehicle-to-vehicle, vehicle-to-everything, vehicle-to-Infrastructure.

of CAV technologies in the region and helps to ensure that the organisation addresses challenges relating to safety, air quality improvements and congestion reduction.

The project aims to:

- Make the West Midlands an attractive and internationally recognised destination of choice for companies looking to research, develop and trial CAVs and supporting technology.
- Accelerate development of SAE Level 4 + CAVs (those that can detect their surroundings and operate without user input).
- Offer insight to TfWM and other public sector agencies about the infrastructure change and investment required for future mobility services on roads.
- Offer insight in policy and regulation that may be required for CAV, ITS and mobility service provision

Challenges and Lessons Learned

On commencement of the WMCA CAV Testbed project, the requirement for a central data-hub was recognised, allowing for the storage, reception and distribution of desired datasets. One of the main challenges associated with the data-hub was the identification of relevant and desirable datasets and receiving them in a suitable format. An example of this is Traffic Regulation Order (TRO) data, which defines any regulations and restrictions on specific roads. Existing means of storing data vary widely across and within local authorities in the region, requiring standardisation to make this data useful to Testbed customers.

The project required extensive collaboration with a range of partners from across the public, private and academic sectors, sharing different areas of expertise and working together towards a common goal. To deliver a project such as this, local authorities need to establish and develop collaborative relationships and sustain these after individual projects have been completed. Local authorities are well placed to act as coordinators and facilitators of consortia, bringing a wide range of parties together.

The support of national government is also crucial for such a large project. The UK Government is committed to positioning the UK as a world leader in the field of CAVs, and as such Government funding has been made available to support projects in this field.

Outcome and Impact

The completion of this project will strengthen the West Midlands as a leader in the provision of innovative technology and infrastructure projects, creating economic growth within the region.

The Midlands Future Mobility consortium has been successful in winning a further £7.9 m of government funding to extend the CAV Testbed into rural and inter-urban roads. This extension will turn the testbed into a 300 km test environment offering vehicle manufacturers and technology providers the chance to test vehicles and technologies in a range of settings.

TfWM is also leading a consortium delivering the ConVEx project, (Connected Vehicle Data Exchange), funded by Innovate UK and running from 2019–21, which will deliver a facility for open and commercial sharing of data resources relevant to CAV development and deployment. This facility will offer data cleansing and analysis, will draw together relevant datasets and explore connections that generate further insights into CAV development, deployment and operation and enable organisations to monetise data resources that may have previously been left dormant.

In 2019, TfWM was awarded £22 m of government funding to act as the pathfinder Future Transport Zone in the UK. This broad programme will build on the legacy of experience in the projects detailed above and deliver a wide range of projects aimed at translating innovations into services that the general public can use. These include an autonomous vehicle trial, trials of demand-responsive transport and MaaS, and improvements to digital infrastructure, such as digitisation of traffic regulation orders that will support future developments of innovative transport services in the West Midlands, [31].

The learnings from the SUITS project have supported TfWM in being able to deliver this programme. For instance, TfWM is delivering a package of transport measures in Coventry in collaboration with private transport service providers such as car hire companies. Furthermore, the experience from the SUITS project in relation to data processing has informed the technology choices that TfWM has made, as it has developed 'big data' processing solutions in collaboration with Amazon Web Services.

Coventry—Strategic Programme: Intelligent Variable Messaging System (iVMS) Smart Tech Traffic Flow

Coventry, like many other cities, has invested substantially in traffic management systems over recent years, including seeking to utilise new technology. Nevertheless, there was little to no active interaction between travel management systems, and 'managed interventions' within travel systems (for example, to respond to major incidents). These tend to be highly reactive, and 'blanket' in communicating to all users in a non-targeted way and requiring high levels of human resource to intervene.

As part of its vision of Integrated Mobility, and alongside its involvement in SUITS, Coventry initiated a project to develop, demonstrate and test new traffic management systems, with innovative iVMS technology, on three main road routes into Coventry; and to demonstrate the potential for a reduction in congestion in one section of the city.

The aim of the project was to receive constant real time traffic and journey data from strategically placed street infrastructure (Automatic Number Plate Recognition, Bluetooth Radar, CCTV, etc.) and 'connected vehicles' (including, potentially, their drivers' / passengers' smartphones), and then to use this data stream to manage the traffic system as a whole through traffic prediction and simulation models, and to provide live in-journey information and guidance to travellers regarding journey conditions, routes and alternatives. The system aimed to enhance travellers' decision-making, with the potential to tailor information to the needs of specific user groups such as freight, commuters, businesses, etc., ultimately leading to reduced congestion.

The key objectives of the iVMS project were to:

- increase the effectiveness of traffic management in Coventry, leading to reduced congestion and associated economic and social benefits.
- encourage behavioural change by individual travellers in support of congestion reduction.
- provide an enhanced testbed environment for future development of vehicle technologies and transport systems.

The project was coordinated by a consortium of partners from key corporate and academic institutions within the region [32].

A sustained period of upgrade and development provided new 'on-street' traffic signalling and communications equipment and capability (ANPR cameras, Bluetooth Radar sensors, loops detecting vehicles, cloud-based technology, etc.) and their integration with existing, and upgraded, communication systems in order to improve the level of communication between on-street devices and the existing Urban Traffic Control (UTC) system.

The system upgrade focussed on making real-time information available to drivers of vehicles and systems, on journeys through selected corridors—entry, exit, route, speed of travel, etc.,—and the capability to manage these [33].

A mobile phone app was developed as part of the project by partner SGIL and underwent a series of 'on-road' trials to ascertain the app's readiness and effectiveness to provide real time information to users during their journey. It was made available to download in November 2017, but due to a number of limitations related to the nature of the project and external factors, was not taken up or tested further.

Challenges and Lessons Learned

The project created the infrastructural capacity and the tool to encourage behavioural change. While the upgrade to infrastructure was a beneficial investment for the future, the iVMS app itself did not have the resource, expertise or time to be tested in a meaningful manner. The app failed to gain usage amongst its target set of drivers, mainly due to a lack of incentivisation and associated marketing. Factors limiting the take-up of the app included:

- the app was only available for Android phones (due to project resources)
- the app's journey planning features were limited to a suggested route and time of departure
- its emphasis on 'peak spreading' objectives acted as a disincentive to individual take-up
- the app was aimed primarily at commuters into Coventry, and the intended take-up incentives (such as prize draw rewards of free parking) were not secured such that the consequent behavioural change was underestimated by the project

- the emergence of other travel-planning apps in the region
- legal changes concerning the use of mobile devices whilst driving, also impacted upon the inclusion of incident reporting.

In conclusion, an app is in place which could change individual journey behaviour, but it has not been launched in a manner to fairly test its ability to do so, and its proposition of behavioural change through gamification and incentivisation remains untested with individual travellers.

Outcome and Impact

Although the behavioural change aspects of the app remain untested, the iVMS project has achieved an upgraded and better integrated traffic management platform that provides the infrastructure and capacity for targeted traffic management on the designated corridors. The capacity now exists, and has been tested, for traffic management strategies and operational decisions to be implemented, in real time, and with subsequent data feedback loops providing 'impact of decision' information—although these traffic management plans and strategies are yet to be written and implemented. Modelling has identified the relatively small scale of driver behavioural change required on the individual corridors to achieve congestion benefits—essentially small numbers of drivers slightly spreading their peak journeys—providing further clarity on the business case for, and desired travel outcomes sought, from any future traffic congestion intervention.

More broadly, the project has substantially informed greater knowledge and understanding of the costs and benefits of scaling the traffic management system to the whole of the city, including the infrastructure, technical and operational challenges of achieving such an outcome.

The most substantial achievement of the iVMS project has been to develop and extend the local test bed environment for vehicle technologies (and related smart city activity) across a number of dimensions. These include:

- An enhanced traffic management platform generated under iVMS as the catalyst, integrator and leverage point for new research and service developments.
- The availability of new data streams is enabling Coventry University's Institute for Future Transport and Cities to undertake 'near blue skies' research into driver behaviour.
- The development of (regional) research partnerships, supporting a deeper and diverse intelligent transport ecosystem for the whole region.
- Further consolidated recognition of Coventry City Council as sitting at the heart of new developments in intelligent transport systems, with involvement in projects such as UK CITE, UK Autodrive, CATCH!, SUITS, PARK-AV and Urban DNA.

Overall, the project has provided an innovative model for traffic management technologies for the city, and a 'test-bed' capacity for future trials of innovative technology, which can be expected to support further local economic benefit to the sectors, businesses, and citizens of Coventry.

9.4   Conclusions

In this chapter, we have tried to give a flavour of how the local authorities in the SUITS partner cities were able to improve their capacity to plan and implement sustainable transport measures using the support of the project team, and the skills and learnings shared within the project. For more about how cities overcame challenges and learned new ways of working, through the organisational change required to deliver projects effectively, see Chap. 6.

Working together with partners in the SUITS project resulted in much shared experience and lessons learned together. A few shared experiences stand out:

- the importance of thinking long-term and across integrated transport measures.
- success through public consultation
- including highly visible transport measures to maximise citizen awareness and acceptance of sustainable goals.
- the need to mix technology innovation with policy innovation.
- working with a range of partners from other sectors, industry, business and academia to maximise and supplement gaps in skills and capacity
- finding new ways of working, innovating financing and procurement
- maximising the use of data through intelligent and smart transport systems.

As a project team, SUITS partners supported the local administration teams with information and mutual exchange of experience. From this SUITS partners learned a lot about the skills and knowledge requirements of the cities/mobility planners, which we were able to take into account in the creation of the Capacity Building Toolbox. The capacity building toolset is introduced in Chap. 7 and can be found at https://cbt.suits-project.sboing.net/.

In Chap. 18 we will examine the Local Authority perspective further with some of the challenges faced while working on EU-funded collaborative projects and lessons learned for the future.

References

1. WHO.: Mortality and burden of disease from ambient air pollution (2016). http://origin.who.int/gho/phe/outdoor_air_pollution/burden/en/ [Accessed April 30, 2020]
2. European Environment Agency.: Briefing, November 2018 (2018). https://www.eea.europa.eu/airs/2018/environment-and-health/outdoor-air-quality-urban-areas [Accessed Feb 22, 2021]
3. WHO.: Global status report on road safety 2018: summary. Geneva: World Health Organization; 2018 (WHO/NMH/NVI/18.20). Licence: CC BY-NC-SA 3.0 IGO) (2018)
4. Eltis.org, Glossary. https://www.eltis.org/glossary/vulnerable-road-users [Accessed Sept 7, 2020]
5. Engels, D.: Topic guide: urban road safety and active travel in sustainable urban mobility planning (2019). https://www.eltis.org/sites/default/files/urban_road_safety_and_active_travel_in_sumps.pdf. [Accessed Mar 9, 2021]

6. EC.: White Paper, Roadmap to a Single European Transport Area – Towards a competitive and resource efficient transport system,/* COM/2011/0144 final */ (2011). https://ec.europa.eu/tra nsport/themes/european-strategies/white-paper-2011_en. [(Accessed Mar 9, 2021]

7. Panagopoulos, T., Tampakis, S., Karanikola, P., Karipidou-Kanari, A., Kantartzis, A.: (2018) 'The Usage and perception of pedestrian and cycling streets on residents' well-being in Kalamaria. Greece', article in Land **7**, 100 (2018)

8. West Midlands Travel Trends FactSheet.: (2016), https://www.tfwm.org.uk/media/2376/tra vel-trends-web.pdf. [Accessed Mar 1, 2021], for active travel see also 2011 Census Cycling figures from: https://www.theguardian.com/news/datablog/interactive/2013/feb/01/cycle-drive-work-map-census-2011?zoom=12&lat=52.48624299999999&lng=-1.890400999999 9971. [Accessed Mar 1, 2021]

9. Pirra, M., Diana, M.: Integrating mobility data sources to define and quantify a vehicle-level congestion indicator: an application for the city of Turin. Eur. Transp. Res. Record **11**, 41 (2019). https://doi.org/10.1186/s12544-019-0378-0

10. Diana, M., Pirra, M., Woodcock, A.: Freight distribution in urban areas: a method to select the most important loading and unloading areas and a survey tool to investigate related demand patterns. Eur. Transp. Res. Rev. (2020). https://doi.org/10.1186/s12544-020-00430-w

11. 2MOVE2. (https://civitas.eu/project/2move2)

12. CARAVEL. (https://civitas.eu/content/caravel)

13. JASPERS is a technical assistance facility for the twelve EU countries that joined the EU in 2004 and 2007. It provides the Member States concerned with the support they need to prepare high quality major projects, co-financed by EU funds. https://ec.europa.eu/regional_policy/arc hive/thefunds/instruments/jaspers_en.cfm

14. For further information about AIM projects see: https://www.apulum.ro/index.php/sit e/en, https://albaiuliasmartcity.ro, https://www.orange.ro/business/industrii/smart-city/alba-iulia/, https://www.youtube.com/watch?v=4BNduu2-mpY

15. 'Rome adopts a new Sustainable Urban Mobility Plan', https://www.eltis.org/in-brief/news/ rome-adopts-new-sustainable-urban-mobility-plan [Accessed Mar 1, 2021]

16. Nussio, F.: 'SUMP in Rome: The process of co-creation and its approval', paper presented at POLIS 2019 (2019). https://www.polisnetwork.eu/wp-content/uploads/2019/11/3G-Fabio-Nussio.pdf. [Accessed Mar 9, 2021]

17. 'The story of Rome's Sustainable Urban Mobility Plan', https://civitas.eu/news/story-romes-sustainable-urban-mobility-plan. [Accessed Jan 25, 2019]

18. 'Progetto BIP4MaaS (BIP for Mobility as a Service)', https://www.regione.piemonte.it/web/ temi/mobilita-trasporti/bip-biglietto-integrato-piemonte/progetto-bip4maas-bip-for-mobility-as-service. [Accessed Mar 9, 2021]

19. Stuttgart.de [online document] Oehler, S., Seyboth, A., Scherz, S., Wortmann, M.: Transport development concept of the state capital Stuttgart. (https://www.stuttgart.de/leben/mobili taet/nachhaltige-mobilitaet/mobilitaets-konzepte/verkehrsentwicklungskonzept-2030.php). [Accessed Mar 9, 2021]

20. Eltis.org: Glossary. https://www.eltis.org/glossary/awareness-campaign. [Accessed Mar 8, 2021]

21. Ec.Europa.eu: 'Road safety: European Commission sets out next steps towards "Vision Zero" including key performance indicators'. https://ec.europa.eu/transport/themes/strategies/news/ 2019-06-19-vision-zero_en.. [Accessed Mar 8, 2021]

22. Coventrytelegraph.net: 'Residents will be given speed guns to shop motorists as drive to become 20mph city accelerates' (2016). https://www.coventrytelegraph.net/news/coventry-news/res idents-given-speed-guns-shop-11543880 [Accessed 07 Sept, 2020]

23. Final Report Summary – PASTA (Physical Activity Through Sustainable Transport Approaches) https://cordis.europa.eu/project/id/602624/reporting

24. "Towards a (+) sustainable mobility in València. Policies and objectives in the area of mobility and public space of the City Council of Valencia" (2019). https://www.valencia.es/documents/20142/628173/ORDENAN_DE_MOBILITAT_valencia__768__BOP.pdf/67d0a569-3e49-d9be-091e-2e4849050de4 (pdf). [Accessed Mar 8, 2021] See English version at Eltis.org: https://www.eltis.org/sites/default/files/case-studies/documents/towards_sustainable_mobi lity_in_valencia.pdf
25. "Keys to the sustainable mobility ordinance for living and co-living in the city of València"; Claves de la Ordenanza de Movilidad Sostenible, and Ordenanza de Movil-idad, (2019). https://www.valencia.es/documents/20142/628173/claves%2520ordenanza.pdf/e787f66c-66f4-7111-6a44-905b6caaa393 (pdf). [Accessed Mar 8, 2021].
26. València City Council: https://www.valencia.es/val/mobilitat/inici. [Accessed Mar 8, 2021].
27. /https://www.dachaudenktweiter.de/aktuelles/; https://cbt.suits-project.sboing.net/case-stu dies; see also https://www.facebook.com/Dachau.Stadt/posts/10158479200202786/
28. Liotopoulos, F.: 'Data-driven, decision-making tools for S-M sized cities', presented at Urban Mobility Days 2020 (2020). https://www.eltis.org/sites/default/files/umd2020_liotopoulos_f otis_1.pdf. [Accessed Mar 10, 2021]
29. SUITS Stakeholder Workshop.: 'Sustainable urban freight transport: addressing modern era challenges', report, and presentations available (in Greek) from Suits website (2019). http://www.suits-project.eu/news/engaging-stakeholders-on-sustainable-urban-freight-transport-challenges-kalamaria-greece/. [Accessed Mar 10, 2021]
30. SUITS Webinar 06/05/2020: Data Management and Exploitation for Sustainable Urban Mobility: https://www.youtube.com/watch?v=i5EA5P6gyok. [Accessed Mar 10, 2021]
31. Midlands Future Mobility Project Website: https://midlandsfuturemobility.co.uk/. [Accessed Mar 10, 2021]
32. The project partners were 7 organisations which each brought particular skills: Coventry City Council; Siemens PLC; HORIBA MIRA; Serious Games International; Institute for Future Transport & Cities, Coventry University; Centre for Business in Society, Coventry University; InfoHub.
33. CWELP.com: (2017). https://www.cwlep.com/news/innovative-app-reduce-congestion-cov entry; and (2018). https://www.cwlep.com/news/innovative-app-leads-upgraded-transport-inf rastructure-coventry. [both Accessed Mar 10, 2021]

Chapter 10
Local Authorities' Perspectives on MaaS Implementation

Andree Woodcock, Sebastian Spundflasch, Frederic Rudolph, Kain Glensor, Keelan Fadden-Hopper, and Katie Miller-Crolla

Abstract Mobility as a service (MaaS) is being developed as a means of providing passengers with a convenient, on-demand, multimodal transport service. Pilot trials claim that MaaS can bring substantial benefits at individual and city level in terms of increases in efficiency, health and environmental factors. The SUITS project hosted a one-day conference in Coventry (UK) on the theme of "stimulating transport innovation through capacity building in small and medium local authorities." The 60 delegates were comprised of transport consultants, academics, local authority representatives, transport stakeholders and representatives from EU transport projects. In a workshop, the challenges that LAs are faced with in setting up MaaS were discussed. The results presented here were collated from moderated roundtable discussions. They reveal considerable concern of local authorities, and the need for consultation and planning on a wider range of issues than those that have been considered so far.

10.1 Introduction

Mobility as a service, or MaaS, is a system which unifies various mobility services, accessible via a single access point, most commonly a mobile app, allowing users to define the way in which they travel and see how much various options will cost [1]. It may form an extension of public transport systems with unified ticketing (as in London and Berlin) to encompass non-public transport modes, such as shared

A. Woodcock (✉)
Coventry University, Coventry, UK
e-mail: A.Woodcock@coventry.ac.uk

S. Spundflasch
Technische Universität Ilmenau, Ilmenau, Germany

F. Rudolph · K. Glensor
Wuppertal Institut für Klima, Wuppertal, Germany

K. Fadden-Hopper · K. Miller-Crolla
West Midlands Combined Authority, Birmingham, UK

© Transport for West Midlands 2023
A. Woodcock et al. (eds.), *Capacity Building in Local Authorities for Sustainable Transport Planning*, Smart Innovation, Systems and Technologies 319,
https://doi.org/10.1007/978-981-19-6962-1_10

vehicles (bicycles, cars or scooters) and taxis [2]. A MaaS system has a number of key components: at it's heart is a "MaaS operator." A public or private body responsible for presenting the MaaS offering to the end consumer, acting as an intermediary between the end consumer and "mobility service providers," also known as transport operators. The MaaS operator would buy capacity from mobility service providers and sell tickets or packages to the end consumer, as well as providing journey planning and fares information [3].

MaaS is seen as having the potential to transform how transport is provided and used. However, in the UK at least, according to a House of Commons Transport Committee report on Mobility as a Service, "the government's current vision focuses too much on the growth of electric vehicles and connected and autonomous vehicles" [4, para 28].

The workshop was held to understand the issues of concern to LAs in relation to MaaS implementation. This understanding was used to develop training material for LAs as part of the SUITS online toolkit and ultimately as part of the wider SUITS training material. The results are timely, given the House of Commons' Transport Committee Report [4] which stressed the need for increased attention to be paid to MaaS and for leadership, practical support and legislative/regulatory changes, and the increasing interest in MaaS reflected in the UK Department for Transport's "Future of Transport Regulatory Review" call for evidence in 2020, which sought views from stakeholders on the regulatory framework required to support MaaS [5].

10.2 Method

The workshop was organised using a version of the world café system [6]. Brief introductory presentations were given by Chris Lane (Transport for West Midlands—MaaS in the West Midlands—Initial Findings), Chris Perry (MaaS Global—implementing a MaaS system in the UK) and Giles Bailey (TravelSpirit—a vision of an open ecosystem MaaS system). Then, the delegates were randomly assigned to one of six moderated tables, each with a different topic (see Table 10.1). After 15 minutes, the delegates moved to another table, and the discussions continued, building upon the previous round of discussions. This cycle was repeated with the combined results of each table summarised, briefly presented in the plenary and feedback provided by the introductory speakers. Following this all material was collated and analysed by the SUITS team.

Table 10.1   Table topic and structure of the workshop

Regulatory framework	
Table 10.1	What are LAs' strengths and vulnerabilities in implementing MaaS measures regarding the regulatory framework?
Table 10.2	What are the opportunities and risks for a LA implementing MaaS measures regarding the regulatory framework?
Financing	
Table 10.3	What are LAs' strengths and vulnerabilities in implementing MaaS measures regarding financing?
Table 10.4	What are the opportunities and risks for a LA implementing MaaS measures regarding financing?
Stakeholder engagement	
Table 10.5	What are LAs' strengths and vulnerabilities in implementing MaaS measures regarding stakeholders?
Table 10.6	What are the opportunities and risks for a LA implementing MaaS measures regarding stakeholders?

10.3   Results

10.3.1   Regulatory Framework

Participants were asked to discuss the two questions shown in Table 10.1. Tables 10.2 and 10.3 show that although the LAs were excited by the opportunities provided by MaaS, there were many unanswered questions and issues in the practicalities of implementation which need to be addressed.

In summary, the participants in each round mentioned that local authorities differ from country to country in the extent to which they can exert control over public transport (PT) or even influence it. In countries where PT is centrally controlled by a national body (e.g. Greece), larger cities have the advantage over smaller cities in that they will have more representation and can therefore influence national regulations. Differences also occur between LAs depending on whether the bus market is regulated or deregulated, with the former having more power to influence changes. Also, many concerns were tabled related to the complicated position of LAs as a (potential) MaaS operator and commercial, regulatory and political actor and the ways in which the widespread adoption of MaaS (especially considering a consolidated multimodal MaaS system) could complicate the myriad—currently separate in many cases—arrangements and agreements with various transport providers.

At the local level, where the LA can influence or control PT, they can encourage integration of services, (e.g. by prioritising or facilitating changes to the regulations) or by forcing providers—including taxi or private hire vehicle operators—to cooperate (e.g. by making it a condition of licensing or a prerequisite for integration with

Table 10.2 Perceived opportunities for MaaS from an LA perspective

Opportunities	
Data	• Generation, control and ownership of data from MaaS would enable LAs to save money through efficiencies and pooling of data, e.g. sociodemographic • Data sourced from MaaS usage could be used to model and deliver new services and provide an evidence base for new tenders • Open data can enable agility and provision of services people want
MaaS offering	Could be incorporated into new tenders and would enable greater competition between service providers
Extended interest	MaaS would provide interest and investment opportunities for a wider range of stakeholders
Innovation	Potential to trial services in cooperation with private companies, leading to new business opportunities and vibrancy in the city
Finances	Reduction in costs of transport service provision, increase in grant availability from national or supranational bodies for development of new services
Regulatory issues	New innovations and different types of service provision will require new regulatory frameworks
Change of focus	• New focus on sustainability and efficient mobility • Equal attention given to MaaS and autonomous and connected vehicles

a MaaS platform, etc.). In the case of taxis, a barrier to MaaS exists in some jurisdictions where regulations control the areas in which passengers can be picked up, so changes in passenger licensing laws are required.

Large cities have an obvious strength in that they more easily can pull together horizontal teams with the necessary skills to implement MaaS—legal, technological, planning, procurement, etc. On the other hand, this may require changes to the way work is traditionally done.

In countries where public consultation is necessary, cities may not be able to move as fast as those where public consultation is not a requirement. Strong political support is necessary for any changes; this may require engaging stakeholders such as mayors or local councillors.

10.3.2 Financing

Participants were asked to consider what were the LAs' strengths, vulnerabilities, opportunities and risks, in implementing MaaS measures in terms of financing. Table 10.4 provides a summary of the main points.

The discussion identified many of the factors outlined in the previous section. Positively, MaaS was seen as an opportunity to increase the number of stakeholders involved as potential funders. However, major problems were anticipated in relation to how existing transport operators would react to new transport services. For

Table 10.3 Perceived risks and unresolved issues relating to MaaS implementation

Risks and unresolved issues	
Market failure	• Extent to which LA would be held responsible (financially or otherwise) for failures in service provision (e.g. vehicle breakdown, weather disrupting services) and bankruptcy/operator failure • Extent to which the LA would be required to take over the service of a failed MaaS operator or mobility service provider • New regulations which need to be developed to mitigate negative effects, including enforcement of service provision
Influence of perceptions of PT	In some countries, public transport (PT) has a very poor image. This may transfer to MaaS and be a barrier to uptake
Mode shift	MaaS may cause people to move to shared vehicles, away from PT, risking PT revenues
Service fragmentation	Funding and operation may be spread across different competing MaaS operators and mobility service providers, increasing complexity and leading to service fragmentation
Finances	• Currently, PT is not fully outsourced, but there is still a lack of service in deprived areas. There is a risk that deprived areas/communities will not be served by MaaS—resulting in new forms of transport poverty. How will guarantees of service/subsidies be worked out to ensure equality and inclusion? • How will fares and benefits to providers (especially for unprofitable routes/services) be distributed to guarantee coverage of services? Would MaaS make it more difficult for LAs to cross-subsidise from more profitable services to less profitable ones? • What effects will MaaS have on revenue streams for LAs (e.g. lost parking and taxi licensing revenue streams)? • There seems to be an implicit assumption that autonomous vehicles will be used collectively, but this is not necessarily the case
Effects on overall quality of service provision	• May lead to the reduction in the need for certain services which are relied on by certain groups of the community, leading to less choice and greater inequalities • How does MaaS fit into the current mix of transport provision, e.g. in terms of taxis and community transport?

(continued)

Table 10.3 (continued)

Risks and unresolved issues	
Competition bias	• That LA may be seen as favouring those providers who take part in MaaS; the LA must be seen to be unbiased towards private operators • Monopoly risk of commercial platforms, especially if driven by mobility service provider themselves—what is the incentive for the MaaS operator to allow competition? • MaaS agreements may limit LAs' future options (e.g. to abolish PT fares)
Change in mind-set/time	• Automotive and mobility sector need to change • LAs have to reach a consensus on MaaS (with open public consultations) • How can a change in mind-set be created to guarantee use of MaaS, what incentives are appropriate?
Data	• How can the information be used to focus on public rather than individual forms of transport? • GDPR may be interpreted in such a way as to undermine MaaS provision (e.g. regular user journey data) • Sharing of open data is an issue, as data have a potential value. How do you value it? Should the aim of LAs be to gain financially by selling data or make the data openly available? • Concerns about handling and security of the data required by MaaS schemes
Regulatory issues	• Concern was raised about the extent to which regulations can keep up with innovations and demand for new services • Need to include regulations which deal with negative/side effects including enforcement of service provision and workers' rights • Need a legal framework for collaborative schemes, liability for non-delivery of services (bankruptcy, etc.) and to encourage competition rather than just favouring MaaS operators • Could create friction in LAs' role, e.g. subsidies could distort the market within the MaaS system. LA might have double role as provider and regulator

(continued)

Table 10.3 (continued)

Risks and unresolved issues	
Additional burdens on LA	• Legal, administrative and regulatory issues will increase • Conflict resolution and disputes over service provision • Management of disruption caused by changing over to MaaS
Inequality and inclusivity	• How can the needs of vulnerable groups be protected? • What happens to hard-to-reach groups, e.g. older people, people without mobile phones?
Predatory competition	• Considerations around the threat from competition, e.g. Uber, especially as they may behave in ways that were not anticipated by the existing regulatory framework • Private operators may concentrate on profitable areas/services, leaving the LA to operate loss-making services • Regulation of their practices, but concern in keeping up and enforcing them

Table 10.4 SVOR analysis in terms of financing for MaaS

Strengths	Vulnerabilities
Transparency and public scrutiny	Limited direct financial opportunities for smaller cities
Central hub for linking suppliers and funding	Lack of knowledge about how to deal with finances
Ability to offer concessions	Lack of knowledge to gain funding/developing financially viable projects/access to crowdfunding/big variability/vulnerability in LAs
Existence of public authority	Big cities' internal competition and lack of prioritisation/competitive funding
	Internal knowledge sharing. Lack of knowledge leads to support in wrong places; high variability
Opportunities	Risks
Backing from political parties	Transparency and public scrutiny
Scale of opportunities for business case. Large city good potential market	Big ideas and too much competition with too little know-how
Better PR and communication to gain public support for taxation system and deliver eco-friendly messages	Cities which already have committed to existing systems are vulnerable
Ability to differentiate volume and destination usage	Ability to offer concessions

example, modal shifts could be away from them, and result in a loss of revenue, if not business closures; consideration is needed as to how revenue streams would be shared between MaaS providers, for example in relation to route sharing, or operating routes with higher proportions of concessionary travel (although data was mentioned as a solution, participants were not able to elaborate how this could be used practically, e.g. for planning of services/distribution of revenues).

10.3.3 Stakeholder Engagement

The development and implementation of MaaS offerings represent a major challenge for LAs as the needs of many stakeholders with different interests must be considered. These interests are sometimes contradictory, difficult to take into account and do not necessarily reflect the interests and ideas of the initiators of such a service. The workshop addressed the question of which stakeholders play an important role in this process and which challenges arise in terms of cooperation. Such multistakeholder negotiation may require different skillsets.

The first step in the group work was to get an overview of who the potential stakeholders are at the suggestion of the participants, a subdivision into internal and external stakeholders was made. Internal stakeholders are mainly initiators or people responsible for the planning, implementation and decision-making of a MaaS undertaking. External stakeholders mostly play an important role in the subsequent operation of the service (Table 10.5).

One major challenge was seen as the need to raise awareness of MaaS within the LA and on the part of policy-makers. The development of a MaaS requires the close cooperation of many departments and a will to implement it. Moreover, it is not easy to get political support, as there is currently little experience on this and

Table 10.5 Overview of internal and external stakeholders	Internal stakeholders	External stakeholders
	Departments within LA • Transport • IT • Technical • Public tender • Highways • Finance • Communication Other • Transport cabinet members • Planning cabinet members • Security board • Local councillors	• Public transport providers • Private mobility service providers • Software developers for a platform • Technology companies • The future users of the service • Infrastructure providers • Press and media

how to manage the uncertainties and risks. The conviction of internal stakeholders is therefore one of the greatest challenges.

In addition, the engagement with external stakeholders is a demanding task. Most will need to invest time and money to get MaaS to work, even if they are not primary beneficiaries. This can become a particular problem for smaller transport service providers.

Key ways in which stakeholders could be encouraged predominantly took a user centred approach, including:

- Promote the possibility for stakeholders to use data (e.g. reason for journey, origin–destination data, frequencies of use)
- Carry out user surveys in advance in order to verify the needs and requirements of the users to the stakeholders. (what might encourage use, what are they dissatisfied with at the moment, what are their expectations?)
- Get people storytelling, e.g. how a nurse gets to hospital at hours when there isn't a commercially viable conventional public transport service
- Show benefits in terms of economics and health
- Show financial benefits (best with concrete calculation examples that compare the ownership of an own vehicle with the use of MaaS)
- What are the benefits for all groups for MaaS?
- Clearly highlight the benefit for the individual user groups

With regard to the opportunities and risks, it was felt important to get the right balance between public and private transport service providers or offerings. A solid competition between participating providers can increase overall quality of the services. Therefore, a danger was seen in putting too much power in the hands of the authority. On the other hand, a completely free market can lead to a situation where MaaS operators and mobility service providers play off each other from positions of power. If the service is offered by a private operator, and the goal is to maximise profits, the main question is as follows: how to create a fair balance between the large and small mobility service providers. Can small providers exist alongside large ones? A situation with just one MaaS operator was also seen as a risk.

LAs could create a fair balance between large and small mobility service providers ensuring that there is a fair distribution of profits between them. Service providers should not simply compete but cooperate to offer the best service to the users. This was seen as a great challenge. However, it was stressed that the user experience should remain the focus. Offerings must be tailored to user needs. If the users recognise that mobility is becoming simpler, more comfortable and above all cheaper, they will rethink their behaviour. Changing mobility behaviour remains one of the greatest challenges, and the LAs need to consider incentivisations and conseqences carefully.

Participants recognised a distinction between small and larger cities, but noted it is very hard to generalise between cities and countries. In terms of strengths, smaller cities may have more flexibility, be more able to identify and reach stakeholders and connect between them. Innovation and research consortia were able to collaborate strongly with LAs, who were better placed to influence the public to adapt to MaaS.

LAs could more easily enforce policy and focus on social visions and had the potential to involve the public through consultations and public enquiry.

However, small cities may lack technical knowledge, know-how and persuasive arguments. They can see complex systems as risky and may be vulnerable to the influence of powerful stakeholders. They were also seen as inefficient with overlapping responsibilities, overly centralised governance and not being able to identify and serve local needs. Clearly, there is some disagreement in the characterisation of small cities, but the overall opinion was that small cities could mobilise and be more receptive to change.

Larger cities were seen as lacking flexibility and being unable to change rules. They were seen as being over-protective and biased, giving preferential treatment to certain stakeholders and incumbents. They found it harder to integrate public and private service providers or transfer services. Additionally, in some cases, the citizens in larger cities were felt to have a weaker knowledge of sustainable mobility.

10.4   Conclusions and Recommendations

The results presented above not only align closely with the views expressed in the House of Commons' Transport Report but go further in expressing the issues each LA will need to contend with. Johan Herrlin was quoted in the report as saying "there is nothing inherently altruistic about MaaS" [4]. Other witnesses noted that MaaS could result in "an increase in use of taxis and private hire vehicles, with negative consequences for road congestion and air pollution" and "the worsening of digital and social exclusion" [4]. LAs have a responsibility to ensure inclusion, and accessibility in transport is maintained: as such, LAs may be required to regulate MaaS operators and/or mobility service providers to ensure this is the case.

The delegates raised many concerns related to the complicated position of LAs in a MaaS deployment. An LA could take on the roles of MaaS operator and/or mobility service provider whilst also being a regulatory and political actor. Which of these roles the LA adopts will depend on local contexts since their role in the current provision of transport also varies substantially. Current arrangements and agreements with various actors, such as incumbent transport operators, could also be complicated by these new roles.

Some of the concerns presented by LAs centre around conflicts between these roles: for instance, the challenge posed by an LA running a MaaS application and simultaneously acting as one of the mobility service providers. The LA taking on a role as regulator of an ecosystem of MaaS operators and service providers may place a burden on LAs, who may not have the appropriate competencies or capacity to do so, particularly in small to medium-sized cities.

LAs' role in MaaS deployments is also likely to influence how LAs can make use of the data available from MaaS deployment. LAs identified opportunities for the use of data, but the availability of these may depend on regulations and local agreements. National regulation may assist LAs in reaching the opportunities presented by data:

for instance, Finland's Act on Transport Services imposes a requirement on MaaS operators and mobility service providers to release a minimum level of data openly [7]. Policy of this nature is likely to aid LAs in reaching the opportunities to use data to improve their own decision-making.

LAs' role in the MaaS ecosystem will likely remain contested as MaaS operators and mobility service providers seek to collaborate with LAs or public transport authorities in order to leverage their brand recognition and perceived service quality. At the same time, other mobility service providers continue to pursue integration of public transport fares data and ticketing in their own offering [8], as well as control of a wider range of transport modes [9]. Technology companies are also integrating public transport ticketing into their own applications.

In the discussions, LAs expressed a great deal of anxiety about the implementation of MaaS and what it would mean for them as an organisation. From the comments, it is quite clear that they are only just starting to grapple with a wide range of issues for which they do not have any answers. There is a clear need to provide guidance, training and a forum for LAs to enable them to discuss such issues and possible solutions as part of existing trials and prior to any stakeholder engagement.

Acknowledgements The SUITS project has been funded by the European Union's Horizon 2020 research and innovation programme under grant agreement no 690650.

References

1. MaaS Alliance website. Accessed 10 Jan 2019. https://maas-alliance.eu/homepage/what-is-maas/
2. Bailey, G.: Presentation to SUITS conference, 21 November 2018. TravelSpirit Foundation: Championing Openness in Mobility (2018)
3. Kamargianni, M., Matyas, M., Li, W., Muscat, J., Yfantis, L.: The MaaS Dictionary. MaaSLab, Energy Institute, University College London (2018). Accessed 11 Jan 2019. Available at: www.maaslab.org
4. House of Commons Transport Committee: Mobility as a Service (2018). Accessed 11 Jan 2019. Available at https://publications.parliament.uk/pa/cm201719/cmselect/cmtrans/590/590.pdf
5. Department of Transport: Future of Transport regulatory review: call for evidence on micromobility vehicles, flexible bus services and Mobility-as-a-Service (2020). Accessed 16 March 2021. Available at https://www.gov.uk/government/consultations/future-of-transport-regulatory-review-call-for-evidence-on-micromobility-vehicles-flexible-bus-services-and-mobility-as-a-service
6. The World Café Community Foundation website. Accessed 10 Jan 2019. http://www.theworldcafe.com/
7. Ministry of Transport and Communications: Act on Transport Services (2017). Accessed 10 Jan 2019. https://www.lvm.fi/en/-/act-on-transport-services-955864
8. Ram, A., Spero, J.: Uber looks to integrate London bus and Tube data into its app, Financial Times (2018). Accessed 10 Jan 2019. https://www.ft.com/content/0f1ee492-fc92-11e8-ac00-57a2a826423e
9. Singh, M.: Ola to invest $327M to set up 'the world's largest scooter factory' in Tamil Nadu, Tec4runch (2020). Accessed 16 March 2021. https://tec4runch.com/2020/12/13/ola-to-invest-327-million-to-set-up-the-worlds-largest-scooter-factory-in-tamil-nadu/

10. Hawkins, A.: Google Maps will now let you pay for public transportation and parking through its app. The Verge (2021). Accessed 16 March 2021. https://www.theverge.com/2021/2/17/222 87043/google-maps-pay-parking-public-transportation-cities
11. House of Commons Transport Committee: Oral evidence: Mobility as a Service, HC 590 (2018). Accessed 17 March 2021. Available at http://data.parliament.uk/writtenevidence/committeeevi dence.svc/evidencedocument/transport-committee/mobility-as-a-service/oral/82485.html

Chapter 11
Social Impact and Hard-To-Reach Groups

Andree Woodcock and Janet Saunders

Abstract This chapter discusses the importance of conducting social impact assessments (SIAs) prior to, during and after the implementation of mobility measures. All transport measures have a direct impact on transport users, but they can also have an indirect impact on users, non-users, and those living further away from the proposed measure. These consequences should be considered as part of the wider cost–benefit/lifecycle of the planned measures. A key aspect of design and implementation of mobility measures is to consider the direct and indirect effects on citizens, in particular vulnerable groups, those who have difficulty accessing transport through lack of finances, poor mobility, ageing or those with dependents/looking after children, or because their needs are not met through current transport provision and are consequently excluded from the opportunities provided by city life. The steps taken to mitigate these effects should feed into planning new transport measures, at all stages from pre-planning through to post-evaluation. This chapter explains the importance of SIA, discusses methodologies, provides a set of factors to be considered in conducting SIAs and provides an overview of groups most vulnerable to negative social impacts, with some suggestions for improving practise.

11.1 Introduction

In any sustainable urban planning process, it is important to consult with and meet the needs of, traditionally disadvantaged and disenfranchised groups. The METPEX project, (part of EC FP7 funding programme 2007–2013), [1] revealed that those engaged in transport and mobility planning found it difficult to engage with members

A. Woodcock (✉)
Coventry University, Coventry, UK
e-mail: A.Woodcock@coventry.ac.uk

J. Saunders
Coventry, UK

© Transport for West Midlands 2023 193
A. Woodcock et al. (eds.), *Capacity Building in Local Authorities for Sustainable Transport Planning*, Smart Innovation, Systems and Technologies 319,
https://doi.org/10.1007/978-981-19-6962-1_11

of 'hard to reach groups'. This led to a systemic under-representation of their needs. Without this engagement, future transport provision will not serve all citizens and may lead to greater inequalities. It is not sufficient to measure transport or mobility measures simply in terms of their performance but as an enabler or barrier to quality of life for all members of society, regardless of their physical abilities, age, sex, gender, ethnic origin, etc.

How can transport planners effectively ensure that mobility measures have a positive impact and meet the needs of all citizens such that no one group is disadvantaged? Impact assessment has traditionally focussed on quantifiable parameters and has prioritised journeys undertaken by (male) car owners or for commuting purposes. This has led to inequalities in transport provision which reinforce other inequalities leading to multiple levels of deprivation, reduction in opportunities (e.g. to access resources such as health, education, employment, and recreational facilities), and poorer quality of life.

In the SUITS approach to capacity building, part of our goal has been to encourage local authorities engaged in the development of SUMPs and implementation of new transport measures, to expand the diversity of those consulted and move away from purely quantitative-based performance indicators to understand how transport-related measures affect mobility and quality of life.

Social impact assessment (SIA) is an approach that aims to improve inclusivity in the design and operationalisation of sustainable transport measures, to ensure that all people's voices are heard and given equal value in planning of mobility services.

In this chapter, we describe the role of SIA in the SUMP development cycle, outline several approaches to SIA and their respective benefits or disadvantages, and identify some of the key parameters necessary to carry out an effective and inclusive SIA.

An important factor in SIA of transport measures is ensuring the inclusion of vulnerable groups, mitigating any potential negative effects, and maximising the benefit to these groups. We outline how 'hard to reach' groups can be engaged in this process.

Research into current good practice was carried out within the SUITS project engaging with a wide range of practitioners in transport planning, and we present our findings here, with implications for improving practice in SIA.

11.2   Social Impact Assessment in the SUMP Cycle

Social impacts relate to the direct and indirect impacts of any policy or intervention on people. Social impacts of transport can be defined as:

'...changes in transport sources that (might) positively or negatively influence the preferences, well-being, behaviour or perception of individuals, groups, social categories and society in general (in the future)', [2].

There are eight crucial principles for successful Sustainable Urban Mobility Planning

Plan for **sustainable** mobility in the entire **'functional city'**

Define a long-term **vision** and a clear **implementation** plan

Cooperate across institutional boundaries

Develop all transport **modes** in an **integrated** manner

Involve citizens and **stakeholders**

Arrange for monitoring and **evaluation**

Assess current and future **performance**

Assure **quality**

Fig. 11.1 '8 crucial principles for successful sustainable urban mobility planning'[1]

Such impacts can be distributed over time and place, discriminatory, cumulative and additive, leading to multiple levels of impact and (dis)advantage with one group benefitting from greater mobility more than another.

Sustainable urban mobility planning (SUMP) in European cities integrates eight crucial principles, in which citizen engagement, cooperation, integration and evaluation are key areas, (Fig. 11.1). Each stage of the development of a SUMP requires attention to be placed on inclusivity and the design of transport systems that meet the needs of *all* citizens—young, old, those from diverse ethnic and gender groups, those with disabilities and from lower socio-economic groups. SIA enables analysis of the effects that transport measures may have on different sectors of society, so it can play a central part in ensuring that the key SUMP principles of citizen engagement and inclusion are planned, implemented and evaluated [3].

The overall aim of an SIA is to evaluate and assess the social impact of transport and mobility measures in order to ensure that they are accessible to all and provide all citizens with access to the range of services and facilities they need for a good quality of life. Considering SIA as part of the SUMP process ensures that no one group is favoured above others (horizontal equity) and that groups which may be at a disadvantage (e.g. with respect to social class, income or disability) receive more opportunities and resources (vertical equity). Transport and mobility measures in the SUMP are measured in terms of their performance as enablers or barriers to improved quality of life.

SIA is not a one-off exercise, but is applicable throughout the SUMP cycle, from pre-planning through innovative financing, procurement and implementation stages to post-implementation phases, as indicated in Fig. 11.2.

---

[1] https://www.eltis.org/mobility-plans/sump-concept [Accessed: Mar 15, 2021].

Fig. 11.2 SIA in the SUMP planning cycle. (*Source* adapted from SUMP Guidelines 2nd edition, [3])

SIA enables analysis of the effects that transport measures may have on different sectors of society. It prioritises the need to recognise and look at intersectional[2] effects of transport, and how transport plans may significantly and differentially affect people's quality of life. For example, a new dual carriageway may speed up flow of traffic into and out of a city, improving the journey time of commuters, but it may cause community severance and significantly reduce the quality of life for those in the immediate vicinity, for example by making it difficult to cross a road, and reach shops and friends.

The benefits of conducting a SIA include identifying affected groups, allaying fears and winning trust, avoiding adverse impacts, enhancing positive impacts, reducing costs, getting approval faster from both decision-makers and the community. SIA is predicated on the notion that decision-makers should fully understand the likely consequences of their decisions before they act, and that people likely to be affected should be notified and have an opportunity to participate in the design of their future [4].

Looking at this in more detail, the starting point for SIA is to engage a wide range of public diversity in the planning process. This should include diversity in terms of gender, ethnicity, income, disability and other characteristics depending on the city or region. This needs to take place at pre-planning and strategy development stages to, to feed into measure planning—in this way, a city can keep in mind the benefits to, and impacts upon, different groups, which can help to shape the measures themselves and how they are implemented. SIA can help to ensure that measures achieve citizen support and take-up, provide actual benefits and avoid poorly-judged investments and procurements.

---

[2] Intersectionality is an approach to people's social and political identities, how they combine and create different and unique modes of discrimination and/or privileges.

In the following sections, we briefly outline how social impact effects can be considered at each phase of the SUMP cycle. However, it should be remembered that SIA should be conducted before any change to transport measures is introduced.

11.2.1 Pre-planning Stage—Needs and Requirements

The development of transport measures should be participatory. The preparation and analysis phase requires mapping and understanding the sociodemographics of those likely to be affected by new transport measures and using this to plan stakeholder and citizen involvement, for example by looking at census returns, conducting surveys and co-creation activities such as interviews and diary studies with those likely to be affected by proposals. It is crucial that the information is understood, used and championed by diverse planning teams sensitive to the needs of all users. Having diversity in the planning team itself, will create a firm foundation for considering inclusivity throughout the project. Diverse teams bring more creativity, depth of knowledge, a wider range of skills and experience increasing the likelihood that inclusive and accessible actions and instruments will be developed. If the team is so small that further diversity cannot be achieved within it, the team should consider its wider network and other strategies to ensure that an intersectional perspective is achievable, such as team training. For example, the TinnGO project, [5], focusses on gender and diversity-sensitive smart mobility and provides a roadmap and tools to support gender action planning, whilst EIGE offers gender mainstreaming training and other toolkits, [6].

11.2.2 Strategy Development

During the strategy development phase, discussions and consultation with citizens and stakeholders should be held to build and assess potential future scenarios, develop a common vision and agree objectives, addressing key problems. Measurable targets and indicators should be agreed considering the needs and requirements of diverse stakeholders, women, people with disabilities and vulnerable groups. SIA is an important tool for evaluating priorities and scenarios and ensuring that all viewpoints are considered in the development of inclusive strategies which can be of benefit to all and maximise improvements in quality of life for previously disadvantaged groups.

11.2.3 Measure Planning

Co-creation activities should continue through the measure planning phase. SIA can be used to assess measures for their impact on sustainability and quality of life. SIA should be included as part of monitoring and evaluation during and post-implementation, for example as part of public procurement and calculation of whole lifecycle costs. Affordability and accessibility are crucial in take-up of measures amongst disadvantaged and vulnerable groups, which will have implications for the social impact of measures as well as their overall take-up and impact on citizens.

11.2.4 Implementation and Monitoring

If plans in earlier phases have included SIA, it will already form part of monitoring the implementation, and analysing successes and failures post-implementation. It is important to continue the dialogue with citizen and stakeholder groups, to share results and lessons learned, and consider emerging challenges and solutions for their social impact.

11.3 Methods for Conducting SIAs

There is a lack of standardisation in methods used to monitor and evaluate transport measures, at regional, national and EU levels, which applies also to SIA. In larger transport measures, SIA may form an integral part of the planning process. In such cases, cost–benefit analysis (CBA), multi-criteria analysis (MCA) or a combination of the two have been used. These are briefly described below and may well be familiar to Local Authority practitioners, but both have serious shortcomings when used in SIA. In the absence of other European examples, desk-based research for SUITS project recommended, Social–Economic Impact Assessment (SEIA) as 'good practice' - described in some detail below in relation to studying the effects of introducing a metro system in Delhi, where the impact of accessibility and mobility on socio-economic well-being (SEWB) of the urban poor was analysed [7].

Cost–Benefit Analysis (CBA) can be used to quantify the costs and benefits of a project (over a certain period) and those of its alternatives, usually in monetary terms, in order to have a single scale of comparison. The economic viability of a project can be assessed and expressed by viability indicators such as benefit to cost ratio (BCR), internal rate of return (IRR) or net present value (NPV). Where environmental and social issues can be monetised, they are also included. However, communication and prioritisation of results can be dominated by a few, easily monetisable indicators and focus only on direct benefits such as measures of reliability and reductions in travel times. The benefits derived from sustainable transport in terms of social equity, urban regeneration and improvements in visual quality require more qualitative approaches, and as such may be overlooked. As such, CBA on its own is not suitable as a means of performing SIA.

Multi-Criteria Analysis (MCA) enables the simultaneous quantitative and qualitative measurement of impact, not necessarily in monetary terms. It is more participatory and holistic but can be subjective leading to bias.

Social–Economic Impact Assessment (SEIA) provides a measure of the expected benefits and costs to different groups, [7, 8]. This approach shows the impact of accessibility and mobility on socio-economic well-being (SEWB) on the urban poor. It uses household survey data to derive indicators of accessibility, mobility and SEWB. The indicators are then aggregated into indices of accessibility, mobility and SEWB. The change in indicators and indices in 'before and after' project scenarios is used to assess the significance of the impact of the project on the urban poor. The approach can be configured to other transport schemes and disadvantaged groups with the incorporation of more qualitative approaches providing richer data.

A comprehensive and worked example of conducting a *Socio-Economic Impact Assessment* was carried out for Delhi (2009) funded by Sida and ADB through the Sustainable Urban Mobility in Asia (SUMA) programme,[3] [8]. This provides a detailed account of the steps that the transport consultants undertook couched within an intersectional framework. This provides transferable definitions and indicators.

This section has been extracted from the SEIA handbook by Arora and Tiwari, in 2007, [8] and the associated document produced by Thynell et al. [9]. These are highly recommended as comprehensive guides for practitioners, providing indicators, equations and a worked example using a case study of the effects on the urban poor of introducing a metro in Delhi. In these examples, Thynell et al. [9, p. 23] consider accessibility; mobility and SEWB:

Accessibility: a description of the proximity of destinations of choice and the facilitation offered by the transport systems (including public transport and non-motorised modes) to reach them.

Mobility: defined as both the ability to travel to destinations of choice and the amount of movement necessary to do so.

Socio-economic well-being: defined as the status of a household where the basic social and economic needs for survival are fulfilled, and the household has the capacity to improve its quality of life.

Arora and Tiwari present SEIA as an 8-stage process, [8]. These eight stages are summarised below for completeness. Thynell et al. document their work as consultants as far as level (stage) V after which they start to define steps they need to complete their work on SIA from a consultant's perspective, [9].

Step I: Problem formulation, in terms of theoretical foundations and key hypotheses. For example, the introduction of a new transport system will change accessibility for a certain target group (in the SEIA example, this was the urban poor),

---

[3] https://www.cleanairnet.org/caiasia/1412/propertyvalue-27072.html, [Accessed: 23/03/2021].

and a change in accessibility changes the mobility profile and the socio-economic well-being of the group in question. Any intervention made in the transport system will have direct and indirect impact on the socio-economic well-being (SEWB) of the group in question.

Step II: Project description including planning history, justification, demand assessment, financial plan, expected usage, expected benefits and identified externalities, if any.

Step III: Identifying the target group including, for example geographic location, time and resource allocation and the population characteristics.

Step IV: Data collection to profile the target group in order to generate a base understanding of the issues in order to estimate values for the indicators or accessibility, mobility and SEWB. This represents a point of departure from more traditional and quantitative methods which rely on census data and secondary data sources. Understanding impact requires a more detailed and nuanced understanding which can be gained from more qualitative methods such as observational studies and face-to-face surveys. Such methods are mentioned in the SUMP2.0 as ways of engaging citizens. Understanding the effects of intersectionality, through the use of, for example gender disaggregated data, provide greater insights into how mobility and quality of life may be affected by new transport measures. Indicators for accessibility, mobility and health are defined, for example in [8, Unit 2, p. 24–27]. Arora & Tiwari provide specimen questionnaires in their Appendices [8, p. 75]. Practitioners may also want to look at the worksheets provided in the UK's WebTAG[4] tool [10].

Step V: Profiling the target group. Collate and analyse results from Step IV to understand the issues and trends shown by the data. The data is used to develop indicators of accessibility, mobility and SEWB, for example, socio-demographic profile of households (e.g. number, age, ethnicity, income, work participation rate, car ownership), accessibility to different forms of transport, travel profile. Changes in accessibility and travel profiles can then be calculated for the introduction of new transport measures. As a simple example, what proportion of the target group will be advantaged/disadvantaged by the rerouting or retiming of a bus service? What effects will this have on their daily activities? What other consequences might this have? Changes to the times of bus services may lead to longer waiting times, which may leave women more vulnerable to attack, rerouting services may mean that older people may no longer be able to access services (and go to shops) as they have further to walk, putting in a new carriageway may cut a community in half, leaving some without access to health care, shops or friendship groups.

---

[4] WebTAG is a comprehensive, mainly CBA approach used in the UK which provides advice on how to set objectives and identify problems, develop potential solutions, create a transport model for the appraisal of the alternative solutions, how to conduct an appraisal which meets the department's requirements. This includes social and distributional worksheets, [10]. The system has been widely used to appraise transport measures. https://www.gov.uk/guidance/transport-analysis-guidance-tag.

Step VI: Estimating the indicators of accessibility, mobility and SEWB. The values for accessibility, mobility and SEWB are calculated using the data collected; these are compared to the values of the indicators due to the projects, enabling a testing of the initial hypotheses. The change in the indicators is the first step towards quantifying the impact of the project.

Step VII: Combining the indicators into indices. In this stage, the indicators of accessibility, mobility and SEWB are aggregated using the principal component analysis (PCA) technique to develop indices of accessibility, mobility and SEWB (Worked examples and equations are given in Arora and Tiwari, [8], Unit 3, p. 51–65). Conducting this step demonstrates how each indicator contributes to accessibility, mobility and SEWB.

Step VIII: Developing the SEIA model. Working at this level will show how changes in accessibility and mobility have changed the SEWB. Describing this in detail is beyond the remit of this chapter, and the interested reader is referred to [8, Unit 3, p. 66–72] for the whole story.

Steps I to V were used by Thynell and Arora [9] as a basis for their final consultant's report and map on to Steps 1 and 2 below. For them, SIA breaks down into the following stages.

Step 1: Scoping

(a) Define the public transport project
(b) Identify relevant government policies and plans
(c) Prepare terms of reference for the SIA

Step 2: Assessment

(a) Determine profile of key interest groups
(b) Identify and prioritise key social issues
(c) Determine indicators for select social issues
(d) Collect data to predicting the impacts
(e) Analyse results

Step 3: Mitigation

(a) Identify possible mitigation measures
(b) Determine the feasibility of mitigation measures
(c) Prioritise and select proposed mitigation measures
(d) Propose compensation measures

Step 4: Reporting

(a) Prepare draft report
(b) Review and discussion of the draft report
(c) Prepare final draft report

Step 5: Decision-making

(a)   Send final report to authorised decision-makers
(b)   Discuss report and make amendments if needed
(c)   Take decisions and make public announcements

Step 6. Monitoring and managing

(a)   Implement the monitoring and management plan
(b)   Conduct an independent evaluation

Step 7: Public Consultation (cutting across all other steps)

(a)   Identify potential beneficiaries and other affected groups
(b)   Decide the approach for public consultation including assessment methods
(c)   Hold the public consultation
(d)   Revise the report based on feedback received

Focussing more on issues around social impact, Stage 2b considered the transport and poverty discourse, efficiency vs equity, access and livelihood of the urban poor, gender bias and health impacts of transport (air and noise pollution, road safety, security and crime) and land use. In Stage 2c, socio-economic well-being indicators related to: social well-being which included indicators of literacy, status of women, infrastructural facilities available, and tenure available to upgrade quality of life and economic well-being (WBE) which included indicators of employment, income and assets.

Basic objectives of a SIA have been usefully outlined by the Council for Social Development (New Delhi), [11] as follows:

• Baseline information about the social and economic conditions in the project area
• Information on potential impacts of the project and the characteristic of the impacts, magnitude, distribution and their duration
• Information on who will be the affected group, positively or negatively
• Information on perceptions of the affected people about the project and its impact
• Information on potential mitigation measures to minimise the impact
• Information on institutional capacity to implement mitigation measures'.

The principles which need to be followed in conduction of an SIA include, diverse public involvement, analysis of impact equity, focussing on the most significant social impacts, transparency in methods, assumptions and definitions of significance, providing feedback in timely and appropriate manner to transport planners so that negative impacts can be mitigated, and positive impacts enhanced, use of trained practitioners, establishment of monitoring and mitigation programmes, identification of data sources, planning for gaps in data.

Following these principles effectively relies on significant collection of data and citizen involvement. SIA relies on both primary and secondary data sources.

• Primary sources of data include quantitative methods (as referenced in SEIA and WebTAG discussed above), census data and socio-economic surveys.

- Secondary data sources include census data, land use records, (including records of land transactions), district gazetteers, administrative records (and previous surveys) and documents from non-governmental organisations.

However, richer insights can be drawn from more qualitative approaches which may be more of a challenge. Typical qualitative methods include: key informant interviews, focus group discussions, co-creation workshops, journey diaries and public hearings. There are numerous tools and methodologies which may be used to harness citizen involvement and maximise the value of their input and feedback at all stages of the process. It is beyond the scope of this chapter to list all such tools and methods. Rather, the intention is to highlight the need for inclusivity in the design and operationalisation of such methods to ensure that the voices of all sectors of society have the chance to be heard and are given equal value. Qualitative methods can be used at a variety of levels of engagement, from informing through to consulting, collaboration and empowerment. For a good overview of methodologies in SUMP, the reader is referred to the SUMP2 guidelines [3, p. 48].

11.4 Factors to Be Included in Social and Distributional Impact Assessments

Social impacts arising from transport interventions can be grouped into 5 broad interconnected categories [12];

1. Accessibility (potential)
2. Movement and activity (realised)
3. Health-related outcomes (road casualties and injuries, air quality, noise, physical activity, intrinsic value, mental health)
4. Finance related (affordability)
5. Community related (social interactions, personal safety and fear of crime and harassment, forced relocation).

The many interconnections make it difficult to measure the impact of individual transport measures, especially as these have distributional, time and group effects, [13]. For example, a change in accessibility brought about, by a change in the operation of a bus line, might have an immediate impact on movement and activity. Some effects may be harmful (e.g. in terms of finances and time), whilst at the same time having beneficial effects (e.g. if it leads to an increase in walking or cycling). Figure 11.3 illustrates how transport-related resources translate into opportunities or risks and ultimately social outcomes and well-being, [14].

Table 11.1 provides an illustration of some of the factors that could be included in a SIA, based on the H2020 CIVITAS SUITS[5] SIA survey [15]. These map closely to the categories used in traditional transport impact assessments and will be

---

[5] https://www.suits-project.eu/.

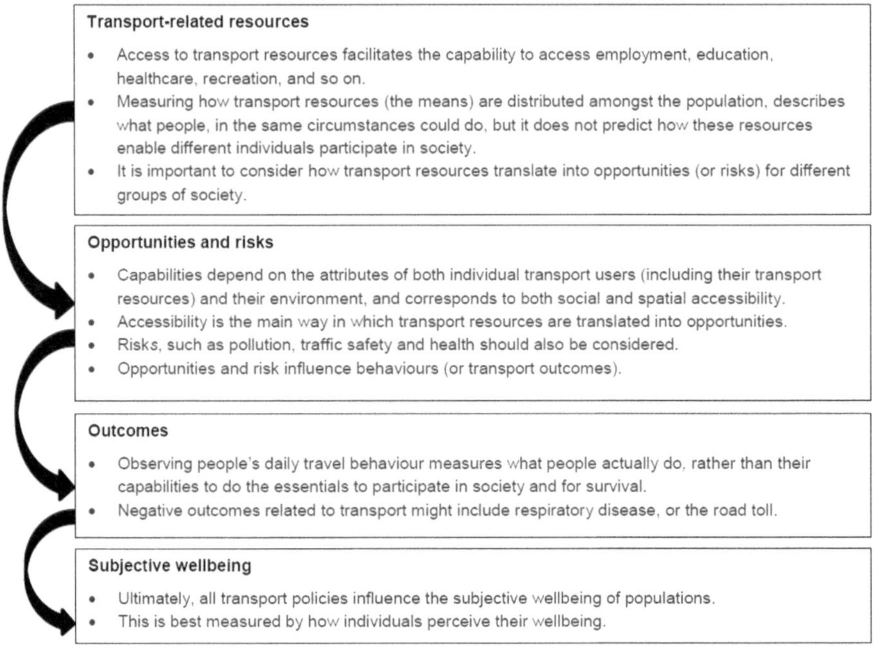

Transport-related resources

- Access to transport resources facilitates the capability to access employment, education, healthcare, recreation, and so on.
- Measuring how transport resources (the means) are distributed amongst the population, describes what people, in the same circumstances could do, but it does not predict how these resources enable different individuals participate in society.
- It is important to consider how transport resources translate into opportunities (or risks) for different groups of society.

Opportunities and risks

- Capabilities depend on the attributes of both individual transport users (including their transport resources) and their environment, and corresponds to both social and spatial accessibility.
- Accessibility is the main way in which transport resources are translated into opportunities.
- Risks, such as pollution, traffic safety and health should also be considered.
- Opportunities and risk influence behaviours (or transport outcomes).

Outcomes

- Observing people's daily travel behaviour measures what people actually do, rather than their capabilities to do the essentials to participate in society and for survival.
- Negative outcomes related to transport might include respiratory disease, or the road toll.

Subjective wellbeing

- Ultimately, all transport policies influence the subjective wellbeing of populations.
- This is best measured by how individuals perceive their wellbeing.

Fig. 11.3 Relationship between transport-related resources and subjective well-being (*Source* Curl et al., 2020, p. 24 [14])

familiar to practitioners. Other categorisations are possible. For example, the INCLU-SION project [16] identifies nine categories: empowerment, empathy, accessibility, affordability, gender equity, safety, convenience and efficiency.

Consideration of social impact should be embedded in the transport planning process as part of increasing inclusivity and accessibility, along with assessments of environmental, health, social and economic factors and new procurement regulations.

The SUITS project also consulted more widely in a second phase of the research (2020), drawing on the experience of 21 practitioners in 7 countries, across a range of organisations including local authorities, transport operators, logistics companies, engineering companies, and NGO's, [17].

From this exercise, we suggest adding or elevating the importance of the following factors:

Environmental factors—this was felt to provide a link between transport and health and providing equity for deprived groups, whether they lived in inner city areas or more remote low-income districts.

Air quality and noise—in light of the growing awareness of air quality during COVID19 lockdowns, overall air quality, with (traffic) noise, was felt to be of increasing importance to quality of life.

Lower income groups in general—possibly considering concessionary travel passes.

Table 11.1 Illustration of factors that should be considered in a SIA

Issue	Factors
Quality of life or liveability issues	Improved accessibility of education, health, employment and other services, ability to take advantage of opportunities
	Overall community and personal satisfaction
	Visual aesthetics of the public realm, streetscape/journey ambience, landscape, effects on historical and heritage resources, property values
	Effects of travel
	Journey quality, transport choice/option value, affordability, travel time, accessibility
	Community cohesion and severance, imposition on physical activity
	Safety and security, casualties and injuries
	Distribution of impacts/amenities amongst vulnerable populations
Environmental issues	Overall quality of the public realm
	Noise, air, soil and visual pollution
Accessibility	Availability and physical accessibility of transport
	Safety and security
	Level of service provided
	Access to spatially distributed services
	Effects of structural issues on pedestrians
	Transportation choice and option values
Economic issues	Connectivity
	Reduction in travel time
	Equity of economic benefits
Social cohesion	Effects caused by reduced opportunities for interaction
	Social isolation and exclusion
	Lack of access to essential services
	(Forced) relocation
Provider and process-based issues	Range and quality of engagement during and after planning
	Poor maintenance and neglect of schemes

Young people on lower incomes—needed better transport equity to allow them to have better employment opportunities.

Deprived communities on the periphery of urban areas—better travel equity for people living in deprived areas, on lower incomes particularly in cheaper housing areas more remote from the city, which were not well-connected by public transport.

Deprived communities related to ethnicity—this reflects a growing awareness that social deprivation is frequently linked to ethnicity, which deprivation has cumulative effects on social well-being.

Greater accessibility is needed for people with disabilities

Enabling independent travel for children and older people

Information accessibility—with more travel information available online and through apps, there is a risk that some users will be excluded.

11.5   Citizen Engagement, Vulnerable and 'Hard to Reach' Groups

To enable collection of adequate data for SIA, with relevant feedback about impact and perceptions of impact, there is potential for citizen engagement throughout the SUMP process, as illustrated in Fig. 11.4.

In the past, research and user engagement has been significantly skewed towards the male car owner or over representation of commuter journeys. This has led to inequalities in transport provision (which may reinforce other inequalities) leading to multiple levels of deprivation, reduction in opportunities (e.g. to access resources such as health, education, employment and recreational facilities) and poorer quality of life.

Fig. 11.4   Citizen involvement in the SUMP process (Sump Guidelines 2019, p. 46) [3]

Consulting with citizens through SUMP development and implementation phases needs to ensure that not only the usual vociferous interests and well-recognised groups are consulted, but also that vulnerable groups are included—and these are often thought of as 'hard to reach'. The groups which need special consideration are those with social vulnerability. The *EMPOWER* project defined this as:

> 'social groups which are disadvantaged in the transport system in general. Generally, this will mean people outside the group of physically and intellectually fit and able employed adults travelling to and from a single workplace on weekdays', [18].

From our desk-based reviews and two surveys (2017 and 2020) when compared with earlier results from METPEX [1], we found evidence from those working in the transport sector, of a broader range of people who are regarded as vulnerable and a more nuanced understanding of intersectional effects including:

- Those from **low-income groups living** in **socially deprived urban areas**. In terms of transport poverty, low car ownership can be detrimental to finding employment; high transport costs may make it difficult to access employment, particularly for young people or those in low-income occupations; poor access to public transport makes it difficult to attend interviews, constrains choice of employment location and makes it harder to maintain employment; long commuter journeys for these groups may also lead to depression; deprived social areas may also have a poor and unsafe urban environment and be exposed to high levels of pollution.
- Those living in **rural or peripheral urban areas** with poor access to public transport may suffer from time, accessibility and affordability poverty. Longer commutes are also detrimental to well-being and mood.
- **Age** has a variety of effects; young men and children from low-income groups are more at risk from traffic accidents; women with young children have a number of mobility restrictions, and older adults (who may have age-related impairments) suffer from a decline in mobility which, may lead to social isolation, loss of confidence, loneliness and dependence on others.
- Little data have been collected on **ethnicity** and transport equity despite over-whelming evidence that people from black, Asian and minority ethnic groups (BAME), or those that appear different from the local norm in terms of facial features, clothing, gait, etc. suffer from many and multiple disadvantages. For 'minority groups', transport is a hostile environment where people face abuse, harassment, crime and experience more traffic casualties and injuries. Public transport may be avoided all together out of fear of harassment. When compounded with low incomes and gender, transport poverty ensues.
- **Gender and diversity**. Research over the last two decades has demonstrated that women's travel patterns are different to men's. Women's mobility is influenced by age, culture and house-care duties. All this affects their access to healthcare, employment, education and recreational facilities. Recognition of the needs of the LGBTQI + community is of central importance here as well. The need for a greater understanding and use of intersectional analysis has become evident during the COVID pandemic.

- Those with a **mobility reduction or hidden disability** (e.g. invisible but chronic conditions, disability in communication, cognitive or social skills) make fewer trips. Transport is a big challenge owing to the design of vehicles, systems, the urban environment and the attitude of drivers, other passengers and road users, [19]. Barriers to travel may include cost (e.g. if special taxis have to be hired to accommodate wheelchairs), design of public transport (in terms of accessibility), road furniture, increasing mixed (and sometimes illegal use) of pavements (e.g. by e-scooters), increased burdens of trip planning, time taken in trips and concerns about safety, control and uncertainty.

With SIA and more intersectional approaches, it is not enough just to regard these as discrete sociodemographic classes. A more nuanced, empathic approach is needed to look at how mobility is affected by transport planning and the effects of transport changes on everyday lives. Factors combine to create levels of multiple deprivation.

However, these groups need to be defined systematically for each SIA, based on a thorough analysis of the populations likely to be affected by a transport measure, e.g. using census returns and/or surveys. Special attention should be placed on inter-sectionality. For example, people on lower incomes may experience poorer health, have lower levels of literacy and may not live in zones supported by regular transport services.

The Covid crisis of 2020–2021 showed how these might change with circum-stances and location. For example, in the UK, the mobility of those with long-term illnesses, the elderly and respiratory problems was curtailed when they were advised to go into social isolation. To reduce the spread of the disease, the country and then parts of it went into lockdown. As the crisis persisted, key workers were considered vulnerable, and then, young people's mobility was curtailed in efforts to stop the rate of infection (with curfews, social distancing and restrictions on the number of people one could mix with). Different countries adopted different practices. The effects of enforced reduced mobility had a number of significant, distributional effects on lifestyle, health and well-being. For example, in the UK, age, gender, ethnicity/culture, housing and employment were found to be key factors in disease risk—showing the importance of understanding intersectional effects.

11.5.1 How to Reach These Groups?

The SUITS project advocates a user-centred, consultative approach to ensure vulner-able users' representation in the planning and construction of new measures and the development of SUMPs. Digital online surveys can reach a wide range of citizens, but may leave out those with poor access to digital devices or the internet (for a host of reasons such as deprivation, cognitive disabilities, age, linguistic ability)—alter-native strategies will be needed to include the views of groups with limited access to or enthusiasm for these channels. Engagement and activities supporting SIA can be achieved through local events, in the community, at a time and place convenient for residents or transport users and should feature active listening/recording of views and follow-ups.

The principles of gender and diversity-sensitive mainstreaming should guide consultation and discussion, ensuring true representation. This simply means considering gender and diversity in everything that we do, working towards becoming fully inclusive, ensuring that all stakeholders are involved in a safe, open and diverse environment which best meets the needs and reality of society and organisations.

To enable the inclusion of vulnerable and 'harder to reach' groups, consideration should be given to composition, timing, location and format of stakeholder and citizen meetings and how groups such as economically disadvantaged, ethnic minorities, women and vulnerable groups will be engaged so that they can share their experiences and provide input in SIA's processes. This may require channelling communication and recruitment to events through social media, in different languages, to various community groups and holding events at times and locations accessible to local communities or groups. For example,

- holding meetings at night, or with crèches so that women with children can attend,
- holding consultation events in the community where there is a large and natural footfall, using accessible buildings,
- levelling the power balance through use of language (avoiding or explaining jargon and acronyms) and material which is readily understandable and attractive,
- using social media campaigns to build up awareness and buy-in of consultation process

Key elements which may be overlooked in the planning of citizen engagement are:

- **adequate preparation, budgeting and planning of events**—allowing time for advertising and recruitment, design of the material, understanding the venue, the audience and how the events and the outcomes fit into the overall strategy.
- **trust and relationship building**—user engagement needs to build throughout planning, development, implementation and evaluation. This means that activities will take place over a number of years. This is not a one-off event, the more people understand about the process, the more they will buy into it.
- **linking SUMP to wider issues of social inclusion** (intersectionality), e.g. in Belgium, [3, p. 59], mobility planning includes consideration of free concessionary travel passes for elders and social fees for taxis for those on low incomes who cannot drive; Greater Manchester's (UK) transport strategy is driven by six societal trials, with an evidence base that is continually updated [3, p. 164].
- **the 'tokenism trap'**, i.e. the apparent involvement of a group or use of a method that hides a lack of commitment to be involved with users or to hear what they are saying. Most often seen at the earlier levels of user engagement (informing and consulting), for example tick-box exercises where engagement targets have been met or the required number of questionnaires/surveys conducted; one person invited on to a committee to represent the voice of their group.

- **understanding the engagement from the perspective of the citizen and wider ethical practices,** e.g. Are the issues framed in ways citizens can understand? Has enough space been left for meaningful interaction? Have they been given an active role? Is the event fulfilling and fun for them? The more someone understands about the proposal, the more they can contribute and the more meaningful and the more rich their input.
- **wider role of consultation at community or individual level**, in terms of capacity building, resilience, and buy-in. Although associated with engagement at levels of empowerment and collaboration, if engagement exercises are framed correctly then citizens could leave with new insights, knowledge and interests. Longer-term engagement can lead to training, e.g. in citizen science, advocacy, expert users and co-creators.
- **need to record, present and act on the results**. This key stage is often overlooked. For face-to-face, qualitative and participatory approaches, this may take longer than the event itself. Without this, the whole exercise has been a waste of time. Key issues include recording and collection of information at the event; collation and analysis of results; format, timing and presentation of key findings; pathway to feed into the overall SUMP process.

Further examples of participatory approaches are provided in the SUMPS2 guidelines [3], and a wide range of support tools can be found on the Eltis website and U4IoT's co-creation tools.[6]

11.6   Improving Practice of SIA

As referred to earlier, SUITS carried out two separate surveys of practitioners over the course of the project (2017 and 2020) to investigate social impact assessment in practise, [15, 17].

The surveys found that SIA could be made more effective through:

- better funding and planning of SIA at the start of the SUMP process so that it can inform design and implementation
- closer cooperation between those conducting the SIA, stakeholders, especially local authorities and users
- employing/working with people who possess good skills to bring groups together and work with disparate groups
- widening of inclusion criteria to those seeking employment, young people and commuters
- making SIA processes, activities and material simple and easy to use and accessible

---

[6] https://www.eltis.org/resources/tools including the *SUMP Participation Kit* See also *U4IoT's co-creative workshop methodology handbook*.

- translating surveys, findings and impacts into layperson's language and different languages of minority groups in the region, bearing in mind levels of understanding and interest in mobility, especially when the scheme may not be in their local area or seem to impact on them
- conducting consultations in safe settings for the group with trusted interviewers, intermediaries and community leaders
- collaborating with established groups, faith communities, women's clubs, migrant groups, social workers, street workers, youth clubs, etc.
- informing key stakeholder groups at each stage
- moving from communicating, to consultation, co-design and empowerment using more creative engagement methods
- extending use of SIA in transport measures to consider, for example

  - how urban transport can be used as a tool for social inclusion of all groups in a society.
  - environmental impact and economic assessment (e.g. motives for buying electrical or hybrid cars).
  - indirect effects of transport measures, e.g. education performance of pupils, effects of cleaner transport on health of citizens, etc.
  - longer vision horizons, not 5 but 15 years ahead.

11.7 Conclusion

A key aspect of the design and implementation of sustainable and innovative transport measures is the direct and indirect effects on citizens, in particular vulnerable groups. The understanding that transport is fundamentally about the movement of people and goods and not vehicles is a relatively recent paradigm shift. This recognition has profound consequences which are, to a certain extent played out in sustainable transport measures.

The initial SUITS survey found, amongst other things that, even when a social impact assessment was conducted, its effectiveness could be reduced by lack of resources, tokenism, poor operationalisation and political issues. This is true for many citizen engagement activities.

Given the potential role of new sustainable transport measures in increasing accessibility and transport equity and reducing transport poverty, together with concerns expressed by LAs that new innovations such as MaaS (and additional concerns noted here about information accessibility) may reduce inclusivity; SUITS would recommend that SIA could be included as a mandatory part of existing regulations such as life cycle costs relating to procurement. This would signal a LAs commitment to ensuring that the health and well-being of all citizens remained at the centre of planning. A holistic appraisal could be undertaken of the 'social' consequences of proposed measures as part of the wider cost–benefit/lifecycle of the planned measures. From this, contingency plans can be developed to address negative impacts such as breaking of communities, displacement of traffic (and its effects) on poorer neighbourhoods.

SUITS approach sought to heighten our partner cities' awareness of social impacts and the value of SIA. Those participants in our survey who were more experienced in a 'hands-on' focus on social impact commented on a growing awareness amongst the general public about 'sustainability and transport' and its social implications. Factors such as a growing awareness of climate change and the additional focus placed on transport impacts during the COVID19 lockdowns may well provide a fertile base within which cities can prove their approach towards SIA and its importance.

References

1. Tovey, M., Woodcock, A., Osmond, J. (eds.): Designing Mobility and Transport Services: Developing Traveller Experience Tools. Taylor and Francis Ltd., Oxon (2016)
2. Geurs, K., Boon, W., Wee, B.V.: Social impacts of transport: literature review and the state of the practice of transport appraisal in the Netherlands and the UK. Transp. Rev. **29**(1), 69–90 (2009)
3. SUMP Guidelines 2nd edn: Rupprecht Consult (editor), Guidelines for Developing and Implementing a Sustainable Urban Mobility Plan, 2nd Edn, p. 30 (2019)
4. IOCPGSIA (Interorganizational Committee on Principles and Guidelines for Social Impact Assessment): Principles and guidelines for social impact assessment in the USA. Impact Assess. Project Appraisal **21**(3), 231–250 (2003)
5. https://www.tinngo.eu/
6. https://eige.europa.eu/gender-mainstreaming/toolkits/gender-equality-training
7. Ph.D. dissertation: Anvita Arora, "Socio-Economic Impact Assessment (SEIA) Methodology for Urban Transport Projects: Case Study Delhi Metro; Thynell, M., Arora, A., Punte, S. Social Impact Assessment of Public Transport in Cities: An approach for people involved in the planning, design, and implementation of public transport systems". ADB/CAI-Asia Center Final Consultants' Report under TA 6291. Pasig City, Philippines (2009). Available http://www.rlarrdc.org.in/images/SIA%20Report%20ADB.pdf
8. Arora, A., Tiwari, G.: A Handbook for Socio-economic Impact Assessment (SEIA) of Future Urban Transport (FUT) Projects, Transportation Research and Injury Prevention Program (TRIPP), Indian Institute of Technology, New Delhi (2007). Available https://www.researchg ate.net/publication/316881853
9. Thynell, M., Arora, A., Punte, S.: Social impact assessment of public transport in cities: an approach for people involved in the planning, design, and implementation of public transport systems. ADB/CAI-Asia Center Final Consultants' Report under TA 6291. Pasig City, Philippines (2009)
10. TAG: Social and distributional impacts worksheets, (last updated 2020). Available https://www. gov.uk/government/publications/tag-social-and-distributional-impacts-worksheets
11. Council for Social Development, New Delhi: Social Impact Assessment, Report of a Research Project on Social Impact Assessment of R&R Policies and Packages in India (2010). Available http://www.indiaenvironmentportal.org.in/content/330023/social-impact-assessment-rep ort-of-a-research-project-on-social-impact-assessment-of-rr-policies-and-packages-in-india/
12. Jones, P., Lucas, K.: The social consequences of transport decision-making: clarifying concepts, synthesising knowledge and assessing implications. J. Transp. Geogr. **21**, 4–16 (2012)
13. Martens, K., Bastiaanssen, J., Lucas, K.: Measuring transport equity: key components, framings and metrics. In: Lucas, K., Martens, K., Di Ciommo, F., Dupont-Kieffer, A. (Eds), Measuring Transport Equity. Elsevier, Amsterdam (2019)

14. Curl, A., Watkins, A., McKerchar, C., Exeter, D., Macmillan, A.: Social impact assessment of mode shift. Waka Kotahi NZ Transport Agency Research Report (2020)
15. Woodcock, A.: Social Impact Assessment Report, Deliverable 7.3, H2020 CIVITAS SUITS (2018). Available https://www.suits-project.eu/reports/
16. Tovaas, K.: The INCLUSION Project, D3.4, Typology and description of underlying principles and generalisable lessons (2020)
17. Saunders, J., Woodcock, A.: Social Impact Assessment Survey 2020 Report, H2020 CIVITAS SUITS (2020). Available https://www.suits-project.eu/reports/
18. Glensor, K.: Development of an index of transport-user vulnerability, and its application in Enschede. The Netherlands. Sustainability **10**, 2388 (2018)
19. Wilkin, D.: Introduction: exploring disability hate crime. In: Wilkin, D. (Ed.), Disability Hate Crime: Experiences of Everyday Hostility on Public Transport. Springer International Publishing, Cham (2020)

Chapter 12
Data Collection and Analysis Tools for Integrated Measures

Mireia Calvo Monteagudo(D), **Miriam Pirra**(D), **Marco Diana**(D), **Fotis K. Liotopoulos**(D), and **F. Tilesch**

Abstract Currently, several data collection tools exist to measure and monitor the daily traffic in urban areas. Such tools should also be used to provide data on passenger and freight movements to inform short-, medium- and long-term mobility plans. One of the objectives of this chapter is to provide an introduction to the current strategies available and used by medium-sized European cities for data collection, including traditional and more technological automated methods. It will provide information on where there are gaps and difficulties in data collection processes, analyse information relating to urban mobility data, draw conclusions and identify possible generic problems of cities by presenting current methods and solutions that have been developed in Europe for data collection. In addition, the results of two actions implemented in the cities of Turin and Kalamaria for the collection of data in a dynamic and innovative way will be shown, along with a tool developed to manage big data applications in the transport sector. As a conclusion, the main benefits of these tools and the traffic data gathering are highlighted.

M. Calvo Monteagudo (✉)
Packaging, Transport and Logistics Research Centre—ITENE, Valencia, Spain
e-mail: mireia.calvo@pnoconsultants.com

M. Pirra · M. Diana
DIATI—Department of Environment, Land and Infrastructure Engineering, Politecnico di Torino, Turin, Italy

F. K. Liotopoulos
SBOING, Thessaloniki, Greece

F. Tilesch
LOGDRILL, Veszprém, Hungary

© Transport for West Midlands 2023
A. Woodcock et al. (eds.), *Capacity Building in Local Authorities for Sustainable Transport Planning*, Smart Innovation, Systems and Technologies 319,
https://doi.org/10.1007/978-981-19-6962-1_12

12.1 Introduction

Currently, urban mobility patterns are characterized by a continuous expansion and a growing dependence on private vehicles. According to a United Nations report, 80% of all European citizens will live and/or work in cities by 2030. Urban transport is producing adverse impacts on sustainable development, affecting the environment, health and safety of the citizens.

The need to change this trajectory has resulted in the increased need to develop and implement sustainable and integrated urban transport systems in recent years. The European Commission adopted an Action Plan on Urban Mobility, which provides a coherent framework for 20 concrete EU-level actions, including the acceleration of Sustainable Urban Mobility Plans (SUMPs) development, the upgrade of data and statistics, the improvement of urban freight transport and the increase of travel information.

According to the recommendations for the development of SUMPs and SULPs [1], it is necessary, as a first step, to understand the city context from the mobility, transport and logistics points of view, in order to identify the specific issues and concerns that need to be tackled in the development of the plan. The development of a baseline scenario requires the collection of transport data. In addition, the use to which the data are put can affect the data collection methodology, and the quantity of data required.

The Urban Freight research roadmap [2] developed in 2014, showed that urban freight flows represented 10–15% of urban traffic and 25% of urban transport-related CO_2; nevertheless, commonly used data collection methodologies do not include these. In this regard, [3] has revealed gaps in data collection, which have implications both for understanding urban freight transport activity patterns and also for developing urban freight models. Issues that have been identified in considering urban freight data gaps include:

- Lack of standardization in the methods used to collect urban freight data. This results in data gaps, making comparisons between datasets impossible.
- Variability in the level of reporting of freight data and analysis between studies.
- Data about light goods vehicle activity are not always available.
- Lack of data about the supply chain as a whole (i.e. the links between urban freight activity and the freight activity upstream in the supply chain)
- Insufficient geographical detail about goods vehicle trips in urban areas.
- Data collection concerning trips carried out by consumers for the purposes of shopping are not considered.
- Insufficient freight data for non-road modes.
- Little information about how data were collected and processed, its reliability and representativeness.

12.2 Context

When defining the most effective data collection methods to use, it is important to consider, at the same time, the analysis strategies which will be used and the capacity of the organization to handle and analyse such data. Large amounts of data can potentially be useful for local authorities and decision-making bodies, in order to implement, assess and compare measures on mobility management. However, this needs to be carefully planned.

One of the problems in collecting massive amounts of data is that much of it has no real value. This requires a selection and pre-filter to get to the really useful data items. In addition, different data collection methodologies result in data gaps when comparisons between datasets are attempted. In this regard, big data systems have been conceived as an opportunity for improving the management of larger amounts of collected data.

One of the most used methodologies for data analysis are key performance indicators (KPIs). KPIs are crucial to assess the current situation and its evolution over time. Although the use of quantifiable passenger-related KPIs is extensive, there are still gaps in the definition of the most useful KPIs for urban freight analysis [4] and more qualitative factors.

Local authorities (LAs) also use decision support systems (DSSs) to make use of the gathered data. These can support decision-makers in understanding and simulating the structure of urban systems and in computing indicators for target setting and benchmarking to identify level of service. Nevertheless, DSS need to be considered as helpful tools, but never as decision systems in themselves. The entire responsibility associated with making a decision using a DSS resides with people who built and use the system.

A survey conducted of the cities participating in SUITS revealed a latent need for increasing information about freight flows, origin–destination traffic matrix and active mobility (especially about bike trips and infrastructures) [5]. Information about traffic flows needs to be addressed from the point of view of better understanding transport patterns and demand characterization (e.g., purpose of journeys, frequency and type of freight loads, vehicle propulsion systems, fleet characteristics or passenger satisfaction data).

Traditionally, this type of data has been gathered via surveys or national statistics. Nowadays, ICT-based technologies also play a relevant role. For example, smartphones or other small devices can act as a data provider for active mobility. In the following SWOT analysis, the main strengths, weaknesses, opportunities and threats to data collection in Europe are explained.

SWOT analysis is an extended methodology that can serve as a precursor to any action, such as exploring new initiatives, making decisions about new policies, identifying possible areas for change or refining and redirecting mid-plan efforts [6].

Strengths
– Creation of SUMPs has increased because of the awareness of cities in relation to their mobility problems
– The European Commission is concerned about the necessity of developing SUMPs and SULPs and pushes local authorities for it
– Greater investments have been made for the development of new systems for data collection and analysis
– Analysis techniques such as decision support systems provide support to decision-makers to understand and simulate the structure of urban systems

Weaknesses
– Current methods such as government surveys do not differentiate between urban and non-urban transport
– Continuous transmission of data generates an important heavy load on the transmission channels
– New technologies do not provide all the data that would have been collected in a traditional survey, so they need to be combined
– There is still some lack of information related to pedestrian and bicycle flows although these modes are increasing in importance in the transport systems
– Commonly used data collection methodologies do not include freight flows, which generates some gaps in freight data collection methods. Sometimes, it is due to the difficulties in accessing private companies' information and confidentiality reasons
– It is common for different data collection processes to use different data collection methodologies. This results in data gaps when comparisons between datasets are attempted
– Data about light goods vehicle activity are not always available
– Trips carried out by consumers for the purposes of shopping usually are under-reported
– There is insufficient freight data for non-road modes
– Often there is relatively little information available about how data were collected and processed
– Although passenger-related KPIs are notably extended, there are still some gaps for the definition of most useful KPIs for urban freight analysis

Opportunities
– The extended use of ICTs in recent years opened new possibilities for high amounts of "real-time" data collection with a relatively low collection
– Local authorities are requiring more information about freight flows, O–D traffic matrix and active mobility
– Potential of new technologies to improve the collection of mobility data
– Long-term economic savings for cities, companies and people
– Real-time traffic information could lead to many improvements such as congestion reduction or dynamic network traffic control
– Active modes relevance in modal share is notably increasing
– Some technologies for collecting traffic information such as pneumatic road tubes or piezoelectric sensors can be also useful for bicycle traffic gathering
– Information about traffic flows needs to be addressed from the point of view of better knowing transport patterns and demand characterization (e.g. purpose of journeys, frequency and type of freight loads, vehicle propulsion systems, fleet characteristics or passenger satisfaction data)
– KPIs are crucial for the assessment of the current situation of urban mobility and to compare the evolution over time

(continued)

(continued)

Threats
– Limitation of traditional traffic collection systems and manual gathering methods for getting the overall picture of the mobility system
– Privacy concerns when using ICT-based solution for data collection (in-vehicle devices, floating car data, smartphones information)
– Combination between ICT-based technologies for automatic data collection and traditional collection methods for the inclusion of some additional factors relevant for the mobility system characterization
– Mistakes in the level of accuracy of data collected
– Public organizations need to work closely with private companies to overcome issues concerned with funding and confidentiality to obtain access to it
– Complexity of methods and models for urban freight data collection and analysis
– Collecting massive amounts can generate data with no real value, which requires a selection and pre-filter to get those really useful data
– Good data analysis strategies are crucial, especially when large amounts of data are collected (e.g. open data)

12.3   Data Collection Methods

Traditional data collection methods are based on manual systems or fixed road sensors, providing specific information about the location where they are installed but not a general overview of the different transport flows. The extended use of ICTs in recent years has opened up new possibilities for high amounts of "real-time" data collection with a relatively low cost.

One problem faced by service providers is the compliance with the specific privacy laws when collecting private information from users (in-vehicle devices, GPS position through smartphone detections, etc.). In addition, ICT-based technologies for automatic data collection need to be combined with more traditional techniques for the inclusion of additional factors relevant for the mobility system characterization, such as land use and qualitative behavioural data necessary for decision-making.

The main data collection methods are summarized below:

12.3.1   Manual Methods

a. Surveys and interviews are used to collect both passenger and freight data. These methodologies provide a variety of useful information about traffic and travel demand patterns. However, they are resource intensive (time, personnel and analysis) and require direct contact with representative users (Fig. 12.1).

b. Manual counts: this usually refers to the practice of counting classified traffic in a "manual fashion." Counting and classification are based on visual examination

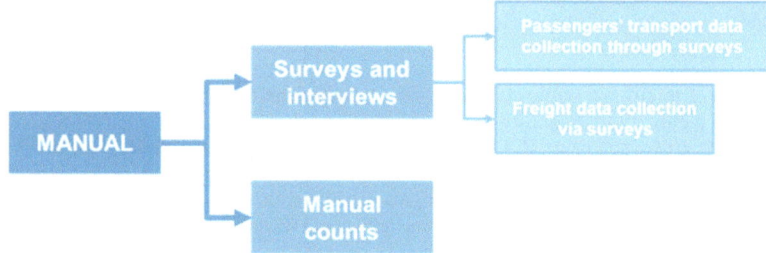

Fig. 12.1 Manual data collection methods

and judgments by individual observers. The data are usually recorded using Tally sheets or mechanical counters. After data have been collected for an interval of time, totals are calculated and registered on a data sheet, which can be input into a computer later.

12.3.2 Automated Methods

See Fig. **12.2**.

Traffic detection systems:

a. Pneumatic road tubes. This is a popular method of vehicle sensing, consisting of rubber tubes placed across road lanes to detect vehicles from the pressure changes that are produced when a vehicle tyre passes over the tube.

b. Piezoelectric sensors: a long strip of piezoelectric material enclosed in a protective casing. Sensors are usually placed in a groove along the roadway surface of the lane(s) monitored.

c. Magnetic loops: one of the most conventional technologies used to collect traffic data. It consists of a coil of wire wrapped around a magnetic core. It measures the change in the magnetic field caused by the passage of a vehicle.

d. Inductive loops: a square of wire embedded into or under the road. The loop utilizes the principle that a magnetic field introduced near an electrical conductor causes an electrical current to be induced. In the case of traffic monitoring, a large metal vehicle acts as the magnetic field and the inductive loop as the electrical conductor.

e. Infrared (passive or active): the sensors are mounted overhead to view approaching or departing traffic or traffic from a side-on configuration. Infrared sensors are used for signal control; volume, speed, and class measurement, as well as detecting pedestrians in crosswalks. Real-time signal processing is used to analyse the received signals for the presence of a vehicle.

f. Microwave radar sensors: used to detect moving vehicles, to determine their speed and to detect motionless vehicles.

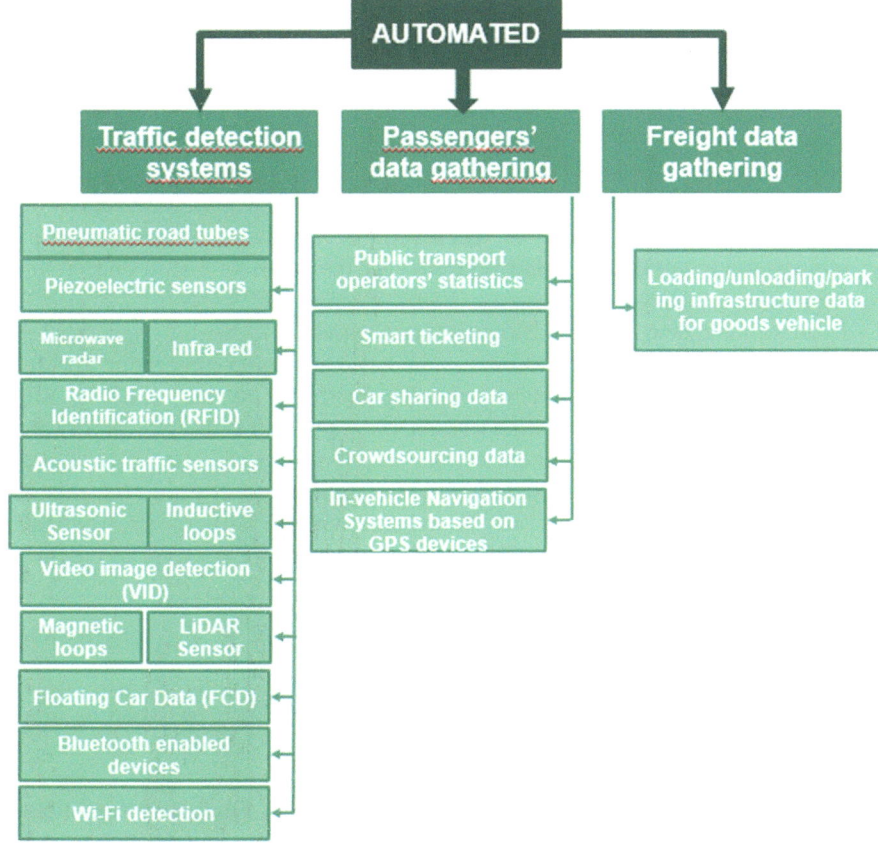

Fig. 12.2 Automated data collection methods

g. LiDAR: "light detection and ranging." Lidar has a wide range of applications; one use is in traffic enforcement and, in particular, speed limit enforcement. Devices are designed to automate the entire process of speed detection, vehicle identification, driver identification and evidentiary documentation.

h. RFID: radio frequency identification refers to small electronic devices that consist of a small chip and an antenna. The chip is typically capable of carrying 2,000 bytes of data or less. The RFID device provides a unique identifier for an object and must be scanned to retrieve the identifying information.

i. Acoustic traffic sensors: measure vehicle passage, presence and speed by detecting acoustic energy or audible sounds produced by vehicular traffic from a variety of sources within each vehicle and from the interaction of a vehicle's tyres with the road.

j. Ultrasonic sensor: traffic detection system that may be used in conjunction with other sensor technologies to enhance presence and queue detection, vehicle counting and height and distance discrimination.

k. Video image detection (VID): system typically consists of one or more cameras, a microprocessor-based computer for digitizing and processing the imagery and software for interpreting the images and converting them into traffic flow data. Cameras can be positioned along streets, on tall buildings or even on drones.

l. Floating Car Data (FCD): collects real-time traffic data by locating some vehicles via mobile phones or GPS over the entire road network. Data generated by the equipped vehicles as a sample are used to assess the overall traffic condition. Some data such as car location, speed and direction of travel are sent anonymously to a central processing centre to get the status of traffic and provide alternative routes.

m. Bluetooth-enabled devices: Bluetooth is a wireless technology standard for exchanging data over short distances. A basic system configuration for data collection via Bluetooth consists of a Bluetooth device that scans for other Bluetooth-enabled devices within its radio proximity and then stores or forwards the data for future analysis and use.

n. Wi-Fi detection: Wi-Fi technology allows the collection of traffic information and can visualize and analyse results to better manage traffic flows, basing the decision on the knowledge of traffic performance and its response to measures implemented.

Passenger data gathering:

a. Public transport operators' statistics: statistics play an important role in data compiling. Public transport operators play an essential part in providing information on KPIs which may be made available through reports or data files on the operators' website.

b. Smart ticketing: system whereby entitlement to travel (or ticket) is stored electronically on a microchip rather than being printed on a paper ticket. As a by-product of smart card use, large volumes of data are automatically captured.

c. Data from shared mobility services: car sharing, bike sharing and micromobility services are increasingly popular means of transport in cities. Local authorities should establish agreements with operators in order to exploit data gathered. Such datasets will provide valuable information about origin–destination journeys, average travel time depending on traffic conditions, users' profiles, etc. and could be linked to socio-demographic variables

d. In-vehicle Navigation Systems based on GPS devices: method used to give precise locations of objects. GPS works by providing information on exact location. It can also track the movement of a vehicle or person.

e. Crowdsourcing data: process through which an entity requests specific resources from a group of people. The objective is to develop new methods of crowdsourcing data and information from citizens over a large scale, via participatory and passive sensing.

Freight data gathering:

a. Loading/unloading/parking infrastructure data for goods vehicles: the urban road freight transport requires specific and significant infrastructure in order to carry and distribute goods to the final destinations. Characterization of vehicle flows and movement offers a general overview of the amount and kind of products delivered in a local area. The analysis of data from loading, unloading and parking areas can provide valuable information for a better understanding of freight operations.

12.4   Key Performance Indicators (KPIs) for Passengers and Freight Mobility Analysis

KPIs provide the basis for a standardized evaluation of mobility systems and the measurement of improvements resulting from the implementation of new mobility practices or policies. The main objectives are to:

- Measure the level (quality) of service
- Diagnose the situation
- Communicate and report on the situation and objectives
- Continual evaluation of progress against benchmarks

12.5   Innovative Data Collection Methods Developed and Tested in SUITS

In the framework of the SUITS project, a deeper test and analyses have been conducted on gathering tools [6]. Two demonstrators were developed. One was based on the freight data gathering in Torino (Italy), and the other focused on ICT and crowdsourcing data in Kalamaria (Greece). The main aim of both demonstrators was to facilitate local authorities in the use of tools that could help them to gain wide information about traffic in their cities. Such data can support the decision planning and the mobility measures to be applied to the cities. In addition, a tool, to store and analyse datasets, (PP4TM) is presented.

12.5.1   Freight Distribution in Urban Areas: Torino Trial

Many cities are observing higher levels of urban freight activities due to the increase of e-commerce in addition to the more traditional distribution to shops. This worsens problems related to congestion and environmental impacts. The analysis performed in the Italian city of Torino illustrates how data can be provided (and combined) to inform the implementation of new measures to meet these problems, and the level of know-how needed to extract useful data.

The trial presented a tool to help cities investigate a specific aspect characterising and influencing their mobility, namely the observation of freight flows from the demand side. The trial consisted of the design, implementation and testing of a spatial cluster analysis approach to understand which are the most important loading/unloading parking spots in an urban setting by processing the GPS traces of a fleet of logistic vehicles. The trial, developed by Politecnico di Torino (POLITO), was performed in three phases.

The first phase involved the exploitation of existing datasets and started with an assessment of the traffic in the city based on Global Position System (GPS) vehicle traces. The information obtained was combined with that available from the city traffic monitoring centre to identify the most congested areas. During this phase, spatial analysis was used though a specific GIS-based data mining technique devoted to finding the most significant clusters (groups) of service stops within the city. This approach helped in seeing where a large number of stops were being made and thus where deliveries were performed.

On the one hand, the results obtained from this analysis provided the opportunity to create a highly disaggregated key performance indicator (KPI) based on the time lost in congestion by each vehicle in each road segment. This KPI can be used to inform a wide range of policy actions within the transport sector, both from the viewpoint of a municipality and from that of an individual actor in the transport system. In the case of the trial, the KPI more specifically helps to understand which vehicles need to be monitored in their monthly deliveries since they lose time in traffic more than the average of the whole fleet (Fig. 12.3).

On the other hand, the analysis provided information about the most congested points of the city (not in terms of levels of service of the roads in the network as is usual, but rather in terms of the total amount of time wasted by vehicles in traffic), the main traffic flows and the most populated zones for loading and unloading goods, among others (Fig. 12.4).

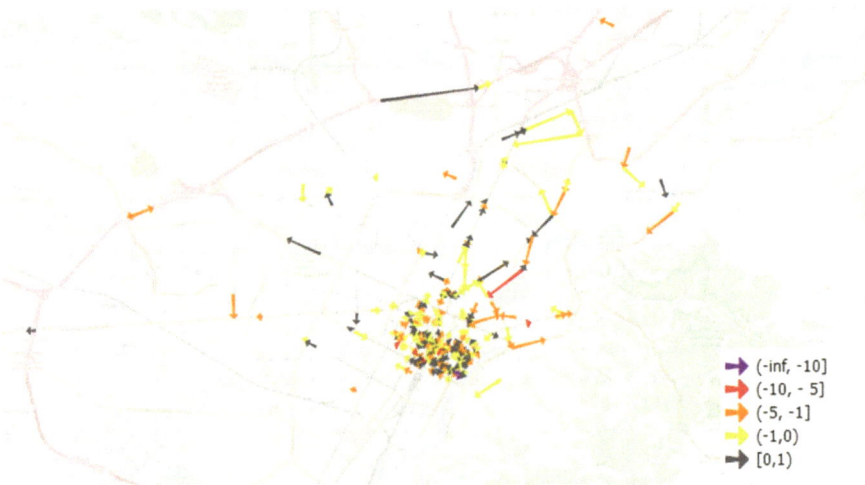

Fig. 12.3 Value of indicator for a selected vehicle, namely vehicle 32 (Reproduced from [7] under the licence https://creativecommons.org/licenses/by/4.0/legalcode)

Fig. 12.4 Value of the mean indicator computed for all the arcs of the road network, zoomed in the city centre (Reproduced from [7] under the licence https://creativecommons.org/licenses/by/4.0/legalcode)

Once the most common clusters of spots for loading and unloading by duty vehicles were identified within the most critical areas in terms of the above-mentioned congestion KPI, the next phase started. One of the areas selected was in the limited traffic zone in the centre of Torino. The aim was to assess the urban deliveries performed in that part of the city through a survey to retailers and shops (Fig. 12.5).

The second phase involved the design of a survey to collect information on the dynamics of freight deliveries (and pickups) at these locations. It was necessary to investigate which operators usually deliver in the selected areas and where the vehicles are parked in order to check the exploitation of the available unload/load areas. Moreover, it was important to check the duration of stops, the number and dimension of packs delivered and the final destination of such deliveries. As express couriers could collect boxes and materials from shops, this operation was recorded too.

The third phase consisted of field activity over 7 days to test the surveying tool at a specific site among the previously identified clusters. This was selected based on two criteria: (1) being inside the limited traffic zone where congestion and pollution problems are relevant and the public space on streets was particularly precious and (2) having load/unload parking spots, the usage of which could be evaluated. Therefore, information gathered in the first phase provided essential guidance in selecting the site(s) on which the survey could be run to maximize the efficiency of the process since surveying activities are very costly and time consuming.

This tool can be used to inform a range of policy actions at the municipality level, mainly on the freight delivery side. The analysis of stops could help in understanding the effectiveness and impact of delivery operations in key areas of the city that could,

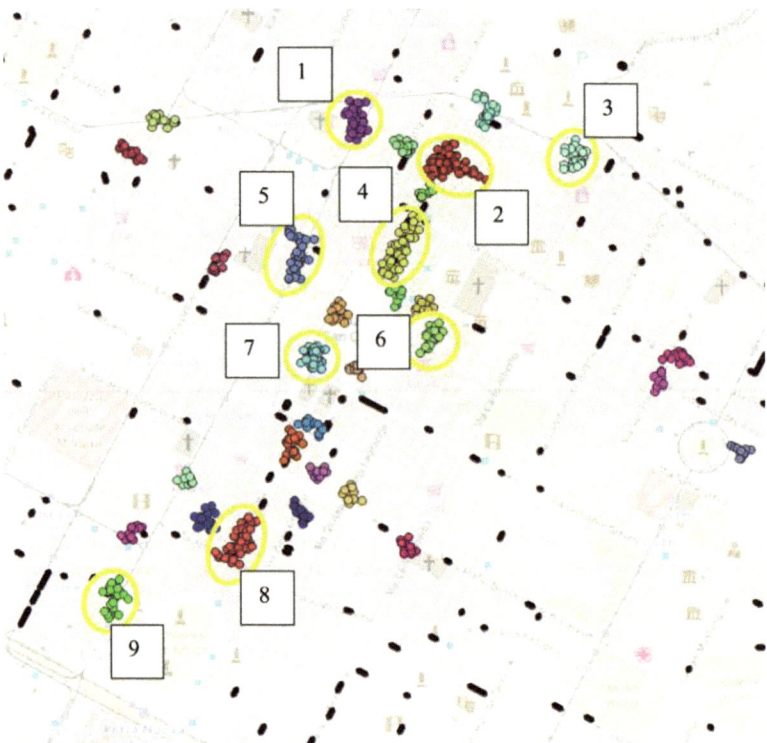

Fig. 12.5 Clusters with at least 10 stops (coloured dots), clusters with at least 20 stops (yellow ovals) and designated load/unload areas (black signs) (Reproduced from [8] under the licence https:// creativecommons.org/licenses/by/4.0/legalcode

for example lead to an assessment of the already available unload/load parking spots. Moreover, the study of the deliveries in a specific street and the investigation of the retailers' and shops' use of express couriers' services helps in understanding which actions the local authorities should address to improve urban freight policies. The methodology can give useful insights to municipalities on a way of monitoring the freight distribution patterns at the more disaggregated individual loading/unloading area. The interested reader is referred for additional details to two openly available technical publications; the first one centred on the congestion KPI [7], and the second one presenting the selection process of loading/unloading areas and the subsequent survey [8].

12.5.2   Urban Traffic Monitoring Through Crowdsourcing: The Kalamaria Trial

One of the biggest gaps in European cities is the inability to identify urban freight flows. The following tool uses crowdsourcing to differentiate such journeys. The trial involved several citizens volunteering to provide their vehicle traces, which were analysed and visualized through the MypolisLive.net platform [9]. This analysis and visualization tool provides real-time traffic data and the opportunity to assess the impact of new mobility measures applied in the city.

The crowdsourcing method of data gathering was tested in Kalamaria (Greece) using several logistics companies, taxi drivers, citizens and city vehicles. A GPS tracker was installed in each vehicle to provide its traces and positions in an anonymous way. Moreover, a navigation app, UltiNavi (and UltiCarNavi for in-vehicle multimedia consoles), developed by SBOING, was used to assist in the mobility data collection, (https://sboing.net/ultinavi).

UltiNavi automatically records the vehicle's route as soon as the driver runs it on the vehicle's console. It operates with or without internet connectivity to record traces. Whenever internet connectivity is available, the user may voluntarily upload real-time or recorded data to the cloud infrastructure. Moreover, the user can "donate" their data (after pseudonymization) to a given stakeholder (e.g., to the municipality or to the city), in which case, the data are forwarded to that party. In this way, a city/municipality may collect (offline) mobility data.

During the 4 months of the Kalamaria trial, the tools collected data and sent it to the platform MyPolisLive.net. The city of Kalamaria was able to visualize the entire map of the town and the traffic information provided by crowdsourcing. The vehicle's location, speed and sensor values were monitored using pseudonymization. The traffic conditions in the city were visualized on a city map based on road colour-coding. Moreover, the average speed, the number of vehicles circulating per road, road name and speed limit were also provided. Based on the datasets obtained, the average speed per road segment was calculated (minimum and maximum), together with the atmospheric conditions. The platform was able to distinguish between the different kinds of vehicles monitored—freight vehicles, citizens, taxi fleet or any other vehicle monitored.

The system has two operating modalities: real-time (live) mode and historic mode. The former provides information in real time about the traffic in the city and is capable of tracking the vehicles. The latter offers, instead, the traffic condition in a particular time-frame and provides a report analysis and graphs based on historical data and a weekly statistical analysis.

The platform is available for any municipality and includes an embedded mapping engine capable of automatically generating the city boundaries and its entire road network. Moreover, complete streets, consisting of multiple road segments, are automatically identified and marked using geospatial queries and algorithms.

The devices and the platform were developed by SBOING, a Greek ICT SME, with the aim of supporting small–medium local authorities and transport stakeholders to gather and control their own data to support integrated transport planning. Based on the crowdsourcing and processing of the vehicles' mobility data, the information obtained is about:

- vehicle tracking
- monitoring the real-time traffic conditions of the city
- collecting and visualizing historical traffic data concerning citizen mobility and freight traffic
- correlation between urban citizen traffic and freight traffic.

Although the above tools appear to provide information already available through other commercial providers, such as real-time traffic conditions, there are a number of advantages in implementing the UltiNavi/MyPolisLive crowdsourcing system. Essentially, all primary data (i.e., individual vehicle traces) rather than simply the statistics are available for analysis and stored in servers under the control of the city. It is then possible, for example to run ex-post analyses on historical data, compute KPIs such as the previously introduced congestion KPI or train traffic models to forecast flows or implement control strategies (signals, real-time information). Indeed, this is one of the reasons why most larger cities still run a traffic operation centre that collects data through extensive infrastructure and hardware deployment, even if we are in the "Google era." Another advantage is that this type of crowdsourcing provides segregation of the source data, i.e., differentiates it by source class (e.g., from taxis, logistics, citizen vehicles, city vehicles, etc.). The use of crowdsourcing tools might enable even smaller cities to do something very similar, with much fewer resources.

A description and resources about the Kalamaria trial and relevant tools are available at the SUITS Capacity Building Toolbox, https://cbt.suits-project.sboing.net/. Open, crowdsourced traffic and mobility data (from the Kalamaria trial and other) are available at the SUITS Open Data Repository: https://cbt.suits-project.sboing.net/suits-tools/dare-tools.

12.5.3 PP4TM: PetaPylon for Traffic Management

The tools presented above illustrate how data can be integrated and used to provide small–medium cities with a low-cost way of gathering data about the flow of traffic, which would enable internal and in-depth big data analysis. However, all the information, which potential is "Big Data," needs to be efficiently stored and analysed. Therefore, an additional tool developed in the framework of the SUITS project was devoted to this task.

The "PetaPylon for Traffic Management" (PP4TM) tool [10] is a fast, versatile and robust big data analytics database solution for urban traffic research and analysis purposes. The system is capable of processing and analysing big data, real-time data and batch (historical, stored) data. The PP4TM high-speed data analyser system is a combination of a unique, non-relational database sub-system and a hardware processing configuration. It provides a multi-purpose, analytical database system, specially designed to perform analysis from semi-structured data and able to process many data modalities in parallel, even in real time.

The first step in using PetaPylon as the base of a traffic management platform is to feed it with the appropriate traffic data. PP4TM is flexible and able to receive and process geographical/numerical/text data, or visual information at the same time, provided the appropriate data format and structure is properly predefined.

The system can perform queries at very high speeds for data sizes as big as petabytes. However, traffic or mobility-related knowledge is not a built-in feature. Only, field experts on traffic and mobility can form meaningful and valid queries to mine interesting results and find the hidden connections between seemingly independent events.

Data collection and data analysis as a business service for municipalities are widely available. One of the favourite service providers is Google who provides a time-limited analytical platform in exchange for local public transport data. (https://cloud.google.com/maps-platform).

However, these tasks, e.g., collecting and analysing data, can also be easily and freely executed by using PP4TM, mainly because the data are already owned by the local authority. There are certain advantages for a municipality to set up its own data analyser infrastructure. First of all because the data are stored and owned locally, there is no burden or limitation on the analytical capabilities. The availability of the historical data allows the performance of wide and deep analysis. Also, by utilizing PP4TM's versatile capabilities, municipalities can also use and exploit datasets from external sources, e.g., from taxi or transport companies or other third parties.

A description, resources and demos of the PP4TM system are available at the SUITS Capacity Building Toolbox, https://cbt.suits-project.sboing.net/suits-tools/pp4tm.

PP4TM is an open-source application, based on Google's Hadoop hardware configuration and Cloudera Impala analytical database. It provides connectors for business intelligence applications, such as QlikView, MicroStrategy, SpotFire, Tableau. Moreover, PP4TM is as follows:

- able to process data in any structure from any source,
- able to provide very fast query response times,
- robust and fault tolerant, linearly scalable hardware system,
- can be attached to most artificial learning/deep learning platforms,
- can be attached to business intelligence tools.

How does PP4TM work?

1. It is necessary to start from an hypothesis about the data that have been gained. For example: "Air pollution is directly proportioned to traffic intensity."
2. Map all the available data from the city.

3. Select the data related to our hyphothesis. Traffic data (real-time traffic situation, traffic intensity per road,…) and environmental data (from air pollution stations, from noise measurement stations…).
4. Connections between the data. Data contain coordinates, date and time fields,… in order to make the connections.
5. The PP4TM tool has all the data imported to it, and it starts the meaningful analysis of the different datasets.
6. Verify if the initial hypothesis is correct or not. In our example, it is necessary to select the environmental and traffic data in a specific date and place through a query in the tool.
7. PP4TM provides a subset of data to analyse as a result. The data obtained can be visualized too for easier and powerful analysis.

This tool is a big data "warehouse" for historical and real-time data. It is oriented towards a data-analytics data centre, enforced with high security and with business and real-time analysis capabilities.

12.6   Data Collection Benefits

The three different tools presented in this chapter support the transformation of mobility and decision-making towards integrated transport planning. Both trials provide a traffic profile of the city, knowledge about the most congested points in the network, the most populated routes, frequent delivery points and main traffic flows.

The analysis of freight distribution in urban areas (Torino pilot) provided information about the use of freight loading and downloading areas which could be used to identify the over and/or underuse of delivery parking spots and assess the impact of the lack or excess of spots. As a result, the usage of the parking spots can be modified according to the city's needs.

The urban traffic monitoring through crowdsourcing (Kalamaria pilot) provides traffic information and visualization tools to enable the assessement of current new measures in order to know what would happen with the traffic when the action is implemented.

The PP4TM tool provides a more in-depth analysis connecting different types of data available and being capable of doing a deep and quick data analysis.

The information collected is useful to [11];

- Test and implement Sustainable Urban Mobility Plans (SUMPs), Sustainable Urban Logistics Plans (SULPs) and mobility measures.
- Evaluate new measures in order to know what would happen with the traffic in the city.
- Assess loading and downloading places (overuse or non-existent) and modify the usage.
- Improve urban accessibility and optimize the distribution of goods.
- Reduce the distances travelled and travel time.

- Reorganize the traffic in case of accidents in the city, when the city holds a big event or in case of street works.

LAs can gain benefit from traffic data because it facilitates the decision-making of the city council and the different transport actors, to make decisions that can result in a reduction of pollutant emissions and energy consumption and consequently an increase in the quality of life with a healthier and more sustainable environment.

However, LAs are not the only agents that can derive benefit from the tools explained here; many other actors involved in the daily traffic of the city can get advantages from them:

- Taxi fleets: the taxi fleets monitoring through crowdsourcing and the analysis of the taxi distribution in the city can provide information about strategic points in the city where a large number of clients take and leave the taxi. Data gathered by the tools can be exploited by taxi fleet companies to gain more passengers and more adequately cover areas in the inner cities. As a consequence, the number of empty runs would decrease, and taxi fleet operations would be more sustainable and efficient.
- Shared services fleet company: the traffic monitoring and the data analysis can provide information about strategic points in the city where a large number of sharing vehicles are needed at certain hours. This information can be used to distribute shared vehicles (especially bikes and mopeds) in the city and provide a better service to the passengers.
- Logistic company: traceability of their vehicles and goods to the client can be monitored, and critical points identified. In addition, they can follow faster routes with less traffic and get benefit from the measures implemented in the city to optimize deliveries.
- IT company: the information gained through the analysis of freight distribution tested in Torino can be used to develop a load/download parking management app. This data can be exploited by an IT company to develop a booking app to provide efficient management of parking lots. This app can be used by the different freight delivery companies to book, for a slot of time, a load/download parking spot and ensure an efficient use.
- Sites attracting a large number of commuter trips: the tools can also be exploited by large employers in their mobility management actions, providing, for example a public transport incentive to encourage the workers to use public transport in the daily travel to work or encourage carpooling and commuting with active transport modes (cycling, walking).

12.7 Conclusions

Small–medium cities are in need of tools for gathering traffic data and most importantly, traffic data analysis in order to get benefits for their transport systems. Different collection tools tailored to contexts where resources are scarce have been presented here, focusing on data gathering from freight distribution in urban areas, urban traffic monitoring through crowdsourcing and a big data storage and analysis database.

The experiences of Torino and Kalamaria show that appropriate tools can be made available appropriate to the resources of small–medium cities with valuable and scalable results.

The first step is to analyse which data are currently gathered and which data the city need. For this step, a SWOT analysis of the methods used for data gathering and an assessment about which data are missing can be a starting point.

The second step is to select the appropriate tool to gather the data needed and improve the current methods. As was stated, there are manual and automated data collection methods that can provide data flows from traffic, passengers or freight data.

The innovative methods tested in SUITS project are suitable for small–medium cities which don't have much budget to invest in data collection. The crowdsourcing traffic monitoring only requires the incentivation of the participant and the installation of the dispositive in the vehicles; then, the monitorization is registered in a platform. The freight data analysis is based on the traffic gained by the city, and it is a methodology to evaluate the flows of different kinds of vehicles, in the case explained the methodology was applied to freight vehicles.

Once the data collection method is selected, the third step is to implement the methodology and gather data.

The fourth step is the data analysis. The assessment of the data gathered is important in order to analyse the situation of the traffic in the city, to apply measures to alleviate it and facilitate the circulation in the city.

The tools and methodologies explained build up the capacity of small–medium cities of monitoring and analysing the different flows of traffic, giving the capacity to apply new measures that alleviate the congestion and reorganize the traffic in the city.

References

1. Wefering, F., Rupprecht, S., Bührmann, S., Böhler-Baedeker, S.: Guidelines: Developing and Implementing a Sustainable Urban Mobility (Logistics) Plan. ELTIS. https://www.eltis.org/sites/default/files/guidelines-developing-and-implementing-a-sump_final_web_jan2014b.pdf (2014)
2. ERTRAC and ALICE: Urban Freight Research Roadmap. https://www.ertrac.org/uploads/documentsearch/id36/ERTRAC_Alice_Urban_Freight.pdf (2015)
3. Browne, M., Allen, J., Woodburn, A., Patier, D., Routhi-er, J.-L.: Comparison of urban freight data collection in European countries. In: Proceedings of the 11thWorld Conference on Transport Research. Berkeley CA (2007)
4. McKinnon, A.: Performance measurement in freight transport. International Transport Forum. https://www.itf-oecd.org/sites/default/files/docs/mckinnon.pdf (2015)
5. Diana, M., Pirra, M., Woodcock, A., Martins, S.: Supporting urban integrated transport systems: transferable tools for local authorities (SUITS). In: 7th Transport Research Arena TRA. Vienna (2018). https://doi.org/10.5281/zenodo.1441138

6. Herrero, D.M., Calvo, M., Martín, N., Rodríguez, J.Á.: Research and gap analysis on data collection and analysis methods. Deliverable 3.1 of the SUITS project. https://www.suits-pro ject.eu/wp-content/uploads/2018/12/Gap-Analysis-on-data-collection-and-analysis-method ologies.pdf. (2017)
7. Pirra, M., Diana, M.: Integrating mobility data sources to define and quantify a vehicle-level conges-tion indicator: an application for the city of Turin. Eur. Transp. Res. Rev. **11**(1), 41 (2019). https://doi.org/10.1186/s12544-019-0378-0
8. Diana, M., Pirra, M., Woodcock, A.: Freight distribution in urban areas: a method to select the most important loading and unloading areas and a survey tool to investigate related demand patterns. Eur. Transp. Res. Rev. **12**, 40 (2020). https://doi.org/10.1186/s12544-020-00430-w
9. SBOING: MyPolisLive.net. https://www.mypolislive.net/. (2018). Last accessed May 2020
10. LOGDRILL: PP4TM: PetaPylon for traffic management. Retrieved from SUITS capacity building toolbox. https://cbt.suits-project.sboing.net/suits-tools/pp4tm. (2018).
11. Woodcock, A., Calvo, M., Diana, M., Liotopoulos, F., Pirra, M.: Demo results, recommenda-tions and guidelines for cities on how to exploit open data and develop business opportunities. Deliverable 3.3 of the SUITS project. (2019)

Chapter 13
Innovative Public Procurement Processes to Implement Sustainable Mobility Policies

Dan Caraman, Ştefan Roşeanu, Isolda Constantin, and Cristiana Dâmboianu

Abstract Investment in goods, services and works is demanding for any local authority in both time and money. The new legislative framework related to procurement across the EU brings a major shift in terms of procedures and requires a reconsideration of public authorities' organisation and capacities. The contracting authorities need to make more strategic decisions in relation to public procurement, to include environmental and social objectives and innovation potential in their selection. This paradigm shift requires changing the mentality and traditional ways of working of managers and professionals in LAs, to enable them to prepare long-term procurement strategies and action plans. Public procurement planning is an on-going process. New procurement procedures should build on the experience gained in the organisation by implementing similar projects, on the experience gained by the market and wider socio-economic changes. This chapter introduces innovative processes in procurement suitable for small–medium LAs wishing to finance sustainable transport measures. These include pre-commercial procurement, procurement of innovative solutions, joint procurement, whole life cycle costs-driven procurement, external transport costs. To further develop capacity in LAs to handle procurement issues, the SUITS Consortium has prepared online guidelines on innovative procurement, which were piloted prior to release in Alba Iulia. The integrated decision support tool is freely accessible from https://www.suits-project.eu/ids-tool/ and provides more detailed information on the contents of this chapter.

13.1 Introduction

As outlined in Chap. 2 (and see [1–8]), European cities are faced with a series of common challenges relating to transport and mobility, which together creates a negative effect on the European economy and on the quality of life of European citizens.

D. Caraman · Ş. Roşeanu (✉) · I. Constantin · C. Dâmboianu
Integral Consulting R&D, Bucharest, Romania
e-mail: stefan.roseanu@integralconsulting.ro

© Transport for West Midlands 2023
A. Woodcock et al. (eds.), *Capacity Building in Local Authorities for Sustainable Transport Planning*, Smart Innovation, Systems and Technologies 319,
https://doi.org/10.1007/978-981-19-6962-1_13

The Transport White Paper [1] proposes strategic objectives to be met by 2050. To reduce congestion and emissions, the paper calls for cities to follow a mixed strategy involving land-use planning, pricing schemes, efficient public transport services and infrastructure for non-motorised modes and charging of clean vehicles. It specifically encourages cities to develop SUMPs bringing all these elements together.

To rise to these challenges, far reaching, city-wide joint actions are needed by local authorities, transport operators, local business, logistic suppliers, landlords, estate developers, stakeholders and citizens. This volume has shown that there are a wide range of tools available to assist cities to develop efficient and effective interventions to benefit citizens. Public procurement reform plays a very important role in reaching these objectives:

- Every year, over 250,000 public authorities in the EU spend around 14% of GDP on the purchase of services, works and supplies. In many sectors such as energy, transport, waste management, social protection and the provision of health or education services, public authorities are the principal buyers. Transparent, fair and competitive public procurement across the EU's Single Market generates business opportunities, drives economic growth and creates jobs [2].
- The estimated value of tenders published in TED excluding utilities and defence 'shows an increase of 9.2%, from 319.66 in 2014 to 349.18 billion in 2015. [...] Excluding utilities and defence, significant increases were in Romania (33%), Estonia (31%), Slovenia (24%), UK (23%) and Malta (21%)' [3].

The principles and legal framework of public procurement within the EU are mainly defined under directives 2014/24/EU [4], 2014/25/EU [5], 2014/23/EU [6] which enhance the efficiency of the public procurement system in Europe and foresee more intelligent rules and electronic procedures. At the same time, these new rules allow the authorities to use public procurement to try to achieve more far-reaching political objectives, such as the social, environmental and innovation objectives. Based on these directives, each member state has implemented a legislative package and methodologies providing a legal framework through which public procurement procedures must be conducted. In brief:

European Directives (17/2004/CE [7] and 18/2004/CE [8]) were developed as the first European regulatory step taken regarding procurement procedures. Research projects financed under European programmes, after 2004, pointed out innovative aspects with a view to a new approach to public procurement and included criteria such as energy consumption, environmental impact and life cycle costs [9–11].

This fuelled the need for reform in public procurement, which was seen as essential in domains of public interest, for instance transport, mobility, energy and innovation. In these areas of long-term development, evolution is rapid and supplies and services have a determining role in sustainable development and in enhancing life quality. As a result of the EC undertakings, as well as of the aspects pointed out through such research works, the following materials were produced:

- Directive 2009/33/CE [12] promotion of clean and energy-efficient road transport vehicles, amended by Directive of the European Parliament and of the Council 2019/1161 [13].
- Regulation (EC) no 1370/2007 [14], public passenger transport services by rail and by road.
- Directive 2014/24/EU on public procurement [4], replacing directive 2004/18/EC [8].
- Directive 2014/25/EU on procurement by entities operating in the water, energy, transport and postal services sectors [5], replacing directive 2004/17/EC [7].
- Directive 2014/23/EU on the award of concession contracts [6], which does not directly replace a previous Directive.

The effect of applying these legal changes is the public procurement reform briefly described in the guidelines [15].

In line with the article referring to the transposition and transitional provisions under the directives, member states were asked to introduce laws, regulations and administrative provisions necessary to comply with these directives by 18 April 2016. The new legal framework contributes to enhancing the efficiency of the public procurement system in Europe and foresees more intelligent rules and a larger number of electronic procedures.

13.2 Context of Procurement Research in the SUITS Project

The overall goal of SUITS is to support cities to improve their capacity to reduce congestion and pollution, to develop sustainable mobility, in order to increase the quality of life of citizens.

One of the key areas of SUITS concerns the gaps in capacity of local authorities in small and medium-sized cities in terms of knowledge and working practices to implement the new EU directives on finance and procurement. To address this, three guidelines have been created to provide decision support for the development of sustainable mobility. Together these aimed to:

- Provide a useful and efficient tool to apply innovative measures in procurement policies and procedures related to sustainable mobility development
- Enhance the administrative capacity of authorities and stakeholders in small and medium-sized cities with a view to facilitating sustainable mobility development.

Three documents have been created, accessible from https://www.suits-projec t.eu/, namely:

1. Guideline for applying innovative and sustainable financing approaches
2. Guideline to innovative public procurement (which is the topic of this chapter) and

3. Guideline for the development of bankable projects, new models and business partnerships

The guideline documents offer a general overview of the respective topic, with a series of annexes providing additional information, examples and selected references useful for understanding concepts (such as life cycle costs and external costs), the way to use them in the process of preparing strategies and public policy documents aimed at developing sustainable mobility.

The guidelines' goals are as follows:

- To provide a useful and efficient tool to apply innovative measures in procurement policies and procedures related to sustainable mobility development
- To enhance the administrative capacity of authorities and stakeholders in small and medium-sized cities with a view to facilitating sustainable mobility development.

In 2004, the European Directives (17/2004/CE [7] and 18/2004/CE [8]) were developed as the first European regulatory step taken regarding procurement procedures. Research projects financed under European programmes, after 2004, pointed out innovative aspects with a view to a new approach to public procurement and included criteria such as energy consumption, environmental impact and life cycle costs [9–11].

13.3 Innovative Public Procurement Processes for Sustainable Mobility

Transport is vital for mobility and economic development, but its development (by transport modes) is still chaotic or short term, and not in line with the White Paper on Transport, COM (2011) 144 final—Roadmap to a Single European Transport Area—Towards a competitive and resource-efficient transport system [1]. This is the case, especially when the external costs are not sufficiently quantified, and agreement has not been reached on their internalisation. External costs (e.g., gas emissions, noise pollution, accident costs, traffic congestion and parking, etc.) are not included in fares paid for travel but are paid by the entire society. EC considered that public procurement procedures are the most important lever that can bring about fast technical progress, in the desired direction, especially in the context of preparing long-term sustainable mobility policy. The following sections introduce some of the more innovative and recent public procurement processes around sustainable mobility.

13.3.1 Pre-commercial Procurement (PCP)

PCP refers to public procurement of research and development services, not including their implementation into the final commercial products. The challenge is to enable public purchasers to collectively implement PCPs to close the gap between supply and demand for innovative ICTs. The objective is to bring radical improvements to the quality and efficiency of public services by encouraging the development and validation of breakthrough solutions through pre-commercial procurement [16].

PCP can be used when the market fails to provide solutions able to meet the purchasers' requirements. Procurement processes are used to get new solutions developed and tested, to meet long-term ambitions of the LA. PCP allows for a comparison across alternative approaches to solutions, by designing the solutions, developing and testing the model or prototype [15].

PCP directs innovative development towards public sector needs, allowing for a comparison or validation of the various approach solutions. PCP achieves market openings for SMEs and an acceptable management of innovation risk through small contracts, gradually growing in size [17].

In PCP, the public purchaser chooses not to keep the research and development (R&D) results exclusively for their own use. This approach is based on:

- Risk-benefit sharing according to market conditions (both for the public purchaser and for the companies involved in PCP),
- Competitive development in phases,
- Separation of the R&D phase from deployment of commercial volumes of end-products.

For instance, the phases may refer to defining the research, exploring the solutions, prototyping, piloting a small series of products or services to prove that they are ready for production and/or delivery at acceptable quality standards. On the other hand, contracted R&D activities do not include commercial development activities such as production in large quantities.

PCP procedures are organised stage by stage, evaluating each R&D stage to progressively select the best solutions. For the last stage, at least, two competitors are selected, to ensure a competitive market.

13.3.2   Procurement of Innovative Solutions (PPI)

PPI can be used when challenges of public interest can be approached through innovative solutions which are almost finalised and do not require research funding to develop new solutions. In this case, suppliers can purchase existing solutions to be delivered and tested at deadlines, integrating them into existing products and services. A PPI may still involve conformance testing before deployment [18].

PPI can thus be used when there is no need for procurement of new R&D to bring solutions to the market. This procedure sends a clear signal from a sizeable amount of early adopters/launch customers that they are willing to purchase and deploy the innovative solutions if they can be delivered to the desired quality, price and time.

The main differences between PCP and PPI processes are summarised below (Table 13.1).

Table 13.1 Main differences between PCP and PPI processes [18]

Category	PCP	PPI
Consortium	3 legal entities, minimum 2 of them public procurers	3 legal entities, minimum 2 of them public procurers
EU grants	90%	35%
When?	Requires R&D to get new solutions developed. Problem clear, but pros/cons of competing solutions not compared or validated yet No commitment to deploy yet	Requires solution which is almost on the market/already on the market in small quantity, but not meeting public sector requirements for large-scale deployment yet. No R&D involved
What?	Public sector buys R&D to steer development of solutions to its needs, gather knowledge about pros/cons of alternative solutions, to avoid supplier lock-in later	Public sector acts as launch customer/early adopter for innovative products and services that are new to the market
How?	Public sector buys R&D from several suppliers in parallel (comparing alternative solution approaches), in the form of competition evaluating progress after critical milestones, risks and benefits of R&D shared with suppliers to maximise incentives for wide commercialisation	Public sector acts as a facilitator establishing a buyer's group with critical mass that triggers industry to scale up its production chain to bring products on the market with the desired quality/price ratio within a specific time. After a test and/or certification, the buyers group purchases a significant volume of products
Eligible activities	• Joint procurement of the R&D services • Implementation of procurement contracts • Assessment of outcomes of the procurement • Confirmation of 'after PCP' strategy for dissemination/exploitation of results	• Joint procurement of innovative solutions • Implementation of procurement contracts • Assessment of outcomes of the procurement • Confirmation of 'after PPI' strategy for dissemination/exploitation of results

13.3.3   Joint Procurement Processes

Another direction that can be followed by public authorities in the preparation of procurement in the field of sustainable urban mobility is to organise public procurement procedures jointly. Joint public procurement means that procurement activities can be shared by two or more contracting authorities (CAs). The key defining feature is that there must be a single offer published by all participating authorities [18].

Joint procurement is not a completely new concept in the European Union legal and regulatory framework. The directive 2004/18/EC [8] and directive 2004/17/EC

[7] included provisions related to joint procurement procedures, and joint procurement agencies have been organised in the past in EU member states, for example public authorities in the UK, the Netherlands and Germany have been buying together for a number of years. However, in many European countries, especially in the South, there is often little or no experience in this area, as we can see in examples selected in Table 13.2 [3, 19]. The 2016 public procurement directives provide better clarifications and rules in setting the procedures.

However, the new directives are not intended to prescribe either joint or separate contract awards, and CAs have to evaluate which is the best way to organise the procurement process, given their past experience and situation [4, 20].

Centralised purchasing activities represent those conducted on a permanent basis, either for the acquisition of supplies and services intended for contracting authorities or for the award of public contracts or the conclusion of framework agreements for works, supplies or services intended for CAs [4].

In this situation, a central purchasing body is a CA providing centralised purchasing activities and, possibly, ancillary purchasing activities [4].

Contracting authorities rarely buy together, with only 11% of procedures carried out through joint procurement. This is a missed opportunity as buying in bulk can result in better prices and higher quality goods and services and help to exchange know-how between CAs [22]. ICLEI considered that joint procurement procedures may represent:

- an entry-door for introducing sustainable procurement
- a launching platform for customers for environmentally innovative solutions
- a solution to reduce the price of environmentally sound products and services
- a way to introduce new products into national markets
- a solution to standardise environmental demands
- a way of pooling environmental expertise
- a means of encouraging suppliers to develop new products [19].

Joint procurement procedures organised by CAs from the same member state can take different forms:

- CAs may coordinate their procurement through the preparation of common technical specifications that will be later procured by a number of CAs, each conducting a separate procurement procedure
- CAs jointly conduct one procurement procedure either by acting together or by entrusting one contracting authority with the management of the procurement on behalf of all contracting authorities.

When implementing joint procurement procedures, CAs are jointly responsible for fulfilling their obligations required by the public procurement directives. This applies also in cases where one contracting authority manages the procedure, acting on its own behalf and on behalf of the other contracting authorities.

The responsibility is linked only to the procedure implemented by the respective CA, and, only where parts of the procurement procedure are jointly conducted by CAs, the joint responsibility is applicable only to those parts of the procedure that

Table 13.2 Joint procurement process approach in EU member states

Countries	Total procurement (Mil. EUR)	Procurement (% GDP)	Joint purchase (%)	Central purchasing
Malta	700	10	71	Yes
United Kingdom	274,600	14	21	Yes
Belgium	52,010	14	15	Yes
Cyprus	1,090	7	15	No
Latvia	2,660	11	15	No
Denmark	33,800	14	14	Yes
Slovenia	4,450	13	12	No
Sweden	68,680	16	10	Yes
Finland	34,460	18	10	Yes
Italy	157,230	10	10	Yes
Hungary	13,730	14	10	Yes
Ireland	15,540	9	9	Yes
Croatia	5,300	12	9	Yes
Slovakia	8,480	12	6	Yes
Lithuania	3,420	10	6	Yes
France	306,980	15	5	Yes
Austria	35,180	11	5	Yes
Czech Republic	21,480	14	5	No
Luxembourg	5,470	12	5	No
Germany	401,730	15	4	Yes
Estonia	2,450	13	4	Yes
The Netherlands	136,320	23	3	No
Poland	46,970	12	3	No
Spain	99,600	10	2	Yes
Portugal	17,290	10	2	Yes
Romania	15,980	11	1	No
Bulgaria	4,810	12	1	Yes
Greece	16,230	9	0	No

Source Directorate-General for Regional and Urban Policy (European Commission), PWC, Stock-taking of administrative capacity, systems and practices across the EU to ensure the compliance and quality of public procurement involving European Structural and Investment (ESI) Funds. Final report–Study, EU publications, Brussels, 2016, 8.2. Appendix 2: Individual country profiles (p. 112) [21]

have been carried out together. For the procedures or part of procedures, one CA is implementing separately; the respective CA is fully responsible [4, 20].

13.3.4   Life Cycle Costs (LCC)-Driven Procurement

The 'Life Cycle Costs (LCC)' concept defines the accumulated costs of a product throughout its whole life cycle (from product design to its disposal).

Lifetime expenses can be assessed on the basis of statistical, probabilistic data and mathematical models, which take into account the components of the LCC: purchase price, energy costs, operating and maintenance costs (labour, consumables, spare parts, repairs, upgrades, training, logistical expenses, etc.), decommissioning at the end of the period, opportunity costs (improving operational safety, reducing consumption and pollution, etc.), inflation, labour costs, fuel, exchange rate, penalties due to unavailability and accidents, etc. As illustrated in Figs. 13.1 and 13.2, acquisition costs play a small ratio during the life cycle of major transport assets (it is less than 23% for railway rolling stock or about 43% for electric buses).

The concept (as well as the modalities of approach, calculation, applications) is defined through standard EN60300-3-3:2004 [23], reviewed through IEC 60,300-3-3 Ed. 3.0 b:2017 [24].

Calculations can be complex and specific to each application. For products and services with a long life having major impacts on the environment, cost and quality of life (such as infrastructures for urban rail transport), the LCC criterion is a more appropriate criterion than the purchase price.

Fig. 13.1 LCC for locomotives used for passenger service [25]

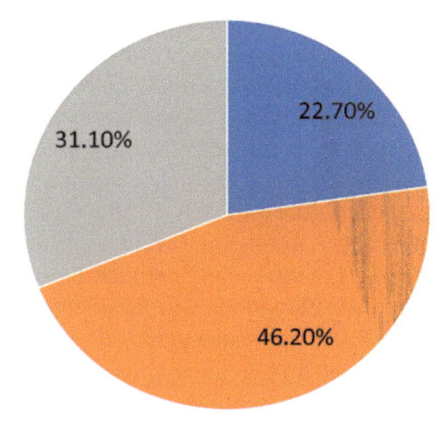

■ Acquisition ■ Energy Consumption ■ Maintenance

Fig. 13.2 LCC for electric
bus with 250 kWh on-board
[26]

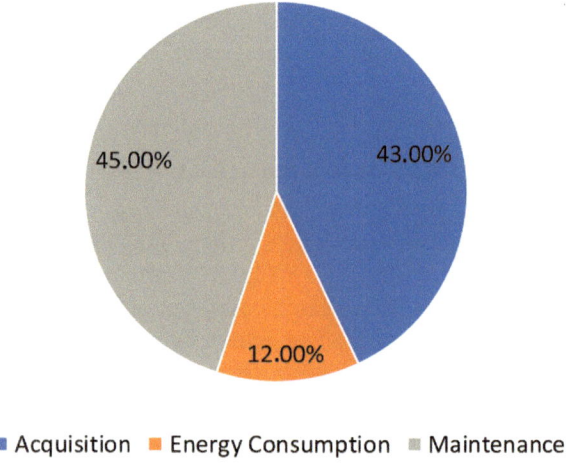

■ Acquisition ■ Energy Consumption ■ Maintenance

When LAs do not have the respective competency to calculate LCC, they can transfer this responsibility to suppliers through the objectives set in procurement documents. It is not necessary for the purchaser to be concerned about their performance and to set this as their objective.

However, it has been found that car transport can be drastically diminished (due to huge external costs, compared to other modes of transport) from the perspective of sustainable development strategies. That is why the EC considered that public procurement procedures are the most important lever for accelerating technical progress in the desired direction.

LCC-driven procurement calls for compliance with the basic principles of any public procurement: non-discrimination, equal treatment, mutual recognition, transparency, proportionality and responsibility.

Given the laboriousness of LCC calculations (Fig. 13.3) and the fact that these have to be done on a case by case basis, as part of the tendering process, the LA should request the LCC results from the tenderers. The tenderers are specialised in the research and design of the respective product, have operational data and are used to making such calculations as part of the product's future development. In the case of a complex product, the suppliers request such data from the sub-suppliers. Meeting the requirements of such a procurement procedure allows competitive suppliers the opportunity to develop innovative products and solutions and show their long-term benefits, which will make them more attractive to investors.

In order to prepare the procurement procedure, the contracting authority should carry out the following specific operations:

- Break down LCC into cost elements—according to EN 60300-3-3/2017, the specific character of the respective product
- Set unit prices to be used in the calculation: electric energy, fuel, emissions, labour, spare parts

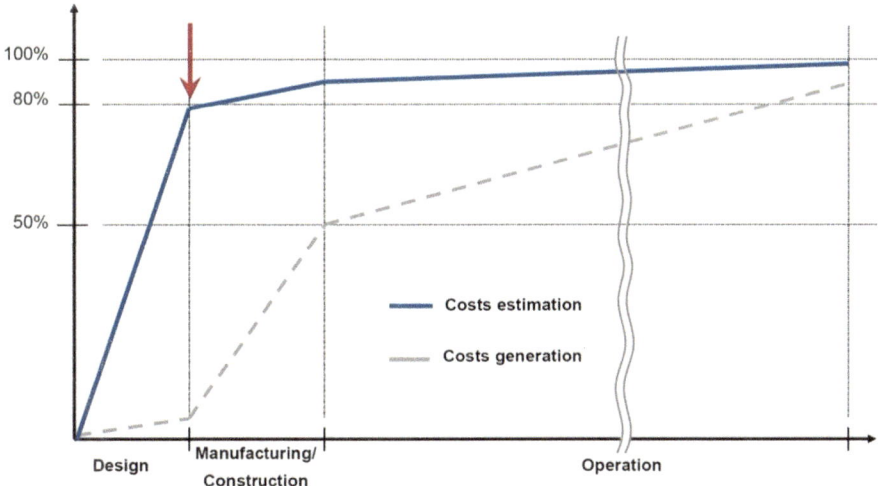

Fig. 13.3 Life cycle of public transport systems and subsystems *Source* Leuven et al. 'Urban Track Project. Final Publishable Activity Report', European Transport Network Alliance 2020, N/A, 2010, p. 70 [18]

- Set a calculation methodology so that LCC may be calculated in the same way by all the tenderers
- Set standardised forms for tendering the required data so that the tenders may be easier to compare to one another
- Set the conditions for data check, with the support of simulation tests, acceptance tests, LCC periodic tests throughout the life cycle or over determined periods of time
- Set the nature of collaboration with the supplier for these checks
- Set the supplier's liability in case of deviations of the data as measured from those provided
- Set, through the award and contract documentation, a bonus–malus policy regarding the results of the periodic checks, the application of subsequent innovative solutions, the collaboration modality, etc.

These activities are quite specialised. If the contracting authority does not have staff with the necessary experience to carry out these activities, they may wish to enlist the help of specialised consultants.

13.3.5 External Transport Costs

External transport costs refer to the transport-generated costs impacting and incurred by the whole society. These include costs such as those caused by traffic congestion, road accidents, air pollution, climatic changes from greenhouse gases (GHG), noise, transport infrastructure wear and tear [27], as well as socio-economic factors related to transport poverty (where life chances are restricted due to inadequacies in transport provision).

The internalisation of external costs is aimed at ensuring fairer competition amongst transport modes [28], with over 93% of the overall external costs are generated by car transport [29, 30], as well as the reduction of the costs incurred by the whole society, in line with the 'polluter pays' and 'user pays' principles [31].

For prices to reflect the internal costs, it is necessary to evaluate the value of each category of external cost [32].

Back in 2011, the European Commission adopted a detailed strategy called 'Transport 2050' transposed through the White Paper on Transports COM (2011) 144 final [31]. The strategy sets a very ambitious framework to support sustainable mobility measures:

- **Creating a fair financial environment: a novel approach to transport taxes.** Transport charges and taxes must be restructured in the direction of wider application of the 'polluter pays' and 'user pays' principle. Infrastructure costs should be applied to motor cars.
- A second stage concerns **internalisation of the costs applicable to all motor cars,** in order to cover both infrastructure costs and the social costs of congestion, CO_2 emissions (if not included in the fuel tax), local pollution, noise and accidents.
- Total transport external costs (all modes) for the EU28, in 2016, were as follows: € 987 billion (6.6% of GDP) [33].
- These costs are distributed across the main categories of external costs as follows: accidents—29%; congestions—27%; air pollution—14%; climatic changes— 14% [33]. Out of the overall external costs, 93% is generated by car transport.
- **Traffic congestion** cost Europe (in 2016) approximately € 270 billion, 1.9% of the gross domestic product per year [33]. Congestion costs will increase by about 50% by 2050 [31]. Apart from social and time costs, traffic congestions significantly increase CO_2 and greenhouse gas (GHG) emissions, noise, harmful effects on health.
- Urban mobility accounts for 40% of all CO_2 emissions of road transport and up to 70% of other pollutants from transport [34].

  - The social costs of accidents and noise would continue to increase. In 2012, the European average was 55 deaths for 1 million inhabitants, with as much as almost 100 deaths per million of inhabitants in some countries. Every year, some 1.3 million people die as a result of road traffic crashes worldwide—more than 3,500 deaths each day [35]. The social cost was estimated at EUR 250 billion in 2012, approx. 2% of the EU's GDP [36].

    69% of road accidents take place in cities [31, 30].

    97% of transport fatalities (all transport modes) occur in road accidents.

- Transport accounts for approximately one quarter of the overall greenhouse gas emissions (GHG) in Europe. Out of the overall traffic emissions in 2014, car transport accounted for **72.8%,** whilst rail transport accounted for 0.6% [37].

Understanding the concept of external transport costs can lead to two types of action:

1. Development of a strategy to develop sustainable mobility based on minimising external costs. Sustainable urban transport will be mainly achieved by developing high performing and attractive public transport based on high-capacity, safe, fast vehicles unlikely to cause traffic congestion and featuring a low emission level [underground trains, light rail, trams]. This principal means of transport will be completed with buses and electric trolleybuses. To achieve this, citizens' awareness needs to be raised about walking, bike and scooter riding for any type of journey. Such policy will result in a reduction of car journeys and a significant reduction in the actual external costs incurred by society. Many studies, projects [38] and European directives [39, 40] set the principles and modalities to develop such strategies.

2. Applying measures which take into account external costs. Guidelines for applying the internalisation of external costs were set through the European Commission Communication COM (2008) 435 final [41], together with the technical annex referring to strategy—SEC (2008) 2207 [32], with the document SEC (2008) 2208 [42]—impact assessment of the internalisation of external costs and SEC (2008) 2209 [43]—summary of the Impact assessment of the internalisation of external costs.

The main economic tools used for the internalisation of external costs are levies, taxes (or utilisation tariffs) and, under certain circumstances [44], emission rights trading.

A good knowledge of the particularities of each external cost, of the parameters it depends on, and of the local conditions are necessary and important in order to analyse the adequate economic tools. The economic tools should be applied in a differentiated way so that cost internalisation may be efficient, support fair competition amongst transport modes and protect the good operation of the domestic market. The general internalisation principle is as follows: 'pricing at the social marginal cost' in keeping with the 'polluter pays' and 'user pays' principles.

Besides the numerous studies dedicated to external costs and their internalisation, we recommend the European Commission's synthesis documents [32, 42, 45, 46], 'Guidelines to Innovative Financing' [46], as well as the other selected documents referred to in the bibliography.

13.4 Conclusions

In most urban centres that manage urban transport systems, the ownership right is exercised by the local public authorities and managed by the transport operator. Due to the restriction of the activity of some of the systems and also in the context of the construction of 'Chinese walls' between the administrative function and the operating one, the technical capacity of the public authority has diminished, which has a negative impact on the preparation and implementation of investment projects in urban road and light rail transport infrastructures.

The increased appetite for new investments, manifested in the post-2010 period, overlaps a reform of the national and European public procurement systems manifested by the package of European directives in 2014. The European Commission has agreed to finance various projects, research that will contribute to the promotion of innovative measures in the field of public procurement with a role in implementing sustainable urban mobility measures. Within the Horizon 2020 research programme, the SUITS research and innovation project represents a research vector in the field, whilst the Guideline on Innovative Public Procurement (https://www.suits-project.eu/innovative-procurement/) will help local public authorities to increase their administrative capacity in the field of sustainable mobility.

In the case of LCC, the correct sharing of responsibility between the contracting authority and the bidders can contribute with significant reductions in financial effort during the 30–50 years of project implementation, thus supporting the applied research in the transport systems investment field.

Using R&D activities in preparing the procurement procedures is the only modality to progress by applying innovative solutions to develop a sustainable modality, to the major benefit of the whole society.

The described procedures can be complemented with local specific elements able to foster the development of partnerships with R&D companies (particularly SMEs and small entrepreneurs), under competitive conditions, which should allow for the main requirements to be identified and carried out in the perspective of future public procurements.

Innovative partnerships and the described procedures include elements which are different from those in the usual procurements. With a view to their efficient and correct development, public administrations should understand their importance and prepare an operational and management team able to carry out the required activities in a creative way and with much determination.

Joint acquisitions can be carried out either through the voluntary association of several local public authorities with common interests or through the organisation of centralised procurement agencies. Such procedures allow standardisation of some key elements of the urban transport system, characterised, otherwise, by excessive localisation and apparently customised technical solutions. This standardisation allows for economies of scale, whilst leaving the responsibility for implementation at the level of each authority participating in the process.

Active cooperation with the research and industrial environment is important for local authorities who want to implement modern sustainable mobility solutions so that the bidding for technical solutions meets the implementation horizon of 30–50 years and avoids obsolete technologies, with negative impact on the public budget.

Such innovative procedures require selection and training of the personnel in charge. This objective can be achieved by means of adequate decision-making support tools in aid of innovative and sustainable financing, the implementation of schemes for innovative procurement and of development of new business models and partnerships.

The SUITS Project Consortium has researched and prepared Guidelines to Innovative Procurements to help decision-makers in S–M cities in relation to the development of sustainable mobility. The results of the guidelines are part of an integrated decision support tool.

References

1. European Commission, White Paper on Transport, COM: 144 final—Roadmap to a Single European Transport Area—Towards a Competitive and Resource Efficient Transport System. European Commission, Brussels (2011)
2. European Commission: DG GROW, Public Procurement. https://ec.europa.eu/growth/single-market/public-procurement_en (n/a). Last accessed 04 March 2018
3. European Parliament, Council of the European Union: Directive 2014/24/EU of the European Parliament and of the council of 26 February 2014 on public procurement and repealing directive 2004/18/EC Text with EEA relevance. Off. J. Eur. Union 57(L94), 65–242 (2014c)
4. European Parliament, Council of the European Union: Directive 2014/25/EU of the European Parliament and of the council of 26 February 2014 on procurement by entities operating in the water, energy, transport and postal services sectors and repealing Directive 2004/17/EC. Off. J. Eur. Union 57(L94), 243–374 (2014b)
5. European Parliament, Council of the European Union: Directive 2014/23/EU of the European Parliament and of the council of 26 February 2014 on the award of concession contracts. Off. J. Eur. Union 57(L94), 1–64 (2014a)
6. European Parliament, Council of the European Union: Directive 2004/17/EC of the European Parliament and of the Council of 31 March 2004 coordinating the procurement procedures of entities operating in the water, energy, transport and postal services sectors. Off. J. Eur. Union 47(L134), 1–113 (2004a).
7. European Parliament, Council of the European Union: Directive 2004/18/EC of the European Parliament and of the Council of 31 March 2004 on the coordination of procedures for the award of public works contracts, public supply contracts and public service contracts. Off J Eur Union 47(L134), 114–240 (2004b)
8. European Commission–Inteligent Energy Europe: Energy efficiency and environmental criteria in the awarding of regional rail transport vehicles and services (ECORAILS). https://ec.europa.eu/energy/intelligent/projects/en/projects/ecorails. Last accessed 2 Oct 2019
9. UIC–International Union of Railways: PROSPER (Procedures for Rolling Stock Procurement with Environmental Requirements). https://www.railway-energy.org/static/PROSPER__Procedures_for_Rolling_Stock_Procurement_with_Environmental_Requirements__32.php (2003). Last accessed 14 June 2019
10. UNIFE–the Association of European Railway Industries: "Railenergy," Adam Foster Web Engineering. http://www.railenergy.org/. Last accessed 14 June 2019
11. European Parliament, Council of the European Union: Directive 2009/33/EC of the European Parliament and of the Council of 23 April 2009 on the promotion of clean and energy-efficient road transport vehicles. Off. J. Eur. Union 52(L120):5–12 (2009)
12. European Parliament, Council of the European Union: Directive (EU) 2019/1161 of the European Parliament and of the Council of 20 June 2019 amending Directive 2009/33/EC on the promotion of clean and energy-efficient road transport vehicles. Off. J. Eur. Union 62(L188), 116–130 (2019)
13. European Parliament, Council of the European Union: Regulation (EC) No 1370/2007 of the European Parliament and of the council of 23 October 2007 on public passenger transport

services by rail and by road and repealing Council Regulations (EEC) Nos 1191/69 and 1107/70. Off. J. Eur. Union **50**(L315), 1–13 (2007)

14. Caraman, D., Constantin, I., Damboianu, C., Roseanu, S.: Guidelines to Innovative Procurement. Integral Consulting R&D, Bucharest (2019)
15. Tsanidis, V.: Policy Framework in Innovation Procurement. European Commission–DG CNECT
16. Leuven, A.V., Corre, F.L., Schnieders, I., Thijssen, D., Schreiter, H., Wragge, V., Vanhonacker, T., Hamoller, G., Janig, N., Rodriguez, N., Jong, M.D., Rupp, T.: Urban Track Project. Final Publishable Activity Report. European Transport Network Alliance (2010)
17. European Commission: Funding & tender opportunities. Horizon 2020 Framework Programme (H2020). European Commission, Brussels (2017)
18. ICLEI–Local Governments for Sustainability, „Joint procurement. Fact sheet," European Commission, DG Environment-G2, Bruxelles (2008)
19. Organisation for Economic Co-operation and Development: Centralised and decentralised public procurement. SIGMA Papers **29**, 2–17 (2000)
20. PWC: Stock-taking of administrative capacity, systems and practices across the EU to ensure the compliance and quality of public procurement involving European Structural and Investment (ESI) Funds. Final report: country profiles. European Commission–Directorate-General for Regional and Urban Policy, Brussels (2016)
21. European Commission–DG GROW: Public procurement strategy. http://ec.europa.eu/growth/single-market/public-procurement/strategy_en. Last accessed 18 July 2018
22. CENELEC–European Commitee for Electrotechnical Standardization: EN 60300–3–3 :2004. Part 3–3: Application Guide. Life Cycle Costing. CENELEC–European Commitee for Electrotechnical Standardization, Bruxelles (2004)
23. IEC–International Electrotechnical Commission: IEC 60300–3–3 Ed.3.0 b:2017. Part 3–3: Application Guide. Life Cycle Costing. IEC–International Electrotechnical Commission (2017)
24. UIC–International Union of Railways: LCC-driven procurement. http://www.railway-energy.org/static/LCC_driven_procurement_87.php (2002). Last accessed 25 Feb 2018
25. O'Donovan, A., Frith, J., McKerracher, C.: Electric Buses in Cities–Driving Towards Cleaner Air and Lower CO2. Bloomberg New Energy Finance (2018)
26. Korzhenevych, A., Dehnen, N., Bröcker, J., Holtkamp, M., Meier, H., Gibson, G., Varma, A.. Cox, V.: Update of the Handbook on External Costs of Transport. Final Report. Ricardo-AEA, London (2014)
27. Commission of the European Communities: COM/95/0691 final–towards fair and efficient pricing in transport policy–options for internalising the external cost of transport in the European Union–Green Paper. Commission of the European Communities, Brussels
28. Essen, H.V., Schroten, A., Otten, M., Sutter, D., Schreyer, C., Zandonella, R., Maibach, M., Doll, C.: External Costs of Transport in Europe. Update Study for 2008. CE Delft, Infras, Fraunhofer ISI, Delft (2011)
29. European Transport Safety Council: 13th Annual Road Safety Performance Index (PIN) Report. https://etsc.eu/13th-annual-road-safety-performance-index-pin-report/. (2019). Last accessed 10 July 2019
30. European Commission: COM/2011/0144 final—WHITE PAPER Roadmap to a Single European Transport Area—Towards a competitive and resource efficient transport system. European Commission, Brussels (2011)
31. European Commission: SEC/2008/2207 final–Stratégie pour une mise en œuvre de l'internalisation des coûts externs–Annexe technique à la stratégie pour une mise en œuvre de l'internalisation des coûts externs. European Commission, Brussels (2008)
32. Essen, H.V: Sustainable Transport Infrastructure Charging and Internalisation of Transport Externalities. CE Delft, Delft (2018)
33. European Commission–DG MOVE: Clean transport, Urban transport. Urban mobility. https://ec.europa.eu/transport/themes/urban/urban_mobility_en (2019). Last accessed 15 Aug 2019

34. Economic Commission for Europe Inland Transport Committee: Informal Document no. 6 Secretariat. Final. (2011)
35. European Transport Safety Council (ETSC): Police Reform and Consequences for Road Safety. European Transport Safety Council (2012)
36. European Commission–Secretariat-General: COM/2016/0501 final–A European Strategy for Low-Emission Mobility. European Commission, Brussels (2016)
37. Rupprecht, S., Brand, L., Böhler-Baedeker, S., Brunner, L.M., Colclough, A., Dragutescu, A., Horvat, M., Durlin, T., Werland, S., Rudolph, F.: Guidelines for Developing and Implementing a Sustainable Urban Mobility Plan, 2nd edn, Final Draft for SUMP Conference. Rupprecht Consult–Forschung & Beratung GmbH (2019)
38. European Commision: COM/2013/0913 final - Together towards competitive and resource-efficient urban mobility. European Commision, Brussels (2013)
39. European Commission: ANNEX 1—Annex. A Concept for Sustainable Urban Mobility Plans to Communication Together towards competitive and resource-efficient urban mobility. European Commission, Brussels (2013)
40. European Commision: COM/2008/0435 final–Strategy for the Internalisation of External Costs. European Commision, Brussels (2008)
41. European Commission: SEC/2008/2208 final–Strategy for an Internalisation of External Costs and the Proposal for a Directive of the European Parliament and of the Council Amending Directive 1999/62/EC on the Charging of Heavy Goods Vehicles for the Use of certain infra. European Commission, Brussels (2008a)
42. European Commission: SEC/2008/2209 Final–Strategy for an Internalisation of External Costs and the Proposal for a Directive of the European Parliament and of the Council Amending Directive 1999/62/EC on the Charging of Heavy Goods Vehicles for the Use of Certain Infra. European Commission, Brussels (2008b)
43. European Parliament, Council of the European Union: Directive (EU) 2018/410 of the European Parliament and of the Council of 14 March 2018 amending Directive 2003/87/EC to enhance cost-effective emission reductions and low-carbon investments, and Decision (EU) 2015/1814. Off. J. Eur. Union 61(L 76), 3–27 (2018)
44. European Commission–DG MOVE: COM/2017/0653 Final–Proposal for a DIRECTIVE OF THE EUROPEAN PARLIAMENT AND OF THE COUNCIL Amending Directive 2009/33/EU on the Promotion of Clean and Energy-Efficient Road Transport Vehicles. European Commission–DG MOVE, Brussels (2017b)
45. European Commission–DG MOVE: COM/2017/0653 Final–ANNEX to the Proposal for a Directive of the European Parliament and of the Council Amending Directive 2009/33/EC on the Promotion of Clean and Energy-Efficient Road Transport Vehicles. European Commission, Brussels (2017a)
46. Lugovoi, A., Parker, A.: Guidelines to Innovative Financing, Appendix C1–Congestion Charge. ARCADIS, UK; SUITS Project (2019)

Chapter 14
Innovative Ways for Financing Transport Infrastructure

Olga Feldman

Abstract Many cities encounter challenges caused by traffic and congestion. Encouraging effective and sustainable transport solutions can help to reduce these issues and to meet objectives in a wide range of policy frameworks, resulting in wider economic and social benefits and leading to better health and wellbeing for citizens. Financing is one of the issues that is faced by many cities and regions across Europe and worldwide. Traditionally, public transportation infrastructure has been financed by public funding via taxation, borrowing, operating profits, or a mixture of these. Following the reduced ability of governments to find funding for mobility and transportation schemes, innovative financing approaches have become increasingly important. This chapter discusses financing mechanisms for sustainable transport and mobility and measures to encourage local, regional, and national authorities to achieve their sustainable mobility objectives. More specifically, it focuses on new and innovative financing approaches which, for a variety of reasons, are not widely used at present. The outcomes of this research will help authorities with the decision-making process of identifying the most appropriate financing approaches to achieve sustainable urban mobility objectives.

14.1 Introduction

Transport accounts for a quarter of the European Union greenhouse gas emissions and plays a crucial role in reducing emission and achieving climate neutrality by 2050 [1]. The European Commission (EC) has been promoting sustainable transport and mobility and is currently working on a strategy for sustainable and smart mobility that will decouple economic growth from resource use and tackle emissions from all transport modes. This new growth strategy seeks to make the European Union a fair and prosperous society with a contemporary, resource-efficient, and competitive economy.

O. Feldman (✉)
Arcadis, London, UK
e-mail: olga.feldman@arcadis.com

© Transport for West Midlands 2023
A. Woodcock et al. (eds.), *Capacity Building in Local Authorities for Sustainable Transport Planning*, Smart Innovation, Systems and Technologies 319,
https://doi.org/10.1007/978-981-19-6962-1_14

Public transportation infrastructure is usually financed by traditional financing sources and methods, including public funding via taxation, borrowing, operating profits, or a mixture of these. Funding shortfalls, making the most efficient use of existing resources, and private sector interest in project development are some of the driving forces behind interest in innovative finance, which is generally defined as a blend of techniques and specially designed mechanisms to enhance traditional financing sources and methods. Following the reduced ability of governments to find funding for mobility and transportation schemes, innovative financing approaches have become increasingly important. They provide opportunities to speed up transportation project delivery process, generate new land value capture strategies and new revenue strategies that better align user benefits and costs, advance new funds management techniques and new types of financial partnerships, and to develop new financing mechanisms that are designed to leverage resources.

Present-day financing systems for sustainable urban transport and mobility vary widely from city to city and include public and private financing systems, as well as Public–Private Partnerships (PPPs). Most cities receive some level of national and EU funding, although small and medium-sized (S-M) cities tend to receive much less funding compared to large cities. This builds the case for helping S-M cities to increase their capacity to take advantage of innovative financing mechanisms that are available to advance their sustainable urban transport and mobility.

Eight cities in seven countries of the SUITS project were selected to demonstrate differences in approaches to developing and financing sustainable urban mobility and have shared case studies and provided the test bed for the development and use of the Guidelines to Innovative Financing [2].[1] These are Rome and Turin (Italy), Kalamaria (Greece), Valencia (Spain), Stuttgart (Germany), Alba Iulia (Romania), Palanga (Lithuania), and Coventry (UK).

This chapter

- Presents some transport regulatory strategies and governance.
- Reports on the current financing in partner cities and gives examples of sustainable transport and mobility funding mechanisms.
- Provides overview of the research methodology and outlines innovative financing principles and the importance of innovative financing approaches.
- Identifies Innovative Financing Mechanisms that S-M cities can utilise to fund sustainable mobility projects, including a summary and general characteristic of each financing mechanism, as well as proven track records, case studies, and guidelines for implementation in S-M cities.
- Outline recommendations.

---

[1] Available from https://www.suits-project.eu/the-cities/.

14.2   Transport-Related Finance Regulations in Europe

Promoting sustainable transport and mobility is important at the EU level, and there is much legislation which focuses heavily on the finance of transport and increasing sustainability. In comparison, at the EU Member State level, there is little legislation and regulation concerning the financing of transport and mobility projects, especially those that are sustainable. However, country-specific examples from Germany, Greece, Italy, the UK, and Romania highlight that there is a large number of legal and regulatory frameworks regarding general finance and investment in EU Member States, as well as various authoritative bodies which control how finance and funding are distributed for transport and mobility projects. In addition, most financing is regulated in accordance with national investment plans, so despite few legal frameworks, there are rules on how different types of funds can be spent, which includes those spent on transport and mobility projects.

Thus, the German government has passed several laws about sustainable transport and mobility financing, including the Municipal Transport Financing Act, which contains provisions for investments in the improvement of passenger transport and infrastructure in cities and municipalities, with a strong focus on achieving a behaviour change which favours local public transport [3]. There is an action plan for sustainable mobility, known as the Urban Transport Development Concept VEK, which started in 2013 and describes the short- and medium-term objectives and measures for sustainable mobility. The action plan serves as a basis for preparing funding for the area, and the Department for Strategic Planning and Sustainable Mobility coordinates the national budget in line with the action plan to support projects such as pedestrian mobility, cycle infrastructure, and corporate mobility measures [4].

The Greece Regulatory Authority for Railways (RAS) was set up in 2011 and has powers which include administrative and financial self-sufficiency, without being subject to control from any government bodies. The RAS regulates how funding and finance for railways are gained and distributed. This example illustrates how finance for transport is often in control of authoritative bodies that are responsible for certain transport sectors rather than provide an overarching legal framework.

The Italian national strategic plan for sustainable mobility schedules a series of priorities for investment. The Italian government also has recently agreed on a series of regulations that should be enacted to give priorities for investment that are set out in Italy's national strategic plan for sustainable mobility [5]. The Italian Transport Authority has financial power over how national budgets for transport are spent and aims to protect and promote competition in the market for railways, airports, ports, toll motorways, and public and passenger transport [6].

In the UK, the Transport Investment Strategy makes the case for continued investment in transport infrastructure and sets priorities and propositions for investment. The priorities tend to be delivered through funds such as the Road Investment

Strategy, local authority block grants, specific funds for large schemes (HS2, Cross-rail), the Transport Technology Research Innovative Grant (T-TRIG), the Innovation Challenge Fund, and other public and private funds [7]. Moreover, the UK's Transport Act 2000 provides several measures regarding transport in the UK and enacts rules concerning transport financing, highlighting that investment in more sustainable transport schemes should be prioritised. Transport in the UK is a 'reserved matter', meaning Scotland, Wales, and Northern Ireland can legislate and create their own policy for transportation.

The General Transport Master Plan of Romania (GTMP) prioritises investments in the field of transport for all transport modes, based on a multicriteria analysis. The Ministry of Transport is responsible for transport finance and coordinates the programmes related to EU cohesion policy through the Large Infrastructure Operational Programme. The priority of the transport infrastructure projects was set based on the methodology set out in the GTMP, approved by the European Commission as correlated with the 2017–2020 Romanian Government Programme. A major financing source for regional and local development by means of which local authorities can make investments in transport and mobility is the Regional Operational Programme 2014–2020.

Spain does not have a specific National Law for the financing of urban collective transport, which prevents the design of a homogeneous model and hinders the planning of services in the medium or long term. Public transport is financed mainly using two variants:

- The programme contracts of Madrid and Barcelona.
- The general operating deficit subsidy model for cities with more than 50,000 inhabitants.

The public passenger transport in Portugal has been regulated by the Regulamento dos Transportes em Automóveis (Regulation of the Transportation in Road Vehicles) from 1948 and by the Lei de Bases do Sistema de Transportes Terrestres (Basic Law on Inland Transportation) from 1990. The public transport provision in metropolitan, urban, and rural areas is regulated by the Regime Jurídico do Serviço Público de Transporte de Passageiros, the Mobility and Transport Authority or by the national government. At the national level, public transportation is subsidised by The Ministry of Finance (Ministério das Finanças) that in cooperation with the Ministry of Planning and Infrastructure (Ministério das Obras Públicas, Transportes e Comunicações) manages and distributes the subsidies and investment resources for transport activities. The municipalities and the communities of municipalities finance the local transport services with local funds. It should be noted that in November 2017, the Public Transportation Service Funds (Fundo para o Serviço Público de Transportes) was approved. Apart from Lisbon and Porto, where the cost of transportation is co-financed by the national government, the financial organisation is spread across different local authorities.

The European Investment Bank is playing an important role in the decarbonisation of transport in Europe by lending to sustainable transport sectors such as rail and urban public transport. Nevertheless, mobility solutions and services are dominated by non-European start-ups due to better access to equity financing for non-European companies and more difficulties in scaling up in Europe due to heterogeneous markets with regards to policies, legislation, and regulation [8].

The European Commission operates closely with national governments across Europe with the aim of developing consistent laws and regulations covering all means of transport.

14.3   Present-Day Financing in Partner Cities

A baseline assessment of each partner city has been carried out regarding the existing funding streams for transport and mobility together with a survey about demographics, socio-economics and the mobility context of each partner city, and the source of the majority of their funding for sustainable transport/mobility schemes (e.g. local/national or international government, loans, taxation, bonds, user fees) and how they decide where this is spent. Examples of how partner cities' sustainable transport and mobility schemes are funded (local, national, or other funds) are shown in Table 14.1.

14.3.1   Cities in the UK

The UK's Department for Transport (DfT) sets the strategic transport and mobility policy for the whole of the UK, including functional and legal constraints for local and regional transport authorities and plays a supervisory role. As a result, small to medium-sized cities have control over local sustainable transport and mobility systems. The central government provides some finance to local transport and mobility authorities in small to medium-sized cities, to help improve infrastructure and develop transport and mobility services. However, local transport and mobility authorities have received 50% cuts in funding from the national level since 2010 [9]. Additionally, most of this funding does not cover sustainable transport and mobility schemes, as the government's current focus is on road infrastructure, and funding tends to be provided only for maintenance of existing roads [10, 11].

To overcome the deficiency in national-level financing, local transport and mobility authorities tend to gain much of their finance for sustainable transport and mobility schemes from mechanisms such as local level taxation (e.g. London Congestion Charge), local motor tax, business rate supplements, community infrastructure supplements and employer/employee taxes (such as Workplace Parking Levy or business rates), user fees (tolls, congestion charges and fares), loans, Public–Private Partnerships (PPPs), Tax Increment Financing (TIF), and innovation funds (e.g. Transport Innovation Fund 2005–2010).

It is important to note that not all public transport and mobility systems in S-M cities are controlled at the local level. National government still has much control

Table 14.1 Examples of sustainable transport and mobility funding mechanisms

City	European funds	National funds	Local funds
Alba Iulia	Passenger transport/adopting SUMP[a]/time schedule for merchandise deliveries in the historic and central areas	Passenger transport/adopting SUMP/time schedule for merchandise deliveries in the historic and central areas	Used for passenger transport/adopting SUMP/time schedule for merchandise deliveries in the historic and central areas
Erfurt		Expansion of tram network (120 million euros) Barrier-free expansion of tram and bus stops/stations	Expansion of tram network (120 million euros) Barrier-free expansion of tram and bus stops/stations Stake in the 'Middle Thuringia' transport association until 2006 (200,000 euros)
Kalamaria	Redesign of Metamorfoseos Street into a pedestrian road and a bike lane (2 million euros)	Redesign of Metamorfoseos Street into pedestrian road and bike lane (2 million euros) Information campaigning, European Mobility Week, and training in a mobility park (10,000 euros)	Upgrading pavements (1.3 million euros) Information campaigning, European Mobility week, and training in a mobility park (10,000 euros)
Palanga	Acquisition of environment and disabled people-friendly buses (0.44 million euros until 2030)		Acquisition of environment and disabled people-friendly buses (0.44 million euros until 2030)
Rome	Smartset (freight)	A cycling plan and its implementation financed by the Ministry of Environment Tumabia	A cycling plan and its first implementation financed by the Ministry of Environment Smartset (freight)
Torino	Novelog (under experimentation)	Metro line (1,046 million euros)	Metro line (1,046 million euros) ZTL (Zona Traffico Limitato)—limited traffic zones (5 million euros) Pedestrian areas in city centre (1 million euros)

[a] Sustainable Urban Mobility Plan.

over rail transport. For example, tracks are owned by state-controlled network rail, and services tend to be provided by private companies at a more regional level, with cities, and especially small and medium-sized cities, having little control. On the other hand, in England, bus services have been deregulated and, in most cities, (other than London, where buses are controlled by Transport for London), private companies operate bus services and have full control over services, fares, and vehicles.

14.3.2 Alba Iulia, Romania

With a population of 63,540 (2011), the Romanian city of Alba Iulia is one of the small to medium-sized SUITS partner cities. The municipality receives little national support and most funding regarding the rehabilitation of the infrastructure comes from Operational Programmes funded by the EU. Currently, the only form of public transport in Alba Iulia is the city's bus network. These services are operated by a private company, Societatea de Transport Public Alba (STP Alba), which owns most of the city's buses, and is subsidised by the municipality. Additionally, the city has seen recent investment in sustainable urban mobility projects and has benefitted from the construction of approximately twenty kilometres of new cycle lanes. The funding for this came directly from the European Union, in the form of grants received through the Regional Operational Programme. For future EU projects addressing mobility at a city level, cycle lanes are seen for the entire length of the city's network of roads.

Overall, Alba Iulia receives most of its finance from private companies at the local level and through international programmes funded by the EU.

14.3.3 Rome, Italy

Rome is one of the larger-sized SUITS partner cities, with a population of about three million inhabitants. The majority of the actual external finance for Rome's sustainable transport and mobility schemes is received from the national level, through the Comitato Interministeriale per la Programmazione Economica (CIPE), the Italian fund for large infrastructures of national interest, which has funded projects such as the Rome's Metro C, and directly from the Ministry of Transport for some other revamping and infrastructure extensions. Additionally, a large amount of finance is received from the Cohesion and Development Fund for the Lazio Region (complementary to ERDF), which is financed by the Italian government as well as the Ministry of Environment for mobility projects linked to transport emissions reduction.

Conversely, some of Rome's financing is internal financing at the local level, derived from fines, fares and parking fees, and the Service Level Agreement for Local Public Transport and Mobility services. Local-level finance tends to fund much smaller-scale schemes such as maintaining public transport services. Additionally, funds for mobility projects are also received from the PON Metro ERDF programme and from CEF, H2020 and INTERREG programmes, but the proportion of funding from the EU received by Rome is small in comparison to other types of finance.

Overall, Rome receives most of its financing for sustainable transport and mobility from public funds at the national level, although some funding is received from public finance at the local level and the EU funding at the international level.

14.3.4 Valencia, Spain

Valencia is another larger-sized SUITS partner city, located in Spain, with a population of about 800,000 inhabitants. Valencia City Council, through the Sustainable Mobility Area, oversees sustainable transport and mobility schemes for the city and manages the public bus fleet (EMT Valencia) and the public bicycle scheme.

Funding for sustainable transport and mobility schemes in Valencia comes from local resources, as funding from the central government is limited. For example, the public bus fleet (EMT Valencia) gets its budget from a combination of funds from the City Council and from ticketing. Valencia City Council attempted to claim a subsidy for metropolitan transport from the central government in the past, however, this was denied. Nevertheless, EMT Valencia still looks for opportunities for regional funding (from IVACE), national funding (such as Subsidy to Collective Urban Transport or MOVEA), and European funding. Valencia City Council also increases opportunities for funding by joining with several municipalities from the metropolitan area, working together as the Mobility Agency of the Metropolitan Area of Valencia, which presents more opportunities to apply for grants.

Overall, most of the finance for Valencia's sustainable transport and mobility schemes is local, derived from funds from the City Council and ticketing.

14.3.5 Stuttgart, Germany

Stuttgart is also a larger-sized SUITS partner city, located in Germany, with a population of approximately 600,000 inhabitants. Since the 1970s, the public transport network in Stuttgart has expanded significantly and comprises of 55 bus lines, 17 light rail lines, and 6 regional metro lines. Verkehrsverbund and Tarifverbund Stuttgart (VVS) regulates all sustainable transport and mobility for the city and includes over 50 public transport companies, the biggest one being SSB, which has more than 500 buses and trains.

Most funding for sustainable transport and mobility comes from the municipal budget at the local level. The Department for Strategic Planning and Sustainable Mobility coordinates the budget, which is then approved by the mayor and City Council. This budget tends to be spent on pedestrian mobility, cycle infrastructure, and corporate mobility measures.

Stuttgart also benefits from some national funding and national sustainable mobility initiatives. For example, there was recently a large nationwide funding programme for e-mobility (to build up and identify charging stations for electric vehicles), although where this money was spent was decided at the national level.

Most of Stuttgart's innovative funding programmes comes from the regional level, through the Baden-Wuerttemberg budget and fund projects such as cycle highways, walking audits, and cycle infrastructure. Additionally, Stuttgart has taken part in several European programmes for research, most recently, as coordinators of CIVITAS CARAVEL and 2MOVE2 projects. However, in Stuttgart, funding for public transport modes tends to be private and operated at the local, regional, and national levels.

Overall, Stuttgart's financial system for sustainable transport and mobility comprises a wide range of financing mechanisms, the majority of which are at local level, from the municipal budget.

14.3.6 Kalamaria, Greece

Kalamaria is a medium-sized SUITS partner city, located in Northern Greece, with a population of approximately 100,000 inhabitants (2011). It is located about 7 kms (4 miles) southeast of downtown Thessaloniki city.

Large projects for sustainable transport and mobility are funded mainly from European funds (PA 2014–2020) at a percentage of 85%. The other 15% comes from national funds. The **PA (Partnership Agreement for the Development Framework) 2014–2020** constitutes the main strategic plan for growth in Greece with the contribution of significant resources originating from the European Structural and Investment Funds (ESIF) of the European Union. The 2014–2020 PA comprises 20 programmes, of which 7 are sectoral and 13 are regional. One sectoral operating programme is dedicated to Transport Infrastructure, Environment and Sustainable Development Operational Programme (YMEPERAA). The objectives of the programme include the development of sustainable and ecological fixed route urban transport to enhance sustainable urban mobility. One of the projects funded by PA is 'Digital services of smart city—quality of life improvement in Kalamaria'. The project refers to the installation of 150 smart parking slots system at 3 roads (on-street) with sensors, around the city's commercial centre.

Studies for projects for urban sustainable transportation are funded from the municipal budget and national funds. For instance, the study of the sustainable urban mobility plan (SUMP) of Kalamaria will be funded from the Green Fund (national funds). The Green Fund is a legal entity governed by public law. Also, studies for parks' redevelopment or other urban facilities/equipment for sustainability are funded from the municipal budget. The municipal budget comes mainly from council tax. The funding source for the supply of new rubbish trucks for 2019, essential for sustainable urban freight transportation, will be from council tax.

'FILODIMOS II' is a new funding programme for urban sustainability. The main objectives of the 'FILODIMOS II' programme are: (a) strengthening balanced, sustainable, and equitable growth; (b) improving infrastructure; and (c) enhancing the competitiveness of municipalities. The purpose of the programme is to finance projects, supplies, services, and project studies of Greek municipalities. The total duration of the programme is from 2018 to 2022, with the possibility of extension. One project under the programme will be the improvement of bus stations in Kalamaria.

At the end of 2021, a new public transport mode—subway—will be operated in Kalamaria, enhancing the sustainability of urban transportation. The source of funding is the European Investment Bank.

In conclusion, Kalamaria's financing mechanisms vary from European funds to national and municipal funds. Large projects are funded from European funds (85%) with a small share from national funds. Smaller projects/supplies are funded from the municipal budget. The new programme 'FILODIMOS II' is intended to increase the total annual budget of the municipality.

14.4   Research Methodology

Innovative approaches to the financing of sustainable transport and mobility in S-M European cities are important, due to increased pressure on traditional sources of financing caused by the reduction of national and local government budgets, increased population growth and urbanisation. In addition, there is a reluctance from traditional transport and mobility investors to invest in sustainable transport and mobility projects, as they do not expect a high enough financial return, as benefits tend to be more diffuse (such as increased city competitiveness, economic growth, or the efficient use of a scarce resource).

The key criteria used to assess suitability and applicability of the innovative financing mechanisms are:

- Environmental and financial sustainability, meaning that the mechanisms can be used to support investment towards low carbon, climate-resilient transport and mobility options without having a negative impact on the ability of S-M cities to cooperate with other public bodies and private organisations or to attract funding from other sources.
- Complementarity, scalability, and efficacy, meaning that the mechanisms can be used with other financing mechanisms (such as traditional funding sources and government revenue) to link funding shortfalls and/or to create new funding streams and can be reproduced and scaled up in other cities in the EU.
- Innovation, meaning that the mechanisms are creative in thinking and feature new methods that are advanced and original.

Innovative financing also comprises financing practices that have not been previously utilised to finance sustainable transport and mobility projects, as well as mechanisms that may be common in some cities, but not generally used in other S-M European cities.

Utilising the tools of innovative finance diverges from the conventional finance in several ways, including new ways for existing revenue to be used to finance sustainable transport, exploitation of financing mechanisms such as debt finance, utilisation of fund management techniques, and formation of new institutional arrangements. Supporting legislation may be necessary for some innovative finance programmes.

The Guidelines for Innovative Financing Mechanisms [2] produced as a result of SUITS research activities provide a useful and efficient tool for local authorities and other stakeholders in small and medium-sized (S-M) cities to apply innovative financing approaches for sustainable transport and mobility projects and schemes. SUITS City Partners provided information regarding present-day financing mechanisms, which they use for sustainable transport and mobility measures, and helped to identify gaps in financing capacity of the cities. Semi-structured interviews and informal conversations with the City Partners were carried out.

Secondary sources included publications from peer-reviewed journals, media publications, and books, as well as open-source reports and papers published by government agencies, universities, not-for-profit entities, private consultancies, and innovation hubs. In some cases, SUITS partners used their professional judgement to determine which data sources and case studies were most appropriate for the guidelines.

Public finance is regulated and allocated by the national, regional, or local government and is usually derived from one of the following sources:

- Taxation at national, regional, municipal, or local level.
- Value capture mechanism by which the agency liable for the development of new infrastructure receives a payment from property owners, land developers, etc., to help fund the infrastructure, while local property owners and developers are expected to benefit from an increase in land and property values in the area surrounding the new infrastructure.
- User fees and charges imposed on users of transport infrastructure and services, or charges for non-compliant use of infrastructure and services.
- EU funding through a wide range of financial instruments, such as grants to support projects, offering public contracts and providing budget and sector support.
- The Public–Private Partnership approach is also well established across Europe and is often used to deliver new investment in the transport sector.

Funding from the central government varies between countries and their geopolitical structures. There is a tendency for the largest and capital cities to receive a larger proportion of finance from the central government due to political attention. On the other hand, public authorities and regional or municipal governments are still actively involved in the financing of urban transport and mobility schemes. Most of the private financing for sustainable transport and mobility schemes in cities comes from banks and institutions through upfront capital investment (loans) to finance the projects that a city needs but expect predictable revenue streams [12].

Innovative financing approaches can be used to fill the gap in funds caused by the reduced funding from government or investors and/or provide more sufficient, effective, and stable funds compared to traditional fiscal investment. Innovative Financing Mechanisms briefs were prepared as part of the guidelines and provide S-M local authorities with targeted information about selected innovative financing approaches, including brief description and background, key characteristics, attractiveness, potential challenges and risks, track record, relevant case studies, together with guidelines for implementation.

In total, 21 briefs were prepared covering the following mechanisms: Congestion Charge, Municipal Green Bonds, Crowdsourcing, Stamp Duty Land Tax (SDLT), Lottery Funding, Voluntary Capture, HGV Charging Schemes, Workplace Parking Levy (WPL), Community Infrastructure Levy (CIL), Advertising, Sponsorship and Naming Rights, Collaborating with other cities, Research Consortia and Private Companies, Citizen Cooperatives, Emission Trading, Planning Obligations/Developer Contributions, Tax Increment Financing, Sales Tax, Toll Roads, Selling Expertise and Technical Know-how, Sale of Land and Property, Donations as Part of Consumer Purchases, and Grants from Private Foundations and Trusts.

In addition to Innovative Financing Mechanism briefs, a Matrix of Financing Mechanisms was developed for users of the guidelines. The matrix shows which Innovative Financing Mechanisms are best suited to specific types of sustainable transport and mobility projects. The matrix was compiled using available information and is intended for indicative purposes only. The matrix does not intend to comprise all sustainable transport and mobility projects that can be financed using the identified financing mechanisms. Furthermore, some financing mechanisms could also be used for other sustainable transport and mobility measures. The applicability of the identified financing mechanisms would depend on local context and would be determined by current legislation and policy, national and local sustainable mobility plans, as well as political will within the S-M city administration [2].

The next section provides a summary of the most popular innovative mechanisms and highlights key challenges and risks. For the full description of the mechanisms, their advantages, and the guidelines for their implementation, please see [2].

14.5 Innovative Financial Mechanisms: Challenges and Risks

14.5.1 Toll Roads and Congestion Charging

Road infrastructures, such as motorways, tunnels, or bridges are very expensive, and governments are often unable or unwilling to commit fiscal spending to build new assets. Many countries have tried to overcome this lack of funds by levying tolls on highway users to generate additional revenue. In fact, toll revenues already provide a much-needed source of revenue to maintain and expand highway networks worldwide, allowing the provision of public infrastructure and services, that otherwise would not be practicable in the same period of time or to the same extent.

There are many examples of the successful implementation of urban toll roads, whereby the revenue generated has been invested into maintaining existing infrastructure, new infrastructure projects, or sustainable transport and mobility projects. For example, the Norwegian cities of Bergen, Oslo, and Trondheim have extensive experience in the implementation of toll roads as a financing mechanism for transport projects and have been applying these policies since the 1980s and 1990s. Other examples of urban road charging in Europe can be found in the UK, Stockholm or Milan, however targeting primarily traffic congestion control, gas emission taxation, or both.

A congestion tax or charge is a financing mechanism and mobility management strategy that surcharges users of public services because of excess demand. A congestion charge for transportation can include higher peak charges for use of public transport or road pricing to manage demand, making it possible to control congestion without increasing supply. The successful implementation of a congestion charge is currently limited to a small number of cities, including Singapore, London, Stockholm, Milan, and a few smaller cities such as Gothenburg in Sweden. New York City has approved a congestion charging scheme that is due to take effect in 2021. Other cities such as Cambridge and Edinburgh were reconsidering schemes as late as 2019/2020.

Challenges and risks:

- The success of urban toll roads is highly dependent on a strong political support and public acceptability for its implementation. Acceptability by citizens is an overriding concern for congestion charging. The way that the impacts, and in particular, the benefits, are perceived by the city authorities and citizens is critical.
- Political and public opposition can occur due to a sense of inequity, for example, the idea that frequent users suffer the most.
- Political and public opposition can also occur if there is a lack of efficient communication of the positive targeted impacts of the tolling system, such as additional revenue for public transport projects. This can be overcome by earmarking toll road revenues for road network investments, as is the case in Bergen.
- Public opposition can also occur due to a user's perception that they are being charged twice, by paying the tolls on top of their taxes.
- There is a cost recovery risk. Traffic and toll tariff levels may not be sufficient to cover all costs, including construction, operation, and maintenance.
- Municipalities and national governments find it difficult to design a scheme that is effective as well as publicly accepted. This can result in original proposals being weakened or watered down.
- In order to generate substantial funds, vehicles need to continue to drive within the congestion charge zone: this contradicts the primary goal of congestion charging, which is to discourage individuals from driving.

- Risk of traffic spilling over into areas outside of the congestion charging zone.
- Implementation causes change in demand of other transport modes; transport planners must ensure that other alternative public transport services are provided and/or improved.
- Without specified projects and funding initiatives, congestion charging provides a general source of funds for public projects, rather than public transport-specific initiatives.
- Requires high capital and operational costs. However, this can be offset by the income generated by the charge.
- Requires national governments to enable congestion charging through legislation.

14.5.2 Municipal Green Bonds

In financial markets, a bond is categorised as a debt instrument which allows the issuer to raise capital from investors. Municipal bonds are debt securities usually issued by governments to finance day-to-day operations and new projects. Green Bonds can only be used to finance projects that support climate change mitigation and resiliency, which includes sustainable mobility initiatives and projects. Toyota's issuance of bonds for financing electric and energy efficient vehicles in 2014 was the first Green Bond in the transportation sector, and subsequently, transport authorities and local municipalities have entered the market.

Challenges and risks:

- Lack of standards and a universal definition for Green Bonds.
- Lack of information and market knowledge for local authorities, national governments, and investors.
- Lack of clear risk profile associated with specific green investments.
- Subnational governments such as city municipalities generally have lower credit ratings than national governments, as their default risk is considered to be higher.
- Local governments often lack legislative power to access the bond market. This could require national governments to establish new legislation to allow the issuance of the municipal Green Bonds.
- Issuance of municipal Green Bonds requires capacity and knowledge which small and medium municipal authorities often lack.
- More suitable for medium and large rather than small cities due to additional costs (the cost of issuance could be high—up to US$50,000), know-how, and demand on resources.

14.5.3 Stamp Duty Land Tax (SDLT) and Tax Incremental Financing (TIF)

Stamp Duty Land Tax is paid to the government by the new owner on most properties above a certain value when the property is purchased, and a set percentage of revenue generated from Stamp Duty Land Tax may be used to finance new transport and mobility schemes within the catchment area, through a concept known as land value capture. Hong Kong and Singapore have both used Stamp Duty Land Tax collection to fund transport projects [13]. There are two schemes of this type in Los Angeles, which have helped reduce traffic congestion by funding improvements to the metro system and the introduction of carpooling lanes [14].

Tax Incremental Financing (TIF) is a way by which local authorities borrow money to build infrastructure. Assuming that improving an area by providing or improving public facilities or infrastructure will increase the value of the properties surrounding those facilities, the local authority can borrow money to build those facilities and pay it using the increased tax revenues related to the increase in economic activity and property values. TIF is reliant on growth in property taxes and assumes property prices will increase as a result of the development of a new facility or infrastructure. In reality, property prices are reliant on a variety of other factors along with the development, only some of which are in the local authority's control. Furthermore, the TIF process may lead to favouritism for politically connected developers, involving attorneys, economic development officials, and others.

TIF has been used to finance many kinds of schemes, not only public transport and mobility. These schemes include public infrastructure, demolition, utilities and planning costs, land acquisition, and improvements such as sewer repair and expansion, pavement and kerb works, traffic control, storm drainage, street construction and expansion, water supply, street lighting, park improvements, environmental landscaping, bridge construction and repair, remediation, and parking structures [15].

Challenges and risks:

- The taxes may place a significant financial burden on a relatively small cohort.
- The success of these types of taxes in funding infrastructure is contingent on a high level of understanding and support for the way the taxes work and a high level of support for the projects themselves.
- Usually, property price increases will occur following the completion of transport infrastructure projects, which means the additional revenue from increased property value cannot be captured until after the project has been completed and the funds spent.
- In all cases where this type of tax has been undertaken, only a percentage of the project's finance was derived from it. It still requires additional funding mechanisms as a top-up.
- The taxes, particularly if too high, risk rendering some redevelopment unviable, resulting in less redevelopment on transit nodes, particularly in the case of development-based land value capture mechanisms.

- Only a small percentage of the total value uplift of properties through the result of infrastructure projects can be captured.
- Most, if not all, examples of this type of finance mechanism have been tried in larger cities: Hong Kong, Singapore, Los Angeles, and London. There are no clear examples yet of this being tried and tested in small to medium-sized cities.
- TIF is reliant on growth in property taxes and assumes property prices will increase as a result of the development of a new facility or infrastructure.
- Although generally sold to legislatures as a redevelopment tool, some TIF districts are drawn up where development would happen anyway, such as ideal development areas at the edges of cities.
- The TIF process arguably leads to favouritism for politically connected developers, implementing attorneys, economic development officials, and others involved in the processes.
- Districts can be drawn excessively large thus capturing revenue from areas that would have appreciated in value regardless of schemes financed through the TIF mechanism.

14.5.4 Lottery Funding

Lottery funding is mobilised through profits generated from lotteries, which are a form of gambling, involving the drawing of lots for prizes, including lotto, electronic terminals, and instant games. In general, lottery funding tends to be used to promote social and environmental outcomes [16]. However, there are some instances where lottery grants have been given for community infrastructure projects such as new bridges, shuttle buses, and easier connections to cycling routes [17–19].

Challenges and risks:

- Lottery funding tends to be a one-off or irregular income source. As a result, it cannot be relied on as a baseline revenue stream to fund all public transport projects.
- Lottery funding budgets depend heavily on ticket sales, retailers' commission, and prize payments and can therefore vary substantially. Consequently, if an organisation relies on lottery funding, any budget cuts could have serious negative impacts.
- The financing mechanism does not incentivise good planning or spending. Funding distributors have no responsibility for the success of the project, and there tends to be a 'fund it and forget it' culture. Therefore, it is solely up to the grant receiver to invest the money in the most efficient way.
- Funding distributors tend to have insufficient skills to identify bankable projects and organisations. Therefore, there is a risk that some bankable sustainable transport and mobility project applications may be overlooked.
- It would only be possible to benefit from this type of funding in EU countries which have lotteries in place.

14.5.5 Voluntary Capture

Voluntary Capture is a deal or partnership between developers or property owners and a local authority, where the developers or property owners offer a voluntary contribution towards the costs of a public infrastructure project. Voluntary contributions tend to be offered when the developer or property owner calculates that the benefits that they will receive from the provision of public infrastructure outweigh the cost of investing in it. Examples of successful implementation of Voluntary Capture to fund sustainable transport and mobility schemes can be found in cities such as Boston, Washington, and Toronto, where voluntary contributions have funded new transit stations and connections with existing transit facilities [20, 21].

Challenges and risks:

- Unlike some other funding mechanisms, Voluntary Capture tends to be a one-off or irregular income source. As a result, it cannot be developed into a reliable and continuous revenue stream.
- It is not a legal requirement, and therefore, individual deals could be hard to come by, as in order for voluntary contribution towards transportation improvement to be viable for developers, a number of conditions need to be meet.
- The areas where public transport and mobility investment are most needed may not be the same as the areas where voluntary contributions are most viable.
- The objectives of a developer who is paying for the transportation improvement do not necessarily correspond with the objectives of the wider community and other stakeholders. Finding a compromise satisfying all stakeholders could be a challenge.

14.5.6 HGV Charging Schemes

As a result of the EU's efforts to internalise the external cost from HGVs' trips across the union's road network, the legislative framework for the HGV charging schemes is based on the 'polluter pays' principle. Most European countries have implemented or are in the process of implementing some form of truck charging scheme, where all hauliers from any country pay for using the roads, either by time or distance. Three different nationwide HGV charging mechanisms are currently used within the EU, namely electronic network-wide toll (distance-based charge), toll with physical barriers (distance-based charge), and time-based charges. Revenue from HGV charging schemes can be used for financing of infrastructure projects, such as the extension and maintenance of the existing road network, and the investment into other more sustainable transport means such as rail and waterways.

Challenges and risks:

- Development of an independent charging policy for HGV tolls that considers all the parameters of the external cost (e.g. pollutant emissions, vehicle gross weight) and traffic conditions could be challenging for S-M cities.
- The introduction of a scheme could be met with resistance from a range of stakeholders, especially from the road freight and haulier industry and local businesses that would be directly affected.
- S-M cities may lack knowledge and capacity to introduce an effective and revenue generating charging scheme on their own without being part of a larger region or nationwide scheme.
- An appropriate national and local legislative framework should be in place, allowing development of an HGV charging scheme and regulating the process of recovering debt from vehicles owners who failed to pay the charge.
- Tolls on the main road network could divert HGV traffic to secondary and local roads.
- The potentially high cost associated with the introduction and operation of the scheme may jeopardise its financial sustainability.

14.5.7   Workplace Parking Levy (WPL)

WPL is a charge imposed by local authorities on employers who provide parking spaces for their employees and may be seen as a type of a congestion charging scheme. The introduction of WPL usually has two objectives: firstly, to be a traffic-demand management measure to deter driving; and secondly, to raise funds for transport and mobility improvement schemes [22]. Under WPL, the employer who is charged and then has the discretion to absorb the cost themselves or to pass on all or part of it to their employees who commute to work by car and use the provided parking. In the UK, by law, any funds generated through WPL must be spent on local transport. Currently, parking levies are used in Australia (Perth, Melbourne, and Sydney), Singapore, and the UK (Nottingham) [23].

Challenges and risks:

- Could lead to potential backlash from local businesses who would be charged extra.
- The WPL could be perceived as unfair by a wide range of individuals, employers, and other organisations, as it does not distinguish between those who travel in congested periods on congested roads and those who do not. It also does not distinguish between those who have practicable public transport options and those who do not. Finally, it places an additional burden on low-income households compared to higher-income households.
- The introduction of a levy requires lengthy and comprehensive consultation with residents and businesses, as well as an audit of the existing workplace parking spaces.

- Finding the right balance for pricing strategy may be difficult. A local authority needs to find a balance between a revenue raising activity, while avoiding setting a charge which is too high, and thus deterring employers from setting up new business.
- Risk of spilling over into surrounding streets unless on- and off-street parking-controlled measures are introduced within the affected areas.

14.5.8   Community Infrastructure Levy (CIL)

CILs are used to raise additional funds for the promotion of existing infrastructure and development of new infrastructures in England and Wales. The levy is placed on new building developers in the local area, as it is expected that the new infrastructure will increase the value of new developments. Revenue generated through a CIL is usually used for funding facilities such as roads and other transportation, schools and other education, medical services and sport/recreation, and open spaces. Case studies include Crossrail 2, which is a proposed new railway in the UK serving London and the wider South East.

Challenges and risks:

- The financing mechanism is dependent on the local authority for pricing and enforcement.
- There is no 'one way' to implement a CIL, thus the local authority needs to be mindful regarding the creation of a CIL charging schedule.
- The charging system may discourage development in areas with high charging rates, in favour of development in areas with either a more affordable CIL or no CIL at all. This can be resolved by implementing a CIL across all local authorities within region.
- Requires an overseer authority (often an authority in addition to the local authority/transport authority) in order to approve the CIL rates and schedule.

14.5.9   Collaborating with Other Cities, Research Consortia, and Private Companies

Research consortia and private companies are interested in working with cities as it offers them an opportunity to test and promote their ideas and new products. In their turn, cities can benefit from additional funding, capacity building programmes, and investments into their infrastructure. Many EU cities have been involved in successful research partnerships which deliver multiple benefits including Dublin, Valencia, Rome, Skopje, Tallinn, Adana, Dubrovnik, Fagaras, Hradec Kralove, Jonava, Kassel, Katowice, Limassol, Lisbon, Ljutomer, Szege, Varna, and Fundao.

Challenges and risks:

- Developing long-term action plans and committing to long-term projects could be challenging due to political constraints and lack of political will.
- Lack of awareness about programmes supporting collaboration between cities and funding available for research consortia.
- Lack of experience, expertise, and relevant knowledge, as well as human resources that need to be dedicated to managing participation in research programmes and partnerships.
- Meeting specified pre-requisites for participation.
- Benefits received from partnering with research organisations and consortia may be limited to technology transfer and training programmes and not necessarily provide additional funding opportunities.
- Even if funding is provided, it is most likely to be tied to specific project(s) or agenda.

14.5.9.1   Emissions Trading

Emissions trading has its origins in economic theories, which assume that if pollution had a price, market forces would eventually deter businesses from polluting the environment because it would become less cost-effective for them to do so [24]. The EU CO_2 Emissions Trading Scheme was introduced by the European Union in 2005. Data in the SLoCaT Partnership Climate Finance Transport Database, which contains information on 277 transport projects covering the period from 1992 to 2016, shows that over \$3 billion of transport-focused investments are by Climate Finance Instruments. Furthermore, a growing number of countries are applying climate finance to sustainable transport projects [25].

Challenges and risks:

- It is a complex mechanism that changes over time due to the conventions on climate change, the state of the market, and other parameters. Therefore, it cannot be relied upon as a continuous revenue stream.
- Even though local governments have strategies, plans, and measures in place in the transport sector to reduce carbon emissions in cities (e.g. if the framework of their voluntary agreements is through the Covenant of Mayors), the buying and selling of carbon quotas for countries take place at national level.
- Many local authorities in S-M cities have limited institutional capacity (e.g. human resources and technical expertise in the area of emissions trading), which may cause them difficulty in identifying viable project options and developing and monitoring them.
- The legislative frameworks that establish and regulate carbon markets have not been designed with city projects in mind.
- Potential overlapping jurisdiction of greenhouse gas (GHG)-emitting sources.

- High transaction costs due to long time frames.
- Risk of projects underperforming due to carbon reductions verified and the amount of carbon credits ultimately delivered.

14.5.9.2 Planning Obligations/Developer Contributions

New residential development could place extra burdens on the existing infrastructure and resources in the local area, such as an increased volume of traffic and congestion. Planning obligations are a way for local authorities to internalise some of the external costs of the development, either through a fixed levy on the development or via direct negotiation between the developer and the local authority. Local authorities can use the money generated from developer contributions to improve existing infrastructure and help minimise the strain which the new development puts on it. Planning obligations are the result of individual, scheme-based negotiations that have proved to be a practical way for local authorities to cover infrastructure costs, as well as capturing some development value. Planning obligations can also be used to restrict or define the way in which the land is used, or in other words, to make acceptable a development which otherwise would be unacceptable to the local authority. Commonly known as 'Section 106' agreements in the UK, or 'public gain' in North America, in many countries planning obligations proved to be successful in making developers contribute funding to offset the site-specific impact of development on public infrastructure.

Challenges and risks:

- Negotiations between developers and local authorities can be lengthy and sometimes difficult, requiring time and effort from both parties.
- Input from multiple neutral surveyors may be required in order to obtain an impartial appraisal of the scheme's viability.
- The outcome of the negotiations remains unclear for the local authority until its end. As a result of this uncertainty, financial planning for potential schemes and projects is more difficult.
- Planning obligations could be perceived as highly subjective when determining the amount to be contributed.
- Without adequate controls or transparency, negotiations between the developer and local planning authority about developer contribution could lead to collusion and corruption.
- Planning obligations may discourage development in areas that ask for planning obligations, in favour of development in areas that are less strict about imposing public obligations.
- Funding for large infrastructure would be potentially dependent on planning obligation payments from many different developers and projects. Due to the one-on-one nature of the planning obligation negotiations, securing the required funding amount could be challenging, as well as time and resource consuming.

- Significant resources could be required to monitor the agreements and conditions pertaining to them after they have been agreed upon, i.e. while and after the land has been developed.
- Negotiations relating to planning obligations, including agreeing viability, can add delays to the planning process.

14.5.9.3   Sales Tax

Sales tax is a tax which is paid to a governing body of a region, such as a local authority, for the sales of certain goods and services. Generally, laws require the seller of the goods or service to collect the funds for the tax from the consumer at the point of purchase. Sales Tax Increase Financing is a mechanism which is sometimes used by the governing body of a region, to generate additional revenue to fund public realm improvements. Where this is the case, a governing body initiates a one-off increase to the current sales tax rate of a region to generate additional revenue. A blanket increase in sales tax is executed, which covers all applicable goods and services sold within the region. The difference between the original tax revenue and newly increased tax revenue is then made available for use in public realm improvements such as the financing of sustainable transport and mobility projects and initiatives [15]. It has been observed that these sales tax increase schemes for funding transport projects are mostly found in the US and Canada and are very rarely seen elsewhere [15, 26] except for India [27], and a proposed scheme in Madrid which failed and was not implemented.

Challenges and risks:

- Local authorities may not have the capacity to undertake the complex adminis-trative processes associated with increasing sales tax in the region.
- The rate of sales tax does not depend on a person's income or wealth, and therefore, an increase in sales tax may be unaffordable for people with lower-income salaries.
- If the tax increase is big enough, it could impact the level of public spending on goods and services, which would impact the local and regional economy [15].
- The public may be resistant to the tax increases if they do not necessarily use the services which will benefit from the tax increase [15].

14.5.9.4   Selling Expertise and Technical Know-How

Selling expertise and technical know-how is a form of collaborative knowledge sharing, where one has an exclusive right to exploit his or her substantive knowledge for economic profit and desires to merchandise it [28]. A local authority may sell its expertise and technical know-how to obtain an additional revenue for its sustain-able transport and mobility projects. An urban local authority can also transfer and share its knowledge through networks, partnerships, and knowledge hubs, for free or for a small price. By doing this, the city can benefit from increased attractiveness and name recognition among the general public and various organisations, including local authority employees from other cities looking for new ideas, private companies looking for a location for their business operation, and potential tourists looking for an exciting destination for their holiday.

Examples of transport authorities selling expertise and technical know-how successfully include the Land Transport Authority (LTA) Academy in Singapore; licencing contactless ticketing technology by Transport for London (TfL); Dutch cities, including City of Nijmegen, City of Amsterdam and Utrecht, sharing their knowledge and expertise about cycling via the Dutch Cycling Embassy.

Challenges and risks:

- The expertise and technical know-how should be transferred to the 'buyer' in a proper manner to avoid misunderstandings, resistance, or pitfalls.
- There could be institutional/organisational challenges within transport authorities and public administrations such as getting people motivated to approach 'selling expertise and technical know-how' as a new financing mechanism.
- The practice of knowledge-sharing should align with local legislation and harmonise with that at global, regional, national, and state levels.
- S-M cities may not have a relevant expertise or knowledge which has a market value.
- Small and medium-sized local authorities may lack capacity and resources to participate in knowledge-sharing networks and partnerships.

14.5.9.5 Selling of Land and Property

Local authorities have publicly owned land and property which in many cases is not being used, especially brownfield land, which lies empty for years. Such government land and property can be used productively to raise funds for sustainable transport projects. This type of initiative will not only generate revenue for housing and jobs and support economic growth but save on the running cost of maintaining such assets. Examples of successful implementation of this form of funding mechanism include Transport for London (UK), which works in partnership with various local authorities to release land for new homes and improved local train stations, and the One Public Estate Programme (UK), which is a national programme delivered in partnership with the Local Government Association and the Office of Government Property.

Challenges and risks:

- Strong political and public support is needed.
- Local authorities may not have the capacity to undertake the complex administrative process requiring strict due diligence and accountability and may require the support of other government bodies.
- There might be planning and land issues that need to be resolved before disposing of assets, and in some case, central government would need to be involved.
- The uncertainties in the economy and, in particular, the property market would pose a risk, as land values can vary a lot depending on many factors.

14.6 Recommendations

Implementation of innovative financing approaches can face many obstacles, including:

- Existing culture (which reinforces the status quo, rather than encouraging innovation).
- Absence of a political champion to take ownership of innovative initiatives.
- Policy and regulation constraints.
- Capacity constraints.
- Lack of appropriate knowledge and experience.
- Lack of awareness and guidance regarding innovative financing mechanisms from national government.
- Risk aversion and fear of perceived and/or real risks.

Local authorities should address these shortcomings by initiating framework conditions and supporting initiatives and actions that provide financial help to innovation and R&D programmes. They should also allocate resources to training and human development and assign staff resources to innovation projects and adopt best practices in innovation policymaking and cooperation between their staff and other national and international local authorities.

Transportation officials should assess each project individually and decide what is the best financing approach for each project as some project types may be more suited for a specific type of financing tool. For instance, the right financial mechanisms, including innovative financial mechanisms, can accelerate projects with the ability to generate revenue and therefore provide a dedicated repayment source and repay project debt over a period of years.

Innovation centres can help to connect academics, city leaders, entrepreneurs, and businesses in specific areas, inspire new smart city solutions, and drive future economic growth. The centres can lead events, networking sessions, training and help oversee innovative projects by providing access to world class expert technical capabilities, equipment, and other resources needed to take innovative ideas from concept to reality. Municipalities and national governments should use innovation centres to pursue sustainable transportation and mobility initiatives and to attract businesses to delve into new and emerging technologies (e.g. blockchain, IoT, and artificial intelligence). Pilot programmes should be created to test new ideas by sponsoring a small, specified number of projects to move forward on a test basis. When pilot programmes are found to be effective, they may be incorporated into law and can be used on a wider basis.

S-M cities should prioritise investment that enhances digital customer services and information provision for transportation services in order to optimise existing networks and resources and to improve conditions for passengers by providing them with live journey and routing information. Open data can help to build creative and useful tools for making more informed travel decisions. Open data is an initiative which is being stimulated throughout the EU.

The Guidelines to Innovating Financing can help European S-M city local authorities to finance and implement sustainable transport and mobility measures and Sustainable Urban Mobility Plans that support mobility transformation. The guidelines can also help to improve administrative capability, increase financial sustainability, and optimise opportunities, such as accessing regional development funds, developing partnerships, and applying new financing approaches. These are freely available at https://www.suits-project.eu/the-cities/.

References

1. Greenhouse gas emissions from transport in Europe. https://www.eea.europa.eu/data-and-maps/indicators/transport-emissions-of-greenhouse-gases/transportemissions-of-greenhouse-gases-12
2. Feldman, O., Lugovoi, A., Parker, A., Farooq, S.: Guidelines to Innovative Financing (2019)
3. International Energy Agency: Municipal Transport Financing Act (GVFG) and Regionalisation Act (RegG) (2017). https://www.iea.org/policiesandmeasures/pams/germany/name-31423-en.php
4. Daude, P.: Corporate Mobility Management in Stuttgart–2MOVE2 (2017). https://www.cities-for-mobility.net/news/corporate-mobility-management-in-stuttgart-2move2/ and interview with Patrick Daude at Stuttgart
5. LCA Law: The 2017 Italian Financial Law: News Concerning Labor. Family, and Transportation, 2016–2018 (2017)
6. Territorio, S.: The New Italian Transport Authority Becomes Operative (2014). http://blogs.dlapiper.com/regulatory-ita/2014/01/17/the-new-italian-transport-authority-becomes-operative/
7. Department for Transport: Transport Investment Strategy (2017). https://www.gov.uk/government/publications/transport-investment-strategy
8. European Commission: Commission Gives Boost to Start-ups in Europe (2016). https://ec.europa.eu/commission/presscorner/detail/en/IP_16_3882
9. Campaign for Better Transport: Buses in Crisis: a report on bus funding across England and Wales (2015)
10. Díaz, R., Bongardt, D.: Financing Sustainable Urban Transport (2013). http://www.sutp.org/files/contents/documents/resources/J_Others/GIZ_SUTP_Financing-Sustainable-Urban-Transport_EN.pdf
11. Feikert-Ahalt, C.: National Funding of Road Infrastructure: England and Wales (2014). https://www.loc.gov/law/help/infrastructure-funding/englandandwales.php
12. BBA: Financing the UK's Infrastructure Needs (2015). https://www.bba.org.uk/news/reports/financing-the-uks-infrastructure-needs/#.WcUccGaWyoW
13. Hui, E., Ho, V., Ho, D.: Land value capture mechanisms in Hong Kong and Singapore: a comparative analysis. J. Prop. Invest. Financ. **22**(1), 76–100 (2004)
14. Los Angeles Metro: Preposition C: The Impact of Preposition C (2018). https://www.metro.net/projects/measurer/proposition-c/
15. King, C., Vecia, G., Thompson, I.: Innovative Financing for Transport Schemes: A European Reference Resource (2015)
16. UNDP: Lotteries (2018). http://www.undp.org/content/sdfinance/en/home/solutions/lotteries.html
17. Big Lottery Fund: Big Lottery Fund website (2018). https://www.biglotteryfund.org.uk/
18. National Lottery: Sustrans Connect2 (2018). http://www.lotterygoodcauses.org.uk/project/sustrans-connect2
19. Ystwyth Transport: Big Lottery Fund (People and Places) Grant (2018). http://www.ystwythtransport.org.uk/general-news/big-lottery-fund-people-places-grant

20. Enoch, M.P., Potter, S., Ison, S.G.: Recapturing value for property owners and developers to finance public transport: a review of possible mechanisms. Publ. Money Manage. **25**(3), 147–154 (2005)
21. Smith, S.: New Balance Buys Boston a Commuter Rail Stop (2013). https://nextcity.org/daily/entry/new-balance-buys-boston-a-commuter-rail-stop
22. Dale, S., Frost, M., Ison, S., Warren, P.: Workplace parking levies: the answer to funding large scale local transport improvements in the UK? Res. Transp. Econ. **48**, 410–421 (2014). https://doi.org/10.1016/j.retrec.2014.09.068
23. Burchell, J.: Investigating the Transferability of the Workplace Parking Levy (2015). https://dspace.lboro.ac.uk/dspace-jspui/bitstream/2134/17192/3/Thesis-2015-Burchell.pdf
24. Kill, J., Ozinga, S., Pavett, S., Wainwright. R.: Trading carbon, FERN (2010). https://www.fern.org/fileadmin/uploads/fern/Documents/tradingcarbon_internet_FINAL.pdf
25. PPMC: SLoCaT Climate Finance Transport Database (2018). http://www.ppmc-transport.org/slocat-climate-finance-transport-database
26. Ubbels, B., Nijkamp, P., Verhoef, E., Potter, S., Enoch, M.: A Case Study Assessment (2000)
27. Dalvi, M.Q.: Financing a metro rail through private sec-tor initiative: the Mumbai metro. Transp. Rev. **19**(2), 141–156 (1999)
28. Crevoisier, O.: The economic value of knowledge: embodied in goods or embedded in cultures? Reg. Stud. **50**, 189–201 (2016)

Chapter 15
New Business Models and Partnerships for Sustainable Mobility and Transport Sector

Iana Dulskaia⬤ **and Francesco Bellini**⬤

Abstract Urban mobility is crucial to European societies in providing access to services for passengers and goods and supporting economic growth. Small and Medium (S-M) European cities are facing similar challenges to larger cities, such as congestion and pollution, and perceive similar trends, such as digitalization, the sharing economy, integrated mobility. To improve urban mobility and the related societal challenges, it requires a wide range of complementary mobility solutions and services adopting innovative user-centric, smart, multimodal, and intermodal approaches. Solving the mobility challenge requires coordinated actions from the private and public sectors. Technological advances and commercialization, funding, intelligent policies, and business model innovation are needed to improve urban mobility and create more sustainable environments in modern cities. The capacity to develop or reshape business models requires organizational know-how and tools. New business strategies enable transportation and mobility organizations to receive investments, whilst well-chosen partners will reinforce the chance of success. This chapter outlines the main results of SUITS' research into developing business models and partnerships. It provides knowledge about innovative business models in urban mobility; addresses existing and new partnership schemes; identifies evolving commercially viable business strategies; introduces the main findings of the research and recommendations.

15.1 Introduction

The situation in European cities regarding the urban environment has reached a critical level. A changing mobility paradigm that properly tackles today's challenges and accommodates current and emerging societal trends will clearly require research into

I. Dulskaia (✉) · F. Bellini
Eurokleis S.r.l., Rome, Italy
e-mail: iana.dulskaia@eurokleis.com

F. Bellini
e-mail: francesco.bellini@eurokleis.com

© Transport for West Midlands 2023   279
A. Woodcock et al. (eds.), *Capacity Building in Local Authorities for Sustainable Transport Planning*, Smart Innovation, Systems and Technologies 319,
https://doi.org/10.1007/978-981-19-6962-1_15

new mobility scenarios, technological innovations, additional mobility services and solutions, as well as new partnership schemes. Over 70% of the EU's population lives in cities (including small and medium-sized cities) and accounts for approximately 85% of the Union's GDP.[1] The present mobility situation has created unsustainable conditions for living: severe congestion, poor air quality, noise emissions, and a high level of CO2. Increasing private vehicle use has caused increased urban sprawl and commuting; however, the expansion of public transport networks has not reached the same level of development. Large European cities are well-known for their critical urban mobility situation, whilst S-M cities are left behind with respect to basic services and lack the necessary institutional capacity to manage their rapidly growing populations and the resulting mobility situation [1]. The European Commission provides measures to address mobility challenges in the S-M cities in the Member States by:

- Facilitating best-practise exchange. Dissemination of experiences and best practises (studies, web portals): Urban Mobility Portal (Eltis)[2]; Platform on Sustainable Urban Mobility Plans; Member States Expert Group.
- Providing platforms for collaboration: Civitas Forum[3] and URBACT.[4]
- Fostering local engagement of citizens and stakeholders: European Mobility Week.
- Providing data and statistics on mobility in Europe.

Lately, new ways to invest in thriving, inclusive and liveable cities have appeared: for instance, non-motorized and electric vehicles are improving local air quality and citizen health; transit-oriented development is optimizing land use, reducing traffic congestion, and tackling urban sprawl. New mobility trends arise, and new business models appear to improve the transport sector situation and make it more sustainable. A key task of the SUITS[5] project has been to research and identify new business models and partnerships in the mobility sector.

15.2 Urban Mobility Solutions

Automobile sales are predicted to increase from about 70 million a year in 2010 to 125 million by 2025, and more than half of these vehicles are foreseen to be purchased in cities. Some automotive analysts predict that today's 1.2 billion strong global car

---

[1] Urban Mobility Package—European Commission https://ec.europa.eu/transport/themes/urban/urban_mobility/ump_en.

[2] ELTIS—http://www.eltis.org/.

[3] CIVITAS—http://civitas.eu/.

[4] URBACT—http://urbact.eu/urban-mobility.

[5] The results of SUITS research on new business schemes are presented in more details in the dedicated Guidelines *"Developing bankable projects, new business models and partnerships"* https://www.suits-project.eu/business-models-for-transport/.

Table 15.1 Traditional mobility schemes versus new mobility trends

Traditional mobility strategies	New mobility solutions
Individual car ownership as a main form of transport	Individual car ownership as one element of a multimodal, on-demand and shared transport offering
Limited consumer choice and poor variety of services	Larger variety of services and service providers
Government-funded public transit	Public–private partnership
Unconnected, suboptimal transportation system	On-demand, connected systems

fleet could double by 2030 [2]. The existing urban infrastructure cannot support this number of vehicles on the road. Congestion has already reached unbearable conditions and can cause such problems as wasted time, wasted fuel, and increased costs of doing business.[6]

Many innovative mobility management strategies may improve the mobility situation such as transportation diversity (travel options available to citizens) and different incentives for citizens to change their way of travelling (frequency, mode, destination, route or timing). Others provide an alternative way of travelling or more efficient land use. Some require policy reforms to develop new transportation planning practises.

Some examples of innovative ways of improving urban mobility are (but not limited to) new "multimodal" services that facilitate everyday journeys combining walking, cars, buses, bikes, and trains, etc. as well as shared transportation services, Mobility as a Service, and Urban Vehicle Access Regulations (see Table 15.1).

A shift towards new urban business strategies can provide such benefits as major savings in public budgets including health, environment, or energy by providing safer transport, less congestion, and a higher rate of employment.[7] In this case, the technological development plays an important role.

15.2.1   Mobility as a Service

The objectives of MaaS are to put users at the core of mobility services, offering them personalized mobility solutions based on their individual needs and enabling easy access to the most appropriate transport mode or service. MaaS has three dimensions that should take place when planning innovation activities and developing new business models:

---

[6] Urban mobility at a tipping point—McKinsey & Company. https://www.mckinsey.com/business-functions/sustainability-and-resource-productivity/our-insights/urban-mobility-at-a-tipping-point.

[7] Eurostat figure. European Commission webpage on mobility facts and figures: http://ec.europa.eu/transport/strategies/facts-and-figures/transportmatters/index_en.htm.

1. *The technological dimension*: data sharing, interoperability, standardization as well as connectivity and built-in sensors of smart devices supporting MaaS.
2. *The behavioural impact*: how do travel and logistics patterns change (e.g. for older commuters); what is the potential for modal shift?
3. *Economic and policy dimensions*, including organizational and regulatory aspects. This might involve a change of roles of different players involved.

15.2.2 Integrated Mobility

Integrated mobility enables connecting commuters from trip origin to their final destination using all transportation modes through the integration of barrier-free planning, design, infrastructure, technology solutions, and personalization. The concept behind integrated mobility means that passengers use more than one mode of transportation. Commuters have different trip needs and may switch modes to get to their destinations.

The benefits provided by integrated mobility are:

- *Combined mobility.* Create a seamless travel experience for the door-to-door journey by integrating public and private transport modes in one single service, guided by an intermodal journey planner.
- *One-stop-shop.* Provide easier travel by combining journey planning, mobile ticketing and fare collection in one single application and perform one single transaction for the whole trip.
- *Personalized solutions.* Every traveller has her/his own travel behaviour that differs from person to person including their travel purpose, final destination and available time. Therefore, each traveller needs the flexibility to choose and adapt her/his individual subscription package.

15.2.3 Shared Mobility

Shared mobility forms part of the wider "collaborative economy" or "sharing economy", in the European agenda collaborative economy defined as *"[a variety of] innovative business models where activities are facilitated by collaborative platforms that create an open marketplace for the temporary usage of goods or services often provided by private individuals"*.[8] Service providers offer their goods, assets, or skills to a variety of users via a platform provided by intermediaries. "Sharing" has also become an urban mobility reality. Shared mobility prioritizes the importance of reaching destinations, often at a smaller individual and societal cost than by using a

---

[8] European Commission, 2016, Communication A European agenda for the collaborative economy http://bit.ly/2cFpEKq.

private vehicle. As shared mobility serves a greater proportion of local transportation needs, multivehicle households can begin reducing the number of cars they own whilst others may abandon ownership reducing future demand.

15.3 Innovative Forms of Partnership in the Mobility Sector

It is crucial to choose a suitable form of partnership that will transform innovation into successful implementation. To develop a sustainable business model of a project, a well-organized partnership can facilitate obtaining investment for the project as different partners can contribute to the project by providing different inputs to ensure financial viability to the investors.

15.3.1 Innovative Public Private Partnerships

The Innovative Public Private Partnership (IPPP) is a new form of partnership where the main actors are public and private organizations and may also include other types of organizations like civil society organizations (CSOs), non-governmental organization (NGOs), or communities. These new forms of collaboration enable identification of opportunities for the design and implementation of long-term strategies for partnership. Each actor of the IPPP has an important role in the alliance.

For instance, public organizations oversee the drawing up, financing and implementation of policies and programmes. In the IPPPs, public organizations are defined as important actors who not only have a key role of supervising, creating incentives and regulatory frameworks, but also developing new opportunities and governance mechanisms to enable sustainable long-lasting collaboration with the private sector and other forms of organization, in order to optimize outcomes, impact and sustainability.

The private sector has a significant role in an IPPP. It contributes by bringing investment and expertise in the alliance from its for-profit business orientation.

Finally, other important actors in this type of partnership such as NGOs, CSOs or communities may bring their expertise and vision of the transport and mobility sector. Establishing an IPPP requires strengthening the capacities of all the actors involved.

The IPPP may provide S-M cities' local authorities with a new mechanism of implementing projects by providing additional value such as:

- Addressing market needs and trends.
- Transferring localized institutional knowledge to public and private organizations.
- Creation of a collective awareness of the innovative solutions developed by the alliance.

- Citizens' engagement.
- Enhancement of the possibility of obtaining investment by involving NGOs, CSOs or communities in the consortium.[9]
- Communities' involvement may bring the innovative vision of urban mobility solutions.
- The CSOs or NGOs may improve social relevance and influence and build capacity for policy monitoring.

Example of the CSO involvement in the transport project:

The Rhein-Main-Verkehrsverbund (RMV) is the largest transport association in Germany. It is responsible for organizing and coordinating public transport in the Rhine-Main region. To improve its services, RMV established a passenger advisory board including members of the general public and a CSO. The advisory board organizes meetings four times a year and has already initiated concrete improvements.[10]

15.3.2 R&D Partnerships

R&D partnerships are strategic alliances between businesses and research organizations capable of developing a new product or service (or improving an old one) and other actors who are economically interested in the development of such innovations. The resource-based view highlights that in order to exploit existing resources and to develop a long-term competitive advantage, organizations need to obtain external knowledge [3]. An organization may benefit from R&D collaboration by coordinating a project with competent R&D partners, sharing risks, resources and expertise and building new knowledge [4].

Depending on the actors involved in the R&D partnership, this form of collaboration can include the following types:

- R&D-Public partnership.

---

[9] Civil Society and Public Private Partnership. Why collaborate? Three frameworks to understand business-NGO partnerships—The World Bank https://blogs.worldbank.org/category/tags/civil-soc iety-and-public-private-partnership.

[10] RMV—http://www.rmv.de/de/Verschiedenes/Informationen_zum_RMV/Der_RMV/Wir_ ueber_uns/Struktur_des_RMV/33022/RMV-Fahrgastbeirat.html.

- R&D-Private partnership.
- R&D-PPP.

Benefits of R&D partnerships for local authorities:

- R&D partners may help to develop new products or services, improve current ones or come up with innovative approaches to operations. R&D partnerships also enable mobility suppliers to remain on the market by monitoring market needs and trends.
- Help public or private organizations to improve their business strategies.
- Local Authorities can share research and development costs and the risks associated with the investment of time, money, and other resources.
- R&D partner may help to provide market analysis or test a prototype.
- R&D partner provides monitoring of the project results.
- The involvement of the R&D partner may provide added value in sourcing investment due to the expertise that this partner can bring.

Example of the R&D institutions involvement in the transport project:

This new form of partnership for transport research was organized in Germany's central region Frankfurt RheinMain by major transport authorities and operators, including partners from industry and consultancy, and supported by the Hessen State Government. Namely, the ZIV Institute was founded at the Darmstadt University of Technology. It enables fostering exchange between research and practise. The Institute conducts research in the sphere of Integrated Traffic and Transport Systems covering the areas of Transport Infrastructure and Traffic Management, Traffic Engineering and Traffic Control, Public Transport, etc. About 25 research associates work together in ZIV on innovative concepts aimed at the optimization of traffic and transport systems. The Institute is funded exclusively through orders for planning and consulting with a focus on application-oriented research and development. ZIV founded a scientific advisory board that has its added value when working on the projects. ZIV has conducted more than 60 projects. ZIV has collaborated in projects with the following organizations: Frankfurt Airport Authority (Fraport AG), German Rail (DB Reise & Touristik AG), Deutsche Lufthansa AG, and the Regional Public Transport Authority (RMV) since 2000.[11]

---

[11] Institute für Verkehr. Transport Planning and Traffic Engineering. http://www.verkehr.tudarm stadt.de/vv/fg_verkehrsplanung_und_verkehrstechnik/forschung_7/profil/index.en.jsp.

15.4 Innovative Business Models

Cities operate in an environment where the mobility sector is highly competitive, and the economic environment is uncertain and rapidly changing. This means that local authorities must take complex and difficult business decisions. Transport and mobility organizations run their businesses in a digital era where new technologies enable innovative business models (BMs) that could solve current mobility problems. Many factors should be taken into consideration when starting a new business such as the mobility business environment, strategic partnerships, technological innovation, market tendencies, revenue streams. A well-developed business model will enable transport and mobility organizations to obtain funds for innovation exploitation, and a well-prepared feasibility study will prove a project's financial viability.

Some researchers in the transport and mobility sector argue that the traditional organizational structure and BMs are no longer viable [5]. Increasing challenges in the mobility sector such as market saturation, environmental issues (poor air quality etc.), congestion, and accelerated urbanization are changing customers' demand and needs, forcing transport authorities to change their BM in order to address these issues. Changing market characteristics and fast-evolving new technologies are leading local authorities to reorganize or even innovate their BM [6]. The evolution of new technologies could enable solutions to some mobility problems, and transport organizations are already implementing them when developing new services.

Technological breakthroughs permit improvements and technological advancements in many areas of transport and mobility, e.g. alternative power trains, digitalization, automotive software and hardware, connectivity and smart device technologies that are further influencing the growth of innovative BM in the transport sector.

> Technology innovations and business model innovations are strongly linked to each other. A business model denotes the way in which companies can make money out of a technology. No matter how the technology is innovative and sophisticated, it will fail, if it is not possible for market players to make profits from it-[7].

Following this statement, it can be argued that emerging technological innovation of the transport industry should be accompanied by BM innovation (see Fig. 15.1).

15.4.1 Car On-Demand

Car on-demand is an innovative, user-focussed approach that leverages emerging mobility services, integrated transit networks and operations, real-time data, connected travellers, and cooperative Intelligent Transportation Systems to allow a more traveller-centric transportation system, providing improved mobility options to all travellers and users of the system in an efficient and safe manner.

- Car sharing
- Bike sharing
- Scooter sharing
- e - Haling/Taxi
- TNC
- Micro mobility
- Smart parking
- Integrated mobility
- Ridesharing

- Public transport
- Car rental
- Parking
- Taxis
- Shuttles

Fig. 15.1 Urban mobility ecosystem (*Source* authors' elaboration)

Car on-demand offers several models:

- **Taxi e-hailing**

Recently, taxi companies have revolutionized their services by applying ICT, providing door-to-door trips, through e-hailing services.[12] E-hailing represents a process of ordering a vehicle (car, taxi or any other form of transport) by using a computer or smartphone. To hail a taxi electronically, a user should provide a taxi company his/her desired or current position, by providing an address or sending a current GPS position (see Fig. 15.2).

- **Transportation Network Companies (TNC)**

A TNC is a corporation, partnership or other types of entity that runs a transportation business using digital technology to connect TNC passengers with TNC drivers. TNC provides "real-time ridesharing", by means of a mobile application that indicates not

---

[12] European Commission: Study on passenger transport by taxi, hire car with driver and ridesharing in the EU. Final report—https://ec.europa.eu/transport/sites/transport/files/2016-09-26-pax-transp ort-taxi-hirecar-w-driver-ridesharing-final-report.pdf.

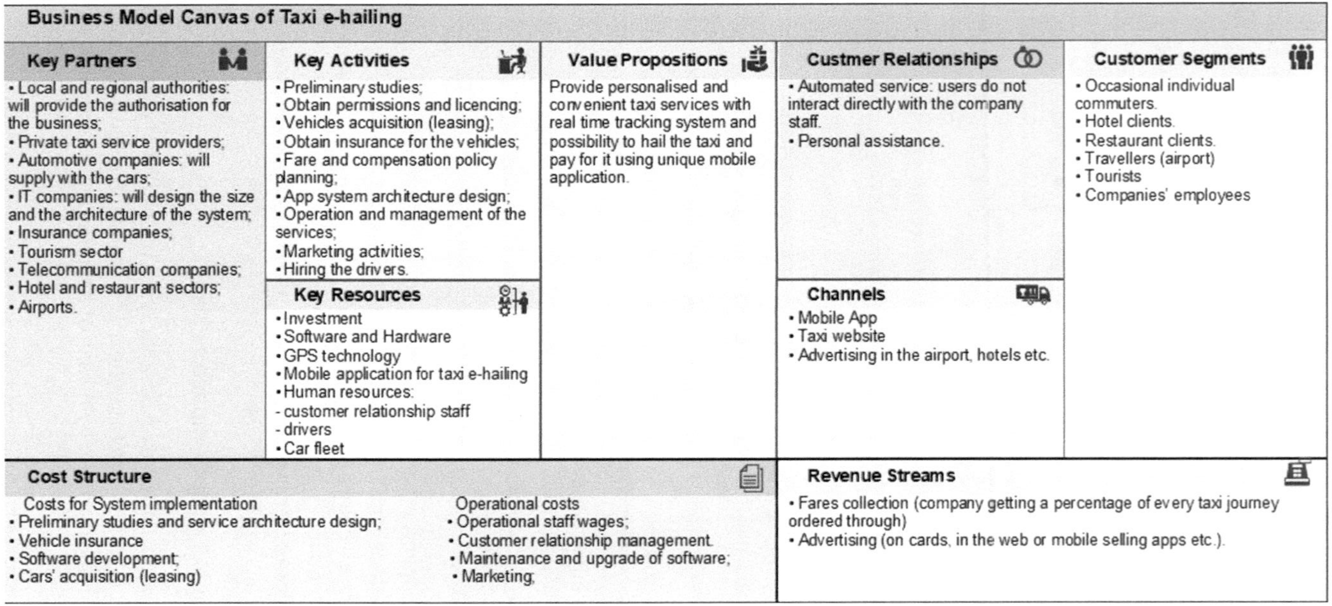

Business Model Canvas of Taxi e-hailing

Key Partners	Key Activities	Value Propositions	Custmer Relationships	Customer Segments
• Local and regional authorities: will provide the authorisation for the business; • Private taxi service providers; • Automotive companies: will supply with the cars; • IT companies: will design the size and the architecture of the system; • Insurance companies; • Telecommunication companies; • Hotel and restaurant sectors; • Airports.	• Preliminary studies; • Obtain permissions and licencing; • Vehicles acquisition (leasing); • Obtain insurance for the vehicles; • Fare and compensation policy planning; • App system architecture design; • Operation and management of the services; • Marketing activities; • Hiring the drivers.	Provide personalised and convenient taxi services with real time tracking system and possibility to hail the taxi and pay for it using unique mobile application.	• Automated service: users do not interact directly with the company staff. • Personal assistance.	• Occasional individual commuters. • Hotel clients. • Restaurant clients. • Travellers (airport). • Tourists • Companies' employees
	Key Resources		**Channels**	
	• Investment • Software and Hardware • GPS technology • Mobile application for taxi e-hailing • Human resources: - customer relationship staff - drivers • Car fleet		• Mobile App • Taxi website • Advertising in the airport, hotels etc.	

Cost Structure	Revenue Streams
Costs for System implementation • Preliminary studies and service architecture design; • Vehicle insurance • Software development; • Cars' acquisition (leasing) Operational costs • Operational staff wages; • Customer relationship management. • Maintenance and upgrade of software; • Marketing.	• Fares collection (company getting a percentage of every taxi journey ordered through) • Advertising (on cards, in the web or mobile selling apps etc.).

Fig. 15.2 Taxi e-hailing Business Model Canvas

only the location of the potential client but also the density of drivers nearby and the waiting time for the closest driver (see Fig. 15.3).

- **Shuttle buses**

Shuttle bus services comprise corporate, regional, and local shuttles that provide limited stops and only pick up passengers at certain points. The final destination may vary depending on the customer segmentation (airports, business centres, etc.).[13]

Shuttle bus business models may differ according to the customer segmentation:

- Tourists and business travellers that commute to/from the airport. In this case, also the type of vehicle can be adapted to this passenger segment. For instance, vehicles can include space for luggage.
- Employees that commute to their organizations. For instance, such vehicles may be provided with Wi-Fi and tables for working.

15.4.2   Micro Mobility

Micro mobility refers to a new category of vehicles that could become an alternative to traditional modes of transportation. Several types of micro mobility vehicles exist, such as scooters/E-scooters and small electric cars with one or two seats.

- **Electric kick scooter sharing**

Electric kick scooter sharing system is a service that makes scooters available for use for short-term rentals. The scooter sharing model is similar to car sharing or bicycle sharing systems. Scooters are normally dockless, which means that they do not have a fixed base location and can be picked up and dropped anywhere in the service area. This business model makes this type of transportation a convenient mobility option for first-/last-mile urban mobility (see Fig. 15.4).[14]

15.4.3   Scooter Sharing

Scooter sharing service provides commuters with access to scooters for short-term use. The vehicles are distributed across a network of scooter sharing spaces within a metropolitan area. Clients can access the vehicles 24/7 with a reservation and are charged by time or by mile. Scooter sharing provides different service models[15]:

---

[13] Stamford Private Shuttle Study, Final report—http://stamfordbusandshuttle.com/documents/Stamford%20Bus%20&%20Shuttle%20Study_Final%20Report.pdf.

[14] McKinsey & Company: Micromobility's 15,000-mile checkup https://www.mckinsey.com/industries/automotive-and-assembly/our-insights/micromobilitys-15000-mile-checkup.

[15] Global Scootersharing Market Report https://www.innoz.de/sites/default/files/howebock_global_scootersharing_market_report_2017.pdf.

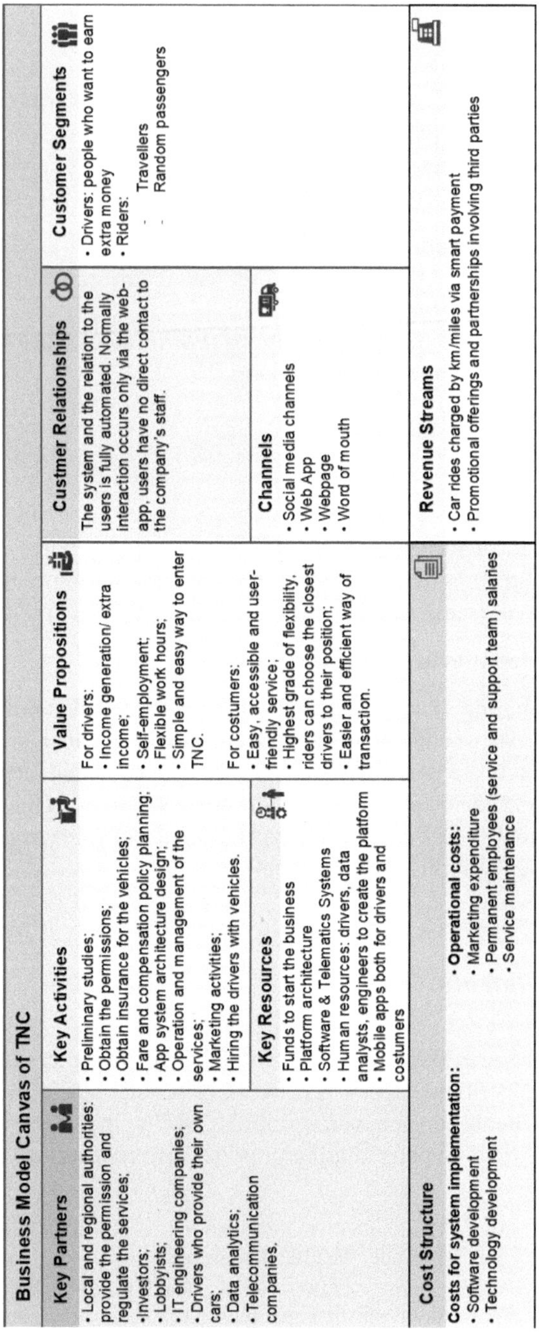

Business Model Canvas of TNC

Key Partners	Key Activities	Value Propositions	Custmer Relationships	Customer Segments

• Local and regional authorities: provide the permission and regulate the services;
• Investors;
• Lobbyists;
• IT engineering companies;
• Drivers who provide their own cars;
• Data analytics;
• Telecommunication companies.
	• Preliminary studies;
	• Obtain the permissions;
	• Obtain insurance for the vehicles;
	• Fare and compensation policy planning;
	• App system architecture design;
	• Operation and management of the services;
	• Marketing activities;
	• Hiring the drivers with vehicles.
	
	Key Resources
	• Funds to start the business
	• Platform architecture
	• Software & Telematics Systems
	• Human resources: drivers, data analysts, engineers to create the platform
	• Mobile apps both for drivers and costumers
		For drivers:
		• Income generation/ extra income:
		• Self-employment;
		• Flexible work hours;
		• Simple and easy way to enter TNC.
		
		For costumers:
		• Easy, accessible and user-friendly service;
		• Highest grade of flexibility, riders can choose the closest drivers to their position;
		• Easier and efficient way of transaction.
			The system and the relation to the users is fully automated. Normally interaction occurs only via the web-app, users have no direct contact to the company's staff.
			
			Channels
			• Social media channels
			• Web App
			• Webpage
			• Word of mouth
				• Drivers: people who want to earn extra money
				• Riders:
				  - Travellers
				  - Random passengers

Cost Structure	Revenue Streams
Costs for system implementation: • Software development • Technology development **Operational costs:** • Marketing expenditure • Permanent employees (service and support team) salaries • Service maintenance	• Car rides charged by km/miles via smart payment • Promotional offerings and partnerships involving third parties

Fig. 15.3 TNC Business Model Canvas

Business Model Canvas of Electric kick scooter sharing

Key Partners	Key Activities	Value Propositions	Custmer Relationships	Customer Segments
• Local and regional authorities; • Local transport operators: public transport services suppliers; • Investors; • IT engineering companies: they will design the size and the architecture of the system; • Scooter manufacturing company: they will provide the operators with vehicles; • Insurance company; • Energy charging companies: these companies will charge the scooter batteries. • Telecommunication companies.	• Preliminary studies; • Introduction of the services to the urban mobility planning; • Obtain the permission and licencing for the service operation; • Kick scooter fleet leasing/acquisition; • Development of the software platform; • Maintenance of the platform; • Scooter fleet management and maintenance; • Charging the e-scooters; • Customer care and feedback; • Marketing campaign. **Key Resources** • Software; • Scooter fleet; • IT infrastructure; • Human recourses (service administration, marketing personnel etc.) • Investments.	• Easy, accessible and user-friendly service; • Pay as you Go approach; • No restriction to stations and dedicated parking areas (pick up and drop the scooter wherever in the service area of the provider); • Covering the main central areas with an elevated number of vehicles in the fleet.	• The relation with the users is fully automated. • Service provider supports the users on online basis. **Channels** • The electric kick scooter sharing platform can be reached by mobile app/ desktop browsers; • Website; • Advertisement at the public places like metro.	• Occasional commuters; • Tourists.

Cost Structure		Revenue Streams
Costs for System implementation: • Preliminary studies; • Insurance; • Scooters acquisitions or leasing; • Software development; • Office equipment and expenses.	**Operational costs:** • Selling, General and administrative expense; • Employees wages (service and support team); • Repairs and Maintenance; • Marketing activities; • Payment for the energy charging companies	• Registration fees • Rental fees (Pay as you Go) • Sponsorship/Commercial (advertisement in the public places)

Fig. 15.4 Electric kick scooter Business Model Canvas

- Round-trip: this type of scooter sharing provides the service where the user must return the scooter to its starting point, at the end of the journey. Round-trip scooter sharing provides the service to the user that demands a long-term journey, in this case, the transport operator offer daily, or day-to-day, charges.
- Point-to-point: (station-based) station-based scooter sharing service permits the users to get a vehicle at one station and return it at different one. Station-based services are considered to be less flexible than free-floating scooter sharing, however, enables more efficient specific trips.
- One-way: (free-floating) scooter sharing enables the users to pick up and leave vehicles at any desired location, within a specified operating area (see Fig. 15.5).

15.4.4   Car Sharing

- Car sharing provides commuters with access to cars for short-term use. The vehicles are distributed across a network of car sharing spaces within a metropolitan area. Commuters can access the vehicles 24/7 with a reservation and are charged by time or per mile/kilometre. Car sharing provides some of the benefits of a personal vehicle without the costs of owning a private one.
- Car sharing provides different service models:

  - Round-trip (membership services, business or institutional fleet, non-membership (e.g. vacation)): This type of car sharing represents the system where the user must return the vehicle to its starting point, at the end of the journey. Round-trip car sharing targets users that have a long-term demand, in this case, the operator is required to offer daily, or day-to-day, charges.
  - Free-floating car sharing enables members of a car sharing programme to pick up and park vehicles at any desired location within a specified operating area (see Fig. 15.6).
  - Station-based car sharing permits the users to get a vehicle at one station and return it at a different station. Station-based services are considered to be less flexible than free-floating car sharing.
  - Peer-to-peer (fractional ownership, P2P Hybrid, P2P marketplace): Individuals provide their private vehicles for other users to rent. In some cases, the vehicles are equipped with telematics devices to provide vehicle-renters with remote access via smartcard, whilst in other systems the car-owner must physically transfer the car's keys to the vehicle-renter.

Business Model Canvas of One-way free - floating scooter sharing

Key Partners	Key Activities	Value Propositions	Customer Relationships	Customer Segments
• Local authorities: public transport policy developers and contracting authority for public transport services; • Local transport operators: public transport services suppliers; • Investors • IT engineering companies: they will develop the application architecture; • Scooter providers: provides transport operators with vehicles; • Insurance companies: will provide the insurance of the vehicles; • Telecommunication companies; • CSO.	• Preliminary studies; • Acquisition of the permission and licensing; • Introduction of scooter sharing to the city plan; • Scooter fleet leasing/acquisition; • Scooter fleet management; • Development of scooter sharing digital platform; • Telematics and web app Management; • Maintenance of the scooter fleet; • Cleaning and refuelling of the scooter fleet; • Marketing activities.	• Urban mobility without transport ownership; • Easy, accessible and user-friendly service; • Highest grade on flexibility (pick up and drop the scooter wherever user wants in the area of the provider); • Pay as you go approach; • No restriction to stations and dedicated parking areas; • High availability of vehicles.	The system and the relation with users are fully automated. Normally interaction occurs only via the web-app. users have no direct contact to the company's staff.	• All citizens with a need for flexible mobility within the city; • Tourists.
	Key Resources • Investments; • Scooter fleet; • Helmets; • Charging stations (for the electric scooters); • Software and Telematics Systems; • Human resources for marketing, scooter maintenance services and cleaning etc.		**Channels** • Web App; • Webpage; • Promotional materials in the public places; • At the charging points.	

Cost Structure	Revenue Streams
Costs for System implementation: • Preliminary studies • Office equipment and expenses • Scooters acquisitions or leasing; • Software development; • Charging infrastructure (for the electric scooters); • Insurance. **Operational costs:** • General and administrative expense; • Employees wages (service and support team); • Repairs and Maintenance; • Marketing activities; • Refuelling and cleaning; • Customer Bonus.	• Rental fees (Pay as you Go); • Sponsorship/Commercial.

Fig. 15.5 One-way free scooter sharing Business Model Canvas

Business Model Canvas of Free - floating car sharing

Key Partners	Key Activities	Value Propositions	Custmer Relationships	Customer Segments
· Local and regional authorities: public transport policy developers and contracting authority for public transport services; · Local transport operators: public transport services suppliers; · Investors; · Insurance companies; · IT engineering companies: they will design the size and the architecture of the system; · Automotive manufacturers; · Telecommunication companies; · CSO.	· Preliminary studies; · Acquisition of the permission/licensing; · Insurance acquisition; · Parking permission from local authorities and · Introduction of carsharing to the city plan; · Car fleet leasing/acquisition; · Car fleet management; · Development of carsharing software; · Telematics and web app Management; · Cleaning and refuelling of the car fleet; · Customer care and feedback; · Marketing activities. **Key Resources** · Investments; · Car fleet; · Charging stations (if the vehicles are electric); · Software and telematics systems; · Human resources for customer support, car services and cleaning.	· Urban mobility without car ownership; · Easy, accessible and user-friendly service; · Highest grade on flexibility (pick up and drop the car wherever user wants in the commercial area of the provider); · Charge by minute with discounts for hourly and daily use; · No restriction to stations and dedicated parking areas; · Covering the main central areas with an elevated number of vehicles in the fleet.	The system and the relation to the users is fully automated. Normally interaction occurs only via the web-app. users have no direct contact to the company's staff. **Channels** · Web App · Webpage · Shops · Pick-up/Drop-off anywhere · Promotional materials in the cars · At the charging points	· All citizens with a need for flexible mobility within the city; · Tourists.

Cost Structure	Revenue Streams
Costs for System implementation: · Preliminary studies; · Office equipment and expenses; · Cars acquisitions or leasing; · Software development; · Insurance. **Operational costs:** · Selling, General and administrative expense; · Employees wages (service and support team); · Repairs and Maintenance; · Marketing activities; · Payment to the energy charging companies; · Customer Bonus.	· Registration fees · Rental fees (Pay as you Go) · Sponsorship/Commercial (advertisement on the car fleet. Promotion material in the cars)

Fig. 15.6 Free-floating car sharing Business Model Canvas

15.4.5   Ridesharing

- **Carpooling**

Carpooling is a way of sharing rides in a private vehicle amongst two or more individuals. It involves the use of the driver's private vehicle to carry one or more passengers [8]. The carpooling platform/app permits quick and easy matching of carpooling users' needs, moreover it helps users to plan itineraries, set prices and pay for journeys [9].

15.4.6   Bike Sharing

A bike sharing system is usually a public service operated by a private company through a public tender. Bike sharing exists in multiple forms, including public, closed community, and peer-to-peer systems. Bike sharing enables users to take short point-to-point trips using a fleet of public bikes distributed throughout a community (see Fig. 15.7).[16]

15.4.7   Smart Parking

Smart parking is a vehicle parking system that generally consists of in-ground smart parking sensors or cameras. Sensors are installed to the parking spots or placed next to them to determine whether the parking space is free or not. Data is collected from parking slots in real-time and is then transmitted to a mobile app or website, which communicates the parking status to users.

Smart parking models:

- Parking Guidance and Information Systems.
- Transit-Based Information Systems.
- Parking app.

- **Parking Guidance and Information Systems**

Assists users in identifying free parking spots and helps drivers in their decision-making process. Occupancy detection of parking spaces is based on vehicle detection technology [10] (see Fig. 15.8).

---

[16] The bike-share planning guide, Institute for Transportation and Development Policy—https://itdpdotorg.wpengine.com/wp-content/uploads/2014/07/ITDP-Bike-Share-Planning-Guide-1.pdf.

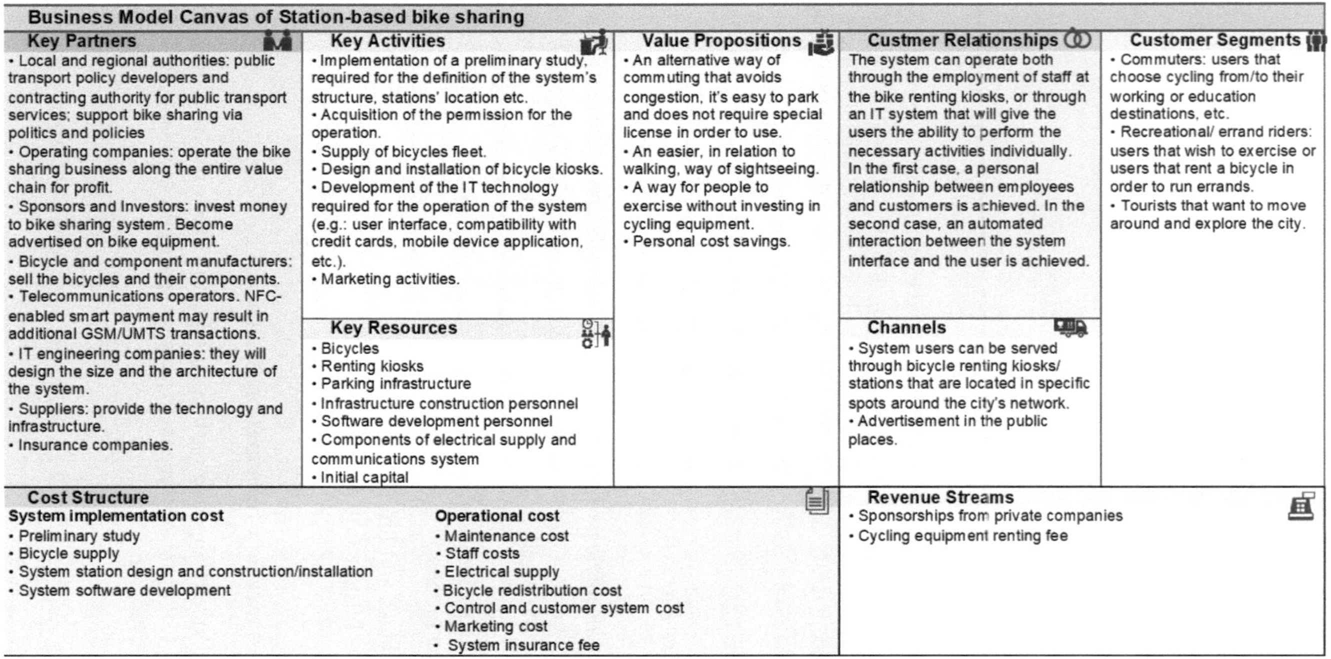

Business Model Canvas of Station-based bike sharing

Key Partners
- Local and regional authorities: public transport policy developers and contracting authority for public transport services; support bike sharing via politics and policies
- Operating companies: operate the bike sharing business along the entire value chain for profit.
- Sponsors and investors: invest money to bike sharing system. Become advertised on bike equipment.
- Bicycle and component manufacturers: sell the bicycles and their components.
- Telecommunications operators: NFC-enabled smart payment may result in additional GSM/UMTS transactions.
- IT engineering companies: they will design the size and the architecture of the system.
- Suppliers: provide the technology and infrastructure.
- Insurance companies.

Key Activities
- Implementation of a preliminary study, required for the definition of the system's structure, stations' location etc.
- Acquisition of the permission for the operation.
- Supply of bicycles fleet.
- Design and installation of bicycle kiosks.
- Development of the IT technology required for the operation of the system (e.g.: user interface, compatibility with credit cards, mobile device application, etc.).
- Marketing activities.

Key Resources
- Bicycles
- Renting kiosks
- Parking infrastructure
- Infrastructure construction personnel
- Software development personnel
- Components of electrical supply and communications system
- Initial capital

Value Propositions
- An alternative way of commuting that avoids congestion, it's easy to park and does not require special license in order to use.
- An easier, in relation to walking, way of sightseeing.
- A way for people to exercise without investing in cycling equipment.
- Personal cost savings.

Customer Relationships
The system can operate both through the employment of staff at the bike renting kiosks, or through an IT system that will give the users the ability to perform the necessary activities individually. In the first case, a personal relationship between employees and customers is achieved. In the second case, an automated interaction between the system interface and the user is achieved.

Channels
- System users can be served through bicycle renting kiosks/stations that are located in specific spots around the city's network.
- Advertisement in the public places.

Customer Segments
- Commuters: users that choose cycling from/to their working or education destinations, etc.
- Recreational/ errand riders: users that wish to exercise or users that rent a bicycle in order to run errands.
- Tourists that want to move around and explore the city.

Cost Structure

System implementation cost
- Preliminary study
- Bicycle supply
- System station design and construction/installation
- System software development

Operational cost
- Maintenance cost
- Staff costs
- Electrical supply
- Bicycle redistribution cost
- Control and customer system cost
- Marketing cost
- System insurance fee

Revenue Streams
- Sponsorships from private companies
- Cycling equipment renting fee

Fig. 15.7 Station-based bike sharing Business Model Canvas

Business Model Canvas of Parking guidance and information system

Key Partners	Key Activities	Value Propositions	Customer Relationships	Customer Segments

• Local and regional authorities: will provide the authorisation.
 • IT engineering companies: will design the size and the architecture of the system.
 • Suppliers: will provide the sensors.
 • Telecommunication companies.
 • Private companies: owns the private parking.
 • Investors.
	• Provide the preliminary studies.
	 • Obtain the permissions from the local authorities.
	 • Create the partnership with all the required partners.
	 • App system architecture design.
	 • Acquisition of the equipment.
	 • Operation and management of the services.
	 • Marketing activities.
	
	 Key Resources
	 • Investment
	 • Software and Hardware
	 • GPS technology
	 • Human resources:
	  – customer relationship staff
	  – marketing personnel
	  – administration and operation personnel
	 • Equipment (sensors, readers)
	 • Office rental and equipment
		• Solving the parking searching process problems by making it faster and more efficient by using innovative technologies.
		 • Reduction of illegal parking.
		 • Better parking management.
			An automated interaction between the system interface and the user
			
			 Channels
			 • Mobile App
			 • Service website
			 • Service promotion in public places
				• Citizens with the need of car parking
				 • Tourists
				 • Companies' employees

Cost Structure	Revenue Streams
System implementation: • Preliminary studies and system architecture design; • Office rental and office equipment; • Software development; • Sensor acquisition and installation. **Operational costs** • Operational staff wages; • Customer relationship management. • Maintenance and upgrade of software and sensors; • Marketing; • Administrative costs	• Fare collection for the parking • Advertising (on cards, in the web or mobile selling apps etc.).

Fig. 15.8 Parking guidance and information system Business Model Canvas

15.4.8 Public Transport

This type of transit comprises buses, trains, ferries, etc., with fixed local routes and express services. It is a core service of shared urban mobility. There is a huge potential for public transport agencies to integrate with or offer shared modes to enhance the access to transport and decrease costs.

- **Bus Rapid Transit**

Bus Rapid Transit (BRT) is fast and flexible road-based public transport that combines stations, vehicles, services, bus lanes, and Intelligent Transport Systems technologies in one system [11] (see Fig. 15.9).

15.4.9 Integrated Mobility

Integrated mobility is a technology-enabled strategic service to ensure that travellers have the most convenient possible transportation journey.

- **Multimodal journey planning**

A multimodal journey planner is a website and/or app that requires and combines the features of a public transport system, forecasting demand and coordinating services having different transport modes and operators as its main elements (Fig. 15.10).

- **E-ticketing and Smart payment**

E-ticketing (or Electronic Ticketing, or Automated Fare Collection, or Smart Ticketing) means, in general, new technologies and integration of services that the user may use to pay by the means of app, smart card.[17] The main features of e-ticketing are:

- Offer related services to users when they buy an e-ticket;
- Offer a new way for public/private transport users to pay for services;
- Improve the overall efficiency and image of the public transport network (see Fig. 15.11).

15.5 Conclusions

Promising urban mobility services that already exist in big cities still need to scale up to their full potential in S-M cities to fully realize the benefits of sustainable urban development. Successful implementation of investment programmes requires

---

[17] Smart ticketing—https://ec.europa.eu/transport/sites/transport/files/themes/its/road/action_plan/doc/2013-urban-its-expert_group-guidelines-on-smart-ticketing.pdf.

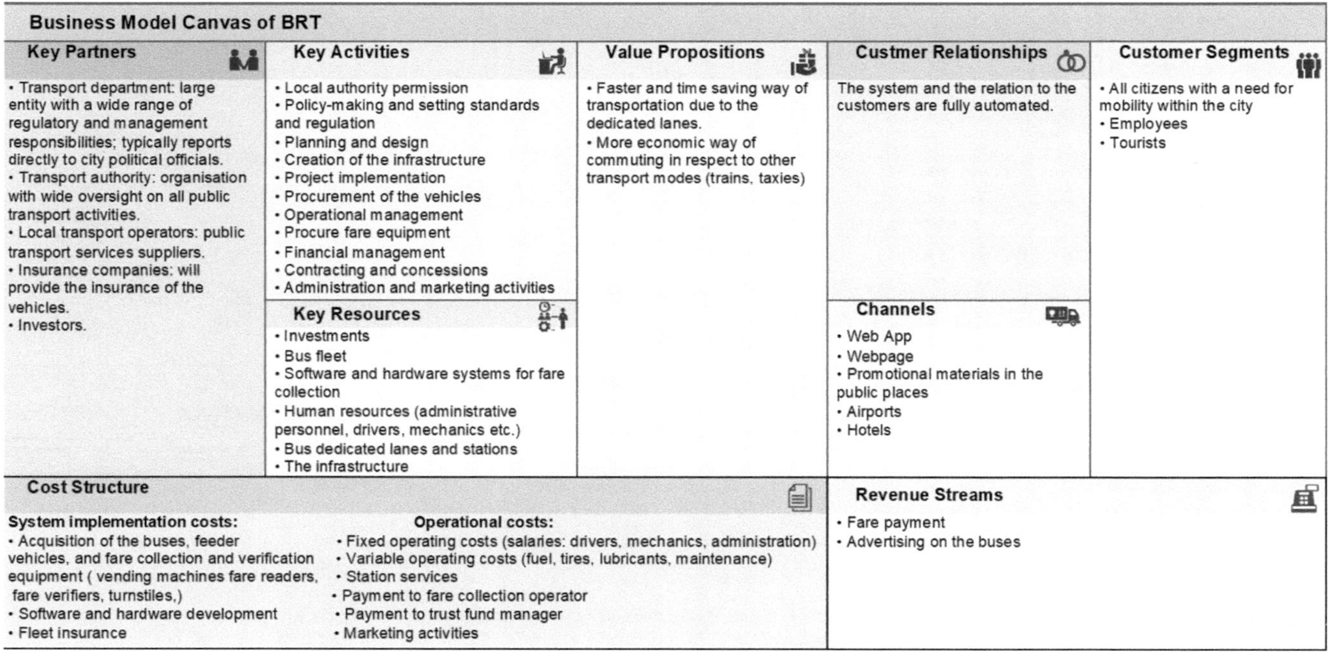

Fig. 15.9 BRT Business Model Canvas

Business Model Canvas of BRT

Key Partners

- Transport department: large entity with a wide range of regulatory and management responsibilities; typically reports directly to city political officials.
- Transport authority: organisation with wide oversight on all public transport activities.
- Local transport operators: public transport services suppliers.
- Insurance companies: will provide the insurance of the vehicles.
- Investors.

Key Activities

- Local authority permission
- Policy-making and setting standards and regulation
- Planning and design
- Creation of the infrastructure
- Project implementation
- Procurement of the vehicles
- Operational management
- Procure fare equipment
- Financial management
- Contracting and concessions
- Administration and marketing activities

Key Resources

- Investments
- Bus fleet
- Software and hardware systems for fare collection
- Human resources (administrative personnel, drivers, mechanics etc.)
- Bus dedicated lanes and stations
- The Infrastructure

Value Propositions

- Faster and time saving way of transportation due to the dedicated lanes.
- More economic way of commuting in respect to other transport modes (trains, taxies)

Custmer Relationships

The system and the relation to the customers are fully automated.

Channels

- Web App
- Webpage
- Promotional materials in the public places
- Airports
- Hotels

Customer Segments

- All citizens with a need for mobility within the city
- Employees
- Tourists

Cost Structure

System implementation costs:
- Acquisition of the buses, feeder vehicles, and fare collection and verification equipment (vending machines fare readers, fare verifiers, turnstiles,)
- Software and hardware development
- Fleet insurance

Operational costs:
- Fixed operating costs (salaries: drivers, mechanics, administration)
- Variable operating costs (fuel, tires, lubricants, maintenance)
- Station services
- Payment to fare collection operator
- Payment to trust fund manager
- Marketing activities

Revenue Streams

- Fare payment
- Advertising on the buses

Business Model Canvas of Multi-Modal journey planning

Key Partners	Key Activities	Value Propositions	Custmer Relationships	Customer Segments
• Local and regional authorities: public transport policy developers and contracting authority for public transport services; • Public transport operators: public transport services suppliers; • Telecommunications operators; • IT Engineering companies: they will design the size and the architecture of the system; • Privat companies	• Develop strategy to collect data and agreements to get access to real time data with regional and stakeholders such as: Authorities. Mobility Agency and Public Transport Companies • Develop web and App with good UX/UI and customization, including alerts and related information to provide best possible experience. • Marketing strategy, especially On-board process for users • Develop strategy for VC and get funds • Marketing activities.	• An alternative way of planning journey, that avoids congestion, taking into account, events, road works, and it will inform commuter on real time of any unexpected issue to save time and money • An easier way to move taking into account personal preferences	• The web/App will communicate automatically. • Personal assistance especially if the Multi Modal Journey App is operated by a Public Transport Company.	• Commuters: users that choose to travel from/to their working or education destinations • Tourists that want to move around and explore the city • Cities and/or Metropolitan Authorities: entities who wish to improve and update old data information system with modern, updated and friendly interface for the user managed by 3rd party.
	Key Resources		**Channels**	
	• Human resources: Sales, Marketing, Design and IT personnel • Initial capital • Cloud • Contact list and relationship with Mobility Agencies and Local Authorities. • Software		• Web and App. • The advertising in the partners transport areas, metro, tram. bus.	

Cost Structure	Revenue Streams
• Marketing and sales cost such as events and trips in order to get agreement on data collection with multiple organizations • System cost using Cloud services such as Amazon services or Google Cloud • Analysis tools for BIG DATA, ML & AI • Staff including Sales, Marketing, Design, IT Systems and software development • Marketing, Sales and Design initiatives	• White label App for local authorities and metropolitan agency • Display advertisements on the site/App • Mobility tools • Geo-Marketing for Local Business • Data & Cross-data generation for 3rd parties

Fig. 15.10 Multimodal journey planning Business Model Canvas

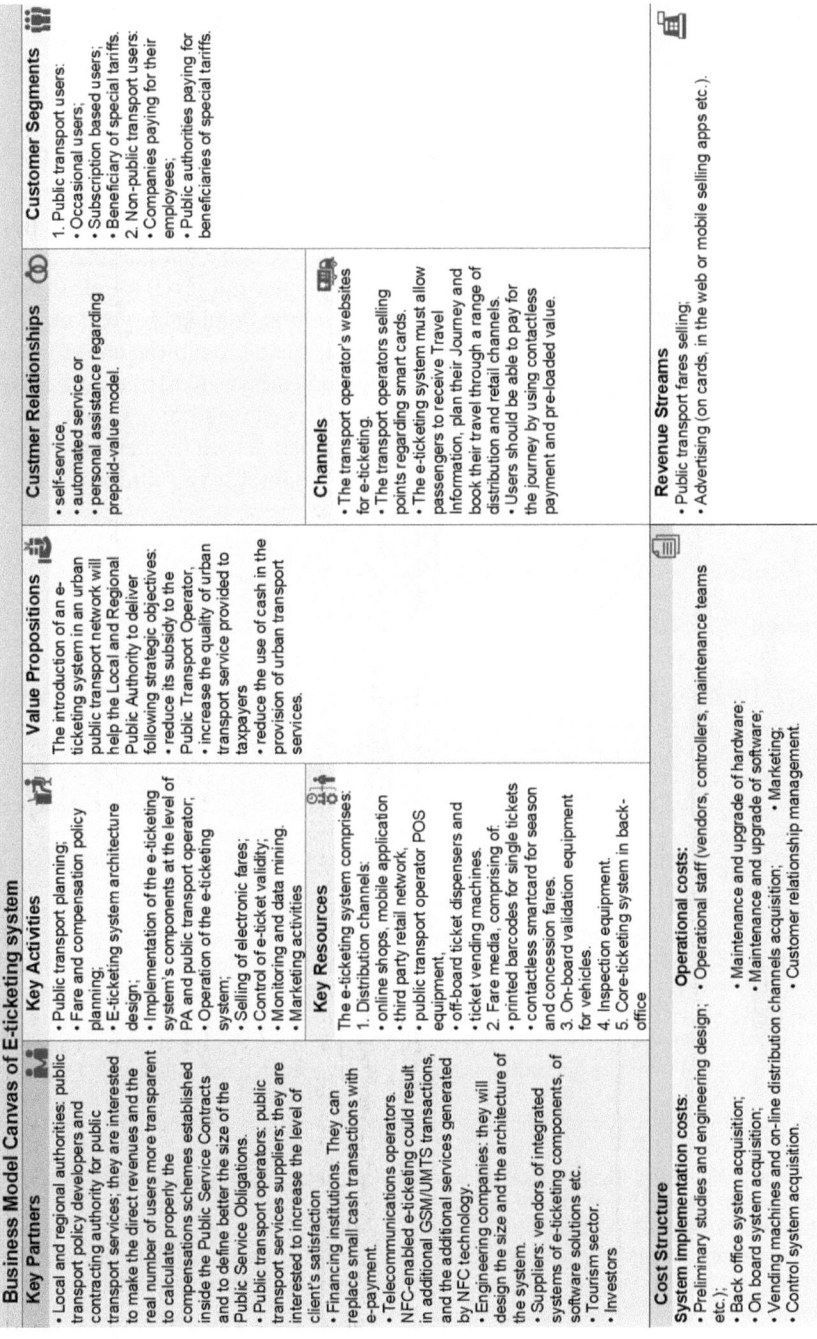

Business Model Canvas of E-ticketing system

Key Partners

- Local and regional authorities: public transport policy developers and contracting authority for public transport services; they are interested to make the direct revenues and the real number of users more transparent to calculate properly the compensations schemes established inside the Public Service Contracts and to define better the size of the Public Service Obligations.
- Public transport operators: public transport services suppliers; they are interested to increase the level of client's satisfaction
- Financing institutions. They can replace small cash transactions with e-payment.
- Telecommunications operators.
- NFC-enabled e-ticketing could result in additional GSM/UMTS transactions, and the additional services generated by NFC technology.
- Engineering companies: they will design the size and the architecture of the system.
- Suppliers: vendors of e-ticketing systems, of e-ticketing components, of software solutions etc.
- Tourism sector.
- Investors

Key Activities

- Public transport planning;
- Fare and compensation policy planning;
- E-ticketing system architecture design;
- Implementation of the e-ticketing system's components at the level of PA and public transport operator;
- Operation of the e-ticketing system;
- Selling of electronic fares;
- Control of e-ticket validity;
- Monitoring and data mining.
- Marketing activities

Key Resources

The e-ticketing system comprises:
1. Distribution channels.
- online shops, mobile application
- third party retail network,
- public transport operator POS equipment,
- off-board ticket dispensers and ticket vending machines.
2. Fare media, comprising of:
- printed barcodes for single tickets and concession fares.
- contactless smartcard for season
3. On-board validation equipment for vehicles.
4. Inspection equipment.
5. Core-ticketing system in back-office

Value Propositions

The introduction of an e-ticketing system in an urban public transport network will help the Local and Regional Public Authority to deliver following strategic objectives:
- reduce its subsidy to the Public Transport Operator;
- increase the quality of urban transport service provided to taxpayers
- reduce the use of cash in the provision of urban transport services.

Custmer Relationships

- self-service,
- automated service or
- personal assistance regarding prepaid-value model.

Channels

- The transport operator's websites for e-ticketing.
- The transport operators selling points regarding smart cards.
- The e-ticketing system must allow passengers to receive Travel Information, plan their journey and book their travel through a range of distribution and retail channels.
- Users should be able to pay for the journey by using contactless payment and pre-loaded value.

Customer Segments

1. Public transport users:
- Occasional users;
- Subscription based users;
- Beneficiary of special tariffs.
2. Non-public transport users:
- Companies paying for their employees;
- Public authorities paying for beneficiaries of special tariffs.

Cost Structure

System implementation costs:
- Preliminary studies and engineering design;
- Back office system acquisition;
- On board system acquisition;
- Vending machines and on-line distribution channels acquisition;
- Control system acquisition.

Operational costs:
- Operational staff (vendors, controllers, maintenance teams etc.);
- Maintenance and upgrade of hardware;
- Maintenance and upgrade of software;
- Marketing;
- Customer relationship management.

Revenue Streams
- Public transport fares selling;
- Advertising (on cards, in the web or mobile selling apps etc.).

Fig. 15.11 E-ticketing Business Model Canvas

shifts in traditional business models that bring public and private interests into alignment. New approaches to developing sound project pipelines are needed to smooth and accelerate the early-stage investment process where knowledge, capacity and interest gaps can exist. Developing new business models to accelerate and scale up investment in sustainable urban mobility will depend on answering the following key questions: what to invest in, how to pay for it, how to mobilize investment capital and how to structure implementation. To best answer these questions, key stakeholders need to develop sustainable solutions together. City decision makers, mobility service providers and investors should build a shared understanding of the challenges and opportunities of different business model choices. New mobility services can have enormous potential for economic development of S-M cities, not just through direct contributions, but also by being a catalyst for innovation in domains beyond transportation, such as technology, communication, procurement. Even if the use of new mobility services is still limited to small and medium urban areas, the concepts that are at the core of new mobility services will serve as an inspiration to improve transportation policy in general and public transit in particular creating new business models and partnerships. New mobility services are an innovative solution for the entire transportation sector and S-M cities in particular.

References

1. Dargay, J., Gately, D., Sommer, M.: Vehicle ownership and income growth, worldwide: 1960–2030. Energy J. **28**(4), 143–170 (2007)
2. Cohen, B.: Urbanization in developing countries: current trends, future projections, and key challenges for sustainability. Technol. Soc. **28**, 63–80 (2006)
3. Hottenrott, H., Lopes-Bento, C.: R&D partnerships and innovation performance: can there be too much of a good thing? J. Prod. Innov. Manag. **33**(6), 773–794 (2016)
4. Caloghirou, Y., Vonortas, N., Ioannides, S.: Research joint ventures. J. Econ. Surv. **17**(4), 541–570 (2003)
5. Chesbrough, H., Rosenbloom, R.S.: The role of the business model in capturing value from innovation: evidence from Xerox Corporation's technology spin-off companies. Ind. Corp. Chang. **11**(3), 529–555 (2002)
6. Holweg, P. (2008) The Evolution of Competition in the Automotive Industry. Judge Business School, University of Cambridge
7. Abdelkafi, N., Makhotin, S., Posselt, T.: Business model innovations for electric mobility–what can be learned from Existing Business Model Patterns? Int. J. Innov. Manag. **17**(1), 1–41 (2013)
8. Dewan, K.K., Ahmad, I.: Carpooling: a step to reduce congestion. Eng. Lett. **14**(1), 61–66 (2007)
9. Furuhata, M., Dessouky, M., Ordóñez, F., Brunet, M.E., Wang, X., Koenig, S.: Ridesharing: The state-of-the-art and future directions. Transp. Res. Part B Methodol. **57**, 28–46 (2013)
10. Mimbela, L.E.Y., Klein, L.A.: Summary of vehicle detection and surveillance technologies used in intelligent transportation systems. Federal Highway Administration's (FHWA) Intelligent Transportation Systems Program Office (2007). https://www.fhwa.dot.gov/policyinformation/pubs/vdstits2007/vdstits2007.pdf
11. Levinson, H.S., Herbert, S., et al.: Bus Rapid Transit: Synthesis of Case Studies. Transportation Research Board, Washington, D.C. Annual Meeting 2003 (2003)

Part III
Reflections and Impact

Introduction

In this, the final part of the volume, consortium partners have been invited to take a more reflective stance on activities and perhaps, with some trepidation look to forthcoming challenges.

As part of the formal requirements of our project, we were asked to conduct an impact assessment. This was led by Eileen O'Connell (Interactions, Ireland) who has conducted a series of evaluations of EU projects. As our project was commissioned in response to the challenge of the need to increase the implementation of sustainable transport measures, in response to the climate (and other crises), impact could not be measured in purely quantifiable terms, e.g. in terms of reduction of vehicle emissions. Such measures are unlikely to be achieved in the lifetime of a project, the main objective of which was to build organisational and indicviudal knowledge and capacity. Rather we have looked at softer measures, working with each partner city to document the impact the project has had on their work practices and capacity to develop transport measures. A fundamental belief in SUITS was that with greater knowledge and understanding of the importance of sustainability and being equipped with the necessary tools and data to understand citizen's needs and mobility patterns of passengers and freight planners are in a stronger position to determine which transport measures are right for their cities. They can lead/discuss options from a position of strength, which recognises whole life cycle and quality of life issues. Chapter 16 is therefore based on the outcome of a structured and continual process and impact evaluation conducted throughout the project.

In contrast to Chap. 16, Chapter 18 was based on a survey of city partners at the end of the project, which allowed them and the project leaders to reflect on their experiences in SUITS, what have been the tangible and intangible benefits. This chapter has been coauthored with the city partners, for other cities who might be considering whether they should take part in EU funded projects. The cities reflect on the steep learning curve, administrative overheads and balancing working on an externally funded project in 'delivery-oriented' environments. Out of their experiences they have been able to develop a list of ways in which small cities can benefit from and maximise the benefits of participation, It is encouraging for all partners, and especially for me, as PI that they have remained enthusiastic and would recommend other cities taking part on future projects.

Chapter 17 acknowledges the impact of COVID-19 on the cities. This has had a profound impact on the whole consortium and our ability to deliver against initial objectives. In the chapter, we have demonstrated how the cities have been effected by COVID, how this has effected transport services and the new challenges this has brought to light.

Andree Woodcock
Principal Investigator of SUITS

Chapter 16
Measuring the Impact of Capacity Development

Eileen O'Connell

Abstract The evaluation work package in SUITS focussed on capacity development in local authorities. It assessed how they adapted and improved their performance on key challenges linked to their sustainable mobility measures, through impact and process evaluation. The main focus of the impact evaluation was on the outputs and outcomes–the changes in performance and capacity in the local authorities in implement sustainable mobility measures. This was done through a self-assessment survey of organisation capacity in addressing key challenges, completed by personnel in each local authority. It was also necessary to link these changes in capacity to the 'inputs' of the project–the Capacity Building Programme and the Organisation Change Programme–and demonstrate the impact of the project, as it is not satisfactory to presume a causal link between the intervention and perceived outcomes. Process evaluation focussed on the change process undergone in the local authorities. The stages of change the organisations went through and, in particular, the barriers and drivers encountered during the capacity development process were assessed. The results of the evaluation programme showed increased capacity in ability to plan and implement sustainable mobility measures. This was directly attributable to the capacity building and organisation change programme of the project.

16.1 Introduction

Evaluation is important when an intervention such as SUITS expects to influence changes at organisational level and have long-lasting effects on policy, strategy and work practises. A capacity building intervention requires a comprehensive evaluation from assessment of the intervention itself, through to outputs and outcomes, to impact on the organisation and its ability to implement plans and strategies, as well as the change processes they have undergone.

E. O'Connell (✉)
Interactions, Wicklow, Ireland
e-mail: eileen@interactions.ie

© Transport for West Midlands 2023   305
A. Woodcock et al. (eds.), *Capacity Building in Local Authorities for Sustainable Transport Planning*, Smart Innovation, Systems and Technologies 319,
https://doi.org/10.1007/978-981-19-6962-1_16

The main goal of the evaluation procedure was to understand factors of success in meeting the objectives of the project. The overall aim of the SUITS project was to challenge and address systemic shortfalls in capacity in local authorities such as lack of transport infrastructure, critical mass, resources, poor data gathering and analysis. This is especially the case in small and medium cities who may be slow adopters, contending with legacy systems, low budgets, and having to deal with rapid changes to mobility patterns and technological advances, as well as new procurement directives. This places them at a significant disadvantage in implementing new sustainable transport measures, to the detriment of the social and economic well-being of their cities.

This capacity gap is not usually recognised in technology-oriented projects, where the focus is on the implementation of measures with little regard to the cultural and behavioural change needed to do this successfully.

SUITS took a **socio-technical** approach, focussing on these capacity gaps and the change needed at individual, organisational and institutional level. Whilst the capacity building programme addressed the technical changes needed, the organisation change programme focussed on the social change–the internal cultural and behavioural processes within the organisation which often inhibit the successful implementation of sustainable, inclusive, integrated and accessible transport strategies.

The evaluation efforts in SUITS were focussed on how the local authority organisations learned, adapted and improved their performance on key challenges linked to their sustainable mobility measures. Local authorities operate in a complex environment with many factors impacting their performance, e.g. budget cuts, innovations, national and international regulations, socio and cultural changes, and health crises. Their operation may also be directly influenced by mayoral directives, changes in higher level management, personnel and different projects/funding priorities. These all influence an organisation's capacities and capabilities. As such, it is difficult to isolate the impacts of one project or intervention from other influences.

The approach to evaluation, therefore, used a number of different methodologies to address these challenges and assess outputs, outcomes, impact and the change process itself:

- Self-assessment of performance on key challenges identified by the project, tracking capacity development over the course of the project
- Assessment of the impact of the changes in capacity–are they reflected in improved ability to plan and implement sustainable mobility measures
- Qualitative assessment of the impact of the project on the capacity development in the organisations. The purpose of this was to determine a link between the inputs of the project's capacity building programme and the changes in capacity
- Finally, a review of the change process itself and the enablers and barriers encountered (process evaluation)

For the purpose of SUITS evaluation and in this chapter, **capacity** is defined as *the ability of people, organisations and society as a whole to manage their affairs successfully* (OECD definition [1]). More specifically for the SUITS project, capacity is the process through which a transport organisation or institution responsible for transport planning and management at the urban level is able to *develop and implement various sustainable transport projects or measures* with the final aim of enhancing integrated transport systems.

This has implications for both technical and behavioural change at the level of the individual, the department and the organisation as a whole.

Capacity development is defined as the process by which individuals, organisations, and countries develop, enhance and organise their systems, resources and knowledge; reflected in their abilities, individually and collectively, to perform functions, solve problems and achieve objectives (OECD definition [1]).

Capacity development in the SUITS project is directed at specific capacity gaps and needs in the local authority organisations, identified during a baseline assessment and from qualitative investigations with the cities in workshops and interviews.

As shown in Fig. 16.1, the result of the SUITS baseline assessments [2] provided a profile of each city, its needs and gaps, the challenges to prioritise, and targets to be achieved. Once the targets were defined, key performance indicators were devised to assess progress towards addressing the challenges and evaluate the outputs and outcomes. See Chap. 4: Sustainable urban mobility plans (SUMPs): setting targets for Local Authorities to increase their capacity to develop and implement sustainable transport measures.

Capacity Inputs: The initial findings from the baseline assessments (Chap. 4) allowed the project partners to create a Change Vision for each city (Chap. 6) and a capacity building programme (Chap. 7).

In turn, the capacity inputs led to:

- **Capacity outputs** which we define as actual changes in individual and organisational capabilities (e.g. staff competencies, new knowledge, access to data)

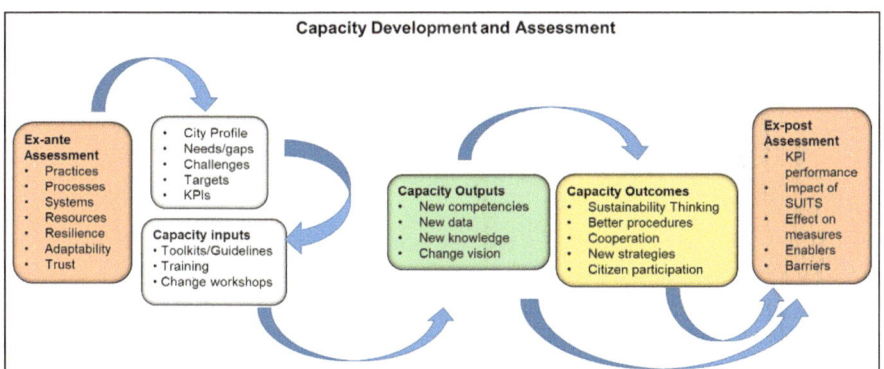

Fig. 16.1 Assessment of capacity development in SUITS

- **Capacity outcomes** which we define as the mainstreaming and transformation of the overall capacity of the institution itself that is necessary for the accomplishment of its mission, beyond the duration of any external intervention (e.g. better procedures, more collaboration)

Capacity development was anticipated in two areas:

- Technical, e.g. skills, knowledge, capabilities, practises and procedures
- Behavioural, e.g. climate, culture, trust, risk, cooperation and collaboration

Increases in **Technical Capacity** were achieved through a Capacity Building Programme devised by project partners including training, guidelines, workshops, products and services (Chaps. 7–12).

Changes in **Behavioural Capacity** as outlined in Chap. 6 were achieved through a behavioural change programme based on the definition of:

- A Change Vision for each city
- The appointment of Change Agents to drive change in the organisation
- The creation of 'Learning Organisations' (whereby the Local Authorities embarked on a journey of organisational change).

16.2   Assessment of Capacity Development

Evaluation is important when a capacity building intervention such as SUITS expects to influence changes at organisational level and have long-lasting effects on policy, strategy and work practises.

In an EU-funded project such as SUITS, the main goal of the evaluation procedure is to assess whether the project has met its objectives and to understand factors of success in achieving this. The overall objective of the SUITS project was to challenge and address systemic shortfalls in capacity in local authorities. The aim of evaluation is to assess the success of this in terms of outputs and outcomes (improved capacity on certain challenges within the local authority), the impact on its ability to plan and implement transport measures, as well as the change processes they have undergone. Finally, evaluation must also demonstrate a causal link with the intervention, as it is not satisfactory to presume a causal link between the intervention and the perceived outcomes.

An assessment was conducted at the beginning of the project using high-level indicators to provide a comprehensive characterisation of each city's socio-economic, demographic, cultural and political context [2] and the capacity of the cities to implement sustainable transport measures. The results revealed the challenges on which each local authority should focus in order to improve its capacity in these areas (Chap. 5).

15 common key challenges that cities encounter during the planning and implementation of mobility measures were identified. The SUITS cities chose 11 of these challenges to address:

- Challenge 1: Institutional cooperation
- Challenge 2: Interaction and cooperation with business partners
- Challenge 3: Citizen participation
- Challenge 4: Use of innovative technologies and data collection methods
- Challenge 5: Understanding political interests and affecting political decisions
- Challenge 6: Understanding and applying innovative financing methods
- Challenge 7: Innovative procurement
- Challenge 8: Estimating the feasibility and acceptance of measures
- Challenge 9: Sustainability thinking
- Challenge 10: Enhancement of knowledge management/knowledge transfer
- Challenge 11: Social impact assessment

The expected outcomes are to show increased capacity in the local authorities' ability to plan and implement transport measures, to demonstrate the link between this increased capacity and the capacity building programme of the project, and to understand the change processes and success factors contributing to this.

In order to achieve this, both impact and process evaluation were undertaken as follows:

1. Impact Evaluation

   o Self-assessment survey of performance on the challenges and indicators
   o A qualitative investigation into whether the changes in capacity can be attributed to the SUITS Capacity Building Programme
   o How this in turn has affected the authority's capacity to implement its transport measures

2. Process Evaluation—a review of the stages of change the organisations went through and, in particular, the barriers and drivers encountered during the capacity development process.

16.2.1 Impact Evaluation

In order to assess the capacity development of the local authorities, an impact assessment framework was proposed [3]. Starting with the mobility measures of the city, specific challenges were linked with them and associated indicators and targets developed. This process is described in detail in Chap. 4. An example of the output is shown in Fig. 16.2.

In order to assess performance, the targets defined for each transport measure were analysed and, in consultation with project partners, refined in the light of the SUITS Capacity Building Programme (CBP) and to reflect the focus and activities of the local authorities.

Project partners worked together to refine these indicators during a dedicated workshop; each of the challenges was addressed and the 'targets' turned into simple, measurable variables. It was important that these indicators were specific

Improve Freight Distribution in the city centre	
Background and goals of the measure City 1 has an historic city centre. In order to save this character and to increase the quality of life and stay in the centre, freight traffic needs to be restricted. This should lead in particular to a reduced pollution and a lower impact of urban freight logistics on circulation of pedestrians and cyclists. For this, the following measures will be implemented: • Increase number of special loading and unloading areas • Access control to restricted areas • Time windows for delivery • Use of new technologies to create efficient distribution and delivery systems that reduce the number of journeys, the length of delivery routes and -time	
SUITS Challenges chosen by the city 1) Institutional cooperation 2) Interaction and cooperation with business partners 3) Citizen participation	Related Indicators 1) Coordination/cooperation between 2) Participatory management 3) None applicable
Targets 1) Organize public meetings with different agents 2) Organize workshop with Local Authorities 3) Conduct surveys, feedback on measures and innovative results	

Fig. 16.2 Development of targets to address challenges

and tangible, making them easy to assess and quantify (e.g. number of partnership agreements with business partners vs. number of citizens participating in sustainable mobility events).

16.2.1.1 Self-assessment Survey of Performance on the Challenges and Indicators

The indicators devised for each challenge were translated into an online survey to assess performance. The assessment was completed by senior personnel in the transport departments of the local authorities.

The initial assessment was conducted in 2019 and a follow-up assessment was conducted in 2020, where the cities assessed their current performance against each indicator relevant to their challenges and measures.

The results showed increased capacity in each local authority in different areas and specific outputs and outcomes were observed. These are shown in Table 16.1 along with the results of the assessment of capacity development below.

Table 16.1 Impact of capacity building programme on measures

City	Project input	Output	Outcome	Measure impact
Challenge 1: Institutional cooperation				
Kalamaria	SUITS capacity building for measures implementation and cooperation between departments	Improved cooperation between departments Increased cooperation with other municipalities	Common spaces for meetings between different municipal staff Online platform for collaboration	More solutions for solving problems with installation of 150 smart parking slots at 3 roads (on-street) with sensors
Palanga	WP5 Capacity Building Workshops to involve politicians	Better interdepartmental cooperation on projects More open communication	Sustainable mobility now involves the economic and architecture department Working groups and workshops organised	n/a
WMCA	Involvement in the project & organisation change models of WP6	Increased internal and inter-municipal cooperation	Stronger links with Coventry City Council MS Teams implemented & now used to communicate with other LAs	Launched a joint procurement for regional e-scooters
Challenge 2: Cooperation with business partners				
Palanga	Guidelines to develop bankable projects, new business models & partnerships, innovative procurement Webinar—'Building S-M LAs capacity to introduce Innovative Transport Schemes'	Stronger relationships with business partners—hotels, restaurants, SMEs More cooperation with business associations	Involve business partners in planning of the measure rather than just at implementation phase Shared vision with business partners 20–50% of projects where business partners are involved (from 0%)	Attracted one of the biggest car-sharing companies to Palanga Prepared joint plan for car-sharing points Private & state-owned electric vehicle charging points opened

(continued)

Table 16.1 (continued)

City	Project input	Output	Outcome	Measure impact
Rome	Exchange of best practise Financial schemes proposed by SUITS	Active participation of stakeholders, associations, academia, etc.	Less conflict to achieve sustainable urban freight distribution	Participation in new EU research project in order to select best practises across Europe and apply them in Rome
Turin	Support from academic partner in the project Learning from SUITS best practise	More cooperation with business partners	More variety of business partners involved	Involving stakeholders in earlier stages of planning Developed the use of Electric Vehicles in urban freight measures
WMCA	Learning from example of Torino's LTZ WP6 organisation change WP4 guidelines on new business models	More intense cooperation with partners in sustainable mobility measures	Involve stakeholders early in the planning process More agreements signed with business partners Collaboration with logistics companies and retailers	

(continued)

Table 16.1 (continued)

City	Project input	Output	Outcome	Measure impact
Challenge 3: Citizen participation				
Kalamaria	SUITS webinars and workshops, emails SUITS capacity building toolbox	Know-how of use of online tools to engage citizens	Survey amongst municipal staff for mobility issues Sending Emails amongst municipal staff, schools in the future Engagement with citizens on social media Advanced know-how	More measures influenced by citizens; More public hearings/open days Use of online platform MyPolisLive.net. High public participation in municipality's platform 4mycity. Results after survey still running till end of October 2020 More crossings and new types (3D crossings) to be scheduled More experience and increased know-how for schedule and design of parking slots More projects to be scheduled and implemented
Palanga	Learning from other cities through partner meetings	Increased awareness of methods of engaging citizens	Using online forums/social media to engage citizens Involvement of citizens in decision-making	Electric vehicle charging network—citizens offered the installation sites they need most
Rome	Exchange of best practise	Increased interaction between local authority, citizens and associations	More active participation of citizens and stakeholders and associations	Making citizens aware of more sustainable forms of transport; improved interest in cycling Involvement of citizens in the planning phase of the cycle network and the drafting of the SUMP

(continued)

Table 16.1 (continued)

City	Project input	Output	Outcome	Measure impact
WMCA	Learning from example of Rome's PUMSRoma	Learned the importance of online as well as face-to-face contact	Procured an online market research community	Consultation with citizens on Ultra Low Emission Vehicle Strategy—crucial for behaviour change for EVs to become common
Alba Iulia	SUITS social media channels	Increased awareness of methods of engaging citizens	More/different communication channel used	More public hearings/open days More use of social media Promotion of cycling
Challenge 4: Use of innovative technologies and data collection methods				
Kalamaria	Access to data platform mypolislive.net	Increased awareness about data reliability and data storage for sustainable mobility	More complete data Data more easily visualised More data now stored in the cloud More real-time data now available	Data from platform MyPolislive.net used in a traffic design near a school in order to undertake mobility measures Data storage for new pedestrian crossings MyPolislive.net in coordination with another project will lead to data integration
Turin	Interaction and support from academic partner Access to data repository		Ability to process own data Ability to analyse and visualise data further	Data used to inform measures—with the result that intermodality has been substituted by MaaS

(continued)

Table 16.1 (continued)

City	Project input	Output	Outcome	Measure impact
WMCA	Learnings from the project about methods/sources & big data approaches Mypolislive.net	Identifying new data collection methods/sources Learned the importance of collecting real-time data for traffic	More complete data Able to handle a much wider array of data	Using big data/processing large datasets Collecting large datasets such as vehicles passing ANPR cameras Able to present information about the effect of COVID-19 on the transport network
Alba Iulia	SUITS facilitated communication with relevant stakeholders in terms of data generation Enabled debate about improving the current status	More aware of the importance of big data in mobility	More admin staff trained in ICT & big data More data now stored in the cloud and more real-time data available Able to process own data more and visualise it	Data used to inform some measures Using real-time data & data from sensors for the first time Making better predictions about mobility to support freight policies and cycling measures & bike lanes
Challenge 5: Understanding political interests and affecting political decisions				
Palanga	WP5 Capacity Building Workshops Innovative financing guidelines Learning from other partner cities	Encouraged to involve politicians Improved dissemination of information to politicians, presentations at meetings, etc.	Greater understanding of sustainable mobility amongst politicians More compromises—looking for common solutions Information taken more seriously	Transport measures now very important in local decision-making Targeting the right political level to advance transport measures Easier to engage with business partners when there is strong support at the political level

(continued)

Table 16.1 (continued)

City	Project input	Output	Outcome	Measure impact
Rome	SUITS examples on how to apply the EU guidelines for Sustainable Urban Mobility Planning (SUMP) Learning from other EU experiences Active modes promotion was supported by SUITS	Consistency between SUITS and SUMP strategies Organisation change to create the necessary background to promote SUMP	Involvement of political interests/consistent with SUMP Increased collaboration and feedback in implementing sustainable mobility measures	Launch of an impressive cycling plan integrated in to the SUMP Approval & financing of 6 public transport infrastructural works included in the SUMP Post-COVID measures included an integrated cycling corridor to complete the SUMP cycling plan SUMP to be included in Recovery Fund request
Alba Iulia	Involvement of decision-makers in SUITS training workshop	Bringing together politicians from all sides to foster novel policies in sustainable mobility—public transport, freight, cycling	More information disseminated at senior level in a user-friendly manner Information taken more seriously	Updating of current SUMP More projects approved and currently implemented ERDF funding applied for and approved
Challenge 6: Understanding and applying innovative financing methods				
Turin	Useful documentation prepared and proposed by the SUITS project guidelines	More understanding and knowledge of innovative financing methods	Congestion charging approved as appropriate Financial strategies in place to implement new funding mechanisms	New agreements with freight operators giving them access to restricted areas in exchange for using more sustainable vehicles As well as access to their data

(continued)

Table 16.1 (continued)

City	Project input	Output	Outcome	Measure impact
Alba Iulia	SUITS guidelines and training on Innovative Financing	More awareness of different financing mechanisms New financing mechanisms were analysed	Financial strategies in place to implement new funding mechanisms Grants, workplace parking levy, municipal green bonds, HGV charging system all approved as appropriate	More collaboration with other cities, private companies and research consortia more funding available for measures, e.g. encouragement of cycling A new parking policy is being implemented including the workplace parking levy
Challenge 7: Innovative procurement				
Kalamaria	SUITS webinars and workshops	Increased awareness of innovative procurement methods	Meetings for using new types of procurement More awareness of impacts and costs	Improved know-how and better schedule for new projects
Alba Iulia	SUITS guidelines and training on Innovative Procurement	More familiarity with Innovative Procurement concepts Cooperation between mobility & procurement departments		Lifecycle costs, pollution reduction and energy consumption were tested and applied as criteria to mobility procurement

(continued)

Table 16.1 (continued)

City	Project input	Output	Outcome	Measure impact
Challenge 8: Estimating the feasibility and acceptance of measures				
Palanga	WP5 Capacity Building Workshop: Building S-M LAs capacity to introduce innovative transport measures Experience gained from project partners and project cities	Started to check interfaces between measures	New thinking at strategic planning stage to integrate measures Using more social instruments to encourage feedback from citizens	Fewer constraints during the implementation of measures Dependencies with other measures are checked Measures are more integrated
Challenge 9: Sustainability thinking				
Palanga	WP5 Capacity Building Workshops Policy Brief 3—Implementation needs behaviour change Learning from other cities through partner meetings	Sustainability is now a top priority	There is a transparent long-term vision for sustainable mobility Training on sustainable mobility topics Employees on-board with vision	Environmental impact added to the focus of measures
Rome	The measures & tools provided by the project giving models for the organisational changes Sharing other EU experiences with on-site visits Participation of the SUMP personnel in project meetings	Developing a change vision Awareness of training needs amongst employees	There is now a long-term vision for sustainable mobility, and employees are on-board and given training More use of participative planning approach Evaluation of short/long-term effects of measures Establishment of a learning process	Providing suitable financing schemes and procurement Development of cost-effective solutions Extension of public transport infrastructure during SUMP—change towards multimodality where public transport can transform the city Minimisation of environmental impacts of mobility

(continued)

Table 16.1 (continued)

City	Project input	Output	Outcome	Measure impact
Turin	WP6 Organisation change models & workshops	Change vision developed	Participative planning Networking with other administrative areas	More focus on electric vehicles compared to the past, evaluation of pollution aspects Minimisation of environmental impacts
WMCA	WP5 & WP 6 workshops/training Knowledge sharing with SUITS partner cities	Increased networking with other administrative areas	Employees more engaged with sustainability vision More training for employees in sustainable mobility topics Enhanced relationship with Coventry City Council	Increased ability to introduce SUMP measures, e.g. scooter trial
Alba Iulia	WP6 Organisation change models & workshops SUITS guidelines	Sustainability is now a top priority Change vision developed; Employees made aware and engaged and trained	Technical & empirical capacity to promote new sustainable mobility at the local level	Sustainability criteria introduced into public procurement SUMP under elaboration Redirection of freight around the city Successful at getting ERDF funding for 3 new mobility projects
Challenge 10: Enhancement of knowledge management/knowledge transfer				
WMCA	WP6 survey of staff to discuss knowledge management processes Looking at the way things are done in other cities	Categorisation of lessons learned now introduced	People are keener to look for lessons learned	Taking a holistic view of MaaS based on lessons learned—better understanding of the issue and hope to deliver a better MaaS offer

(continued)

Table 16.1 (continued)

City	Project input	Output	Outcome	Measure impact
Challenge 11: Social impact assessment				
Kalamaria	SUITS dissemination in Municipality's magazine and website	More events and increased awareness New surveys	SIA more robust and inclusive SIA influences development of measures SIA now conducted for some measures	Innovation in smart mobility systems Participation in European mobility week events Installation of pedestrian touch devices as additional equipment in traffic lights with pedestrian affected phases and for use by visually impaired people
Palanga	Policy brief 2 SIA of transport measures & systems	Greater awareness of SIA	SIAs more robust & inclusive	SIAs conducted more frequently SIAs have slightly more influence on the development of measures
Rome	SUITS guidelines on suitable financing schemes and procurement	Integration of SUITS guidelines into the LA process	SIA more robust and inclusive	SIA now has more influence in the development of measures MCA rather than CBA Evaluation of measures according to SIA results

(continued)

Table 16.1 (continued)

City	Project input	Output	Outcome	Measure impact
Turin	Some suggestions coming from SUITS capacity building programme		More robust and inclusive SIAs conducted	
WMCA	SUITS project emphasised the importance of SIA		SIAs more robust & inclusive	More SIAs now conducted

16.2.1.2 Link to CBP and Impact on Measures

The next step in evaluation was to demonstrate the impact of the SUITS project, i.e. to link the performance outcomes found in the first phase to the capacity building inputs of the project.

Local authorities operate in a complex environment with many factors impacting their performance, e.g. budget cuts, innovations, national and international regulations, socio and cultural changes and health crises. Their operation may also be directly influenced by mayoral directives, changes in higher level management, personnel and different projects/funding priorities. These all influence an organisation's capacities and capabilities. As such, it is difficult to isolate the impacts of one project or intervention from other influences.

The approach to evaluate the impact of SUITS is based on the Rapid Assessment of Capacity Development (RAC) methodology of the EU. This complementary approach to traditional evaluation methodologies was developed to demonstrate causal links between the intervention and capacity development at institutional level.

We adapted and piloted the Rapid Assessment of Capacity (RAC) tool from Capacity4dev [4], the European Union's knowledge sharing platform, created and managed by the Directorate General for International Cooperation and Development (DEVCO).

The RAC is based on a thorough evaluation methodology, according to the EU evaluation guidelines. It shares many features with similar methodologies recently elaborated and tested by different international agencies, such as the United Nations Development Programme (UNDP), the World Bank Institute (WBI), and particularly the five capabilities (5Cs) methodology developed by the European Centre for Development Policy Management (ECDPM) [5].

The purpose of the RAC tool is to assess the impact that the interaction with an intervention generates at individual and at institutional level in terms of capacity development and/or strengthening. It is a qualitative approach, using individual and collective interviews with key stakeholders, using specific data collection tools and specific questions to be followed.

The RAC methodology advocates a 4-step approach:

- Step 1: Identification of:

  - Facilitating factors that positively influence the capacity development process and therefore also the development intervention
  - Factors that tend to hinder or limit the effectiveness of the capacity development process

- Step 2: Assessment of the capacity outputs
- Step 3: Assessment of the capacity outcomes
- Step 4: Verification of the assessments and results

Steps 2–4 were used for the evaluation of the impact of the SUITS project on the capacity development of the cities. Step 1 is dealt with under Process Evaluation below.

Step 1: Establish the capacity outcomes
Present the results of the KPI surveys to establish the extent to which there have been changes in the skills, procedures or organisational set up in the Local Authorities
Step 2: Input Assessment
Identify which inputs of the project (if any) have contributed to those changes
Step 3: Links to mobility measures
Establish whether these changes have had an impact on the sustainable mobility measures in the city, getting examples where possible
Step 4: Summary of findings
The relationships between the capacity inputs and capacity outcomes are summarised and the contribution of the project is established

Fig. 16.3 RAC adapted for SUITS evaluation

The RAC process was adapted and piloted in 2019 in the SUITS cities, and implemented in 2020 for final evaluation. Site Evaluation Assistants (SEAs) held face-to-face interviews with relevant personnel in the local authority, including the Change Agent and the technical department/mobility department, and/or individual measure leaders. The process was as follows Fig. 16.3.

For this assessment, cities are presented with the results of their self-assessment surveys from 2019 and 2020, showing progress (or lack of it) on performance in relation to each of the indicators and challenges over time (Step 1).

They then discuss each change in capacity, whether it could be attributed to the inputs of the SUITS project, and if so, which aspect of the SUITS project contributed to the change (Step 2).

The last step is to determine if the changes in capacity have directly influenced the ability of local authorities to plan and implement sustainable mobility measures. They are asked for specific examples of measures that have improved or been implemented more successfully (Step 3).

In Step 4, the results are summarised.

16.2.1.3 Results of Impact Evaluation

The results are summarised for each challenge in turn, linking the capacity outputs and outcomes to the project inputs that effected those changes, and demonstrating the impact on the transport measures in the city.

The impact of the SUITS project on capacity development in the local authority organisations has been demonstrated using the methodologies described above. The evaluation framework shows that the Capacity Building Programme and the Organisation Change Programme were successful in developing the capacity of the cities to implement their measures. The SUITS Capacity Building Programme delivered webinars and workshops which were highly rated by participants, who stated that they felt their ability to implement change had increased and that they would put the learnings into effect (see Chap. 7). Cities specifically mention knowledge sharing and learning from each other through attendance at workshops.

16.2.2 Process Evaluation

The final aspect of evaluation is to assess the change process itself–to understand the context in which the capacity development programme was implemented and the factors that affected the implementation.

The initial focus of process evaluation in SUITS was on the building of trust and rapport with the local authorities as the concept of cultural change was new, often distrusted and misunderstood. In the latter stages of the project, the focus was on the internal processes of change being undertaken in the organisations as they engaged with the change vision, created quick wins and built long-term capacity development.

Two approaches were used to evaluate these processes:

1. The SATELLITE Process Evaluation methodology was adapted for the first phases of process evaluation [6].
2. The Eight-Step Process for Leading Change outlined by Kotter [7] was used in the latter phase of the project. This is described in detail in Chap. 6 in small-to-medium local authorities.

The CIVITAS SATELLITE methodology is concerned with the process of how initial proposals for a sustainable mobility measure are developed into a feasible design, how the measure is then constructed or implemented, and focussing on the 'stories behind the figures', recording the drivers and barriers to success under various headings:

- Society—People
- Society—Governance
- Transport System
- Economy
- Environment
- Energy

As SUITS was a Research and Innovation project with a focus on capacity development in Local Authorities, the usual SATELLITE approach to process evaluation was not suitable. The SUITS objectives relate to the capacity of municipalities to implement sustainable mobility projects. Rather than 'measures', we are dealing with 'challenges' and evaluating the impact of the SUITS Capacity Building Programme on how cities deal with these challenges as well as the processes of change that were undertaken.

Therefore, the SATELLITE process evaluation methodology was adapted to analyse.

1. the drivers or facilitators which enabled the process of capacity development in the local authorities
2. the barriers to success—those factors which hindered change along the path to becoming learning organisations with increased capacity to implement sustainable mobility initiatives.

Phase 1: Process Evaluation 2017.

A process evaluation form (Appendix) was developed based on the CIVITAS SATELLITE process evaluation methodology [7]. This was used 6 months into the project by the Site Evaluation Assistants (SEAs) and the Change Agents to reflect on achievements to date, the initial barriers encountered and how they could overcome them.

The main barriers were institutional and cultural.
Institutional barriers consisted of:

- impeding administrative structures which meant that there had been a delay in the appointment of personnel to work on the SUITS project in a number of Local Authorities
- not enough staff/staff did not have the time to work on the project
- rigid hierarchical structures in some cities
- bureaucratic procedures

Cultural barriers included:

- lack of experience in EU-funded projects

Drivers that have been or could be capitalised on included:

- political support for projects such as the SUITS project
- openness to new concepts
- interest in best practise and in gaining new skills and knowledge
- public pressure and media interest
- good fit with other projects going on in the city.

Actions taken by the SEAs included:

- widening the scope of involvement beyond transport departments to include environment, finance, etc.
- raising awareness of sustainable mobility and the advantages offered by the project
- demonstrating the 'fit' with existing SUMP or other projects
- face-to-face meetings.

Phase 2: Process Evaluation 2018

A year later, at the second Process Evaluation workshop, SEAs reported on progress over the past year using the same methodology. Whilst the barriers encountered remained the same, some activities had been implemented to overcome them:

- A dedicated person within the city allocated to the SUITS project—in some cases this was an external expert
- Developing a closer relationship between the SEAs and the city mobility planner or department
- Getting input and feedback from the cities in relation to the guidelines and training courses being developed by the project
- Supporting them on technical and planning issues, or providing detailed information specific to their measures
- Involving a wider audience of local stakeholders relevant to the project objectives, e.g. logistics operators.

As a result, trust and openness increased as the cities worked on the technical and cultural changes needed to overcome capacity gaps.

Phase 3: Process Evaluation 2019

In the third year of the project, process evaluation was visited again. This time the focus was on the internal processes as the local authorities engaged with the change process.

The Eight-Step Process for Leading Change outlined by Kotter [7] was used to guide the local authorities through the implementation of organisational change (Chap. 6). This model is very transparent, clearly structured and practically oriented and thus, facilitates creation of impactful change.

The model comprises the following eight steps:

1. Create urgency—the local authority has to understand why the organisational change is needed
2. Form a powerful coalition (build a winning team that is responsible for the implementation of the organisational change)
3. Create a vision for change
4. Communicate the vision—so that everybody knows about the long-term target
5. Empower action: removing barriers such as inefficient processes and hierarchies provides the freedom necessary to work across silos and generate real impact
6. Create quick wins: drivers and enabling factors must be recognised, collected and communicated—early and often—to track progress and energise volunteers to persist
7. Build on the change—think about long-term wins, and finally
8. Make it stick—make the new structure and behaviours a part of your daily business

It incorporates process evaluation into the 8 steps of change, enabling organisations to reflect on the barriers and drivers to date, and encouraging them to choose those barriers to focus on to generate impact (Step 5—Empower Action) and those drivers which enable them to create quick wins (Step 6).

Organisational change needs time with few rewards visible at the start of the process. Therefore, it is most important to create opportunities for quick wins and the celebration of these to keep employees' motivated to support the ongoing change processes. Empowerment contributes to increased levels of employee engagement, and more specifically their engagement in the change process by addressing barriers to action.

Since all local authorities had passed the first four steps and had managed at least half of the organisational change process, the workshop focussed on steps 5 and 6—identifying the barriers so they could be addressed and identifying the drivers so they could be capitalised on. Working in groups, the participants discussed and wrote down key obstacles and key enablers they experienced so far during the organisational change. They then grouped and ranked them from the highest to the lowest in terms of relevance for successful organisational change.

The key drivers to promoting organisational change were identified as follows Table 16.2.

In identifying the barriers, participants were asked to choose the barriers that were the easiest to overcome to be tackled next in their organisation. The local authorities identified the following barriers to be addressed in order to keep the momentum going in the change process (Table 16.3).

Following on from this process evaluation workshop, the cities focussed on the drivers and quick wins to advance the organisation through the remaining stages of change. Many organisational change programmes fail because they lack an end point, continuation plans, or they do not reflect on their process. Organisational change is

Table 16.2 Key drivers

1. Sharing best practises across local authorities in order to get support/endorsement from top management
2. Having the SUMP implementation already in process & a sustainable mind-set developed within the organisation
3. Having experts identified, e.g. change agents
4. Setting up working groups + stakeholders + steering committees to generate political influence/power
5. Guidelines developed by the project help to promote collaboration
6. Having a Think tank—pool of experts + local authorities (combined resources)

Table 16.3 Barriers to be addressed

1. Lack of engagement from other departments
2. Lack of experts
3. Political barriers
4. Understanding relevant innovative technology
5. Processing change and coming to terms with the changes within the organisation

a continual process in which, in this case, the local authorities kept analysing what could be done better for continued improvement. During the lifetime of the project, all cities had shown further progress on the Kotter organisation change steps–this is reported on in Chap. 6.

16.3 Conclusion

Whilst many projects focus on the implementation of imported measures, SUITS has shown that it is first necessary to build up capacity within local authorities so that they can plan, fund and implement sustainability strategies, appropriate to their current and future needs. This requires buy-in, trust and openness to change.

This chapter shows the importance of adopting a multi-tiered approach to evaluation of project interventions. This provides a comprehensive methodology for the assessment of, in particular, a capacity building intervention:

• from direct outputs and outcomes
• through to impact on the organisation and its ability to plan and implement initiatives
• the processes they have undergone
• as well as demonstrating the causal link with the intervention.

Traditional evaluation methodologies focus on changes in performance and capacity—the outputs and outcomes. Identification of indicators is essential in enabling evaluators to assess these changes. It is important that the indicators are easy to evaluate in a self-assessment format, are relevant to the activities in the organisation, and reflect the objectives of the intervention.

In addition to the assessment of outputs and outcomes, the link to the intervention itself must be determined, as it is not satisfactory to presume a causal link between the intervention and perceived outcomes. The Rapid Assessment of Capacity (RAC), which was adapted, piloted and simplified for the SUITS project, provides a qualitative methodology for evaluating the causal links between a capacity building intervention and any changes in capacity.

It is also important to evaluate the process by which capacity is built. Process evaluation helps the learning organisation to identify drivers and opportunities to achieve quick wins which in turn motivate the capacity building team. It also enables them to identify barriers, decide which can be overcome and focus their efforts on those. The Kotter model not only assesses and evaluates processes but also enables change by encouraging organisations to reflect and empower action to remove barriers, achieve notable results in increasing of their capacity, and thus, became learning organisations which are more flexible and resilient than other organisations.

Appendix: Process Evaluation Form

Phase of project	Objectives	Short description of what you have been aiming to achieve in this period
	Activities	Short description of activities you have carried out in this period
Barriers	Brief and clear description of barriers encountered—what happened? What is the negative impact on this phase of the project?	
Drivers	Brief and clear description of the drivers encountered—what happened? How did it occur? What is the positive impact on this phase of the project?	
Activities	Brief and clear description of the actions you undertook to overcome the barriers and make use of the drivers	
Risks	Estimation of risks to achieving the project objectives Score 1—Low risk Score 5—High risk	
Changes since last period	Have you noticed any improvement or deterioration since last reporting period? Note any positive or negative changes	
Other comment	Other relevant information, e.g. extraordinary conditions, etc.	

References

1. The Enabling Environment for Capacity Development: Perspectives Note. http://www.oecd.org/dac/accountable-effective-institutions/48315248.pdf
2. Kalakou, S., Spundflasch, S.: SUITS D2.1: contextualisation of project cities SUITS project (2018). https://www.suits-project.eu/reports/
3. Dolores Herrero, M.: SUITS D6.1: mapping of transport measures against capacity building and development of KPIs to measure changes in capacity with respect to sustainable transport decision making (2019)
4. Methodology for Rapid Assessment for Capacity Development (RAC). http://capacity4dev.ec.europa.eu/public-cd-tc/minisite/rac

5. https://ecdpm.org/publications/5cs-framework-plan-monitor-evaluate-capacity-development-processes/
6. Engels, D., Van Den Bergh, G.: CIVITAS SATELLITE D2.1: optimised CIVITAS process and impact evaluation framework (2016)
7. Kotter, J.P.: Leading Change. Harvard Business School Press, Boston (1996)

Chapter 17
Meeting the Covid Challenge–Agility and Resilience of SUITS Local Authorities

Ann-Marie Nienaber, Andree Woodcock, and Kat Gut

Abstract The Covid-19 pandemic has created immense social, economic, and political disruption around the world. It has shown the importance of agile, functional and resilient cities. In the fight against Covid-19, public life has been reduced to a minimum, but local authorities (LAs) have had to continue to satisfy existing and emergent citizens' needs and implement, sometimes at very short notice, extreme measures to restrict movement, commerce, education and leisure activities. This poses significant challenges as they have had to not only enforce and communicate government plans but also be proactive and respond to local needs of their cities whilst learning to work in new ways and support the health and well-being of employees. The Covid-19 pandemic has forced organizational change. In many instances, this has accelerated the rate of change, proving that new ways of working are effective and has led to a (temporary) advancement of sustainable transport. This chapter focuses mainly on the experience of SUITS LAs in the final year of the project.

17.1 Introduction

In the last year of the project, it was planned that cities test out and embed the result of the project in their everyday practise. Since March 2020, all our authorities were affected by the pandemic. We remained in contact with them and most of this chapter is based on their experiences during the first two lockdowns in 2020. Given the remit of the project the main focus of the chapter is on organizational change and sustainable transport measures, moving to a consideration of wider emergent issues, that have been brought to light, such as the need to address more holistically inequalities (such as health and transport poverty) through an intersectional lens and the need to preserve support for active forms of travel and rebuild trust in public transport.

A.-M. Nienaber (✉) · A. Woodcock · K. Gut
Coventry University, Coventry, UK
e-mail: ann-marie.nienaber@coventry.ac.uk

© Transport for West Midlands 2023
A. Woodcock et al. (eds.), *Capacity Building in Local Authorities for Sustainable Transport Planning*, Smart Innovation, Systems and Technologies 319,
https://doi.org/10.1007/978-981-19-6962-1_17

17.2   Covid-19 and European Cities

Cities have become epicentres of the pandemic, with high population densities and transport networks accelerating the spread and transmission of the virus at local, national and international levels [1]. Although urban areas are associated with economic growth, lower unemployment and good social infrastructure—the gains and services are not equally distributed. Restrictions and lockdowns have slowed down the economy and led to economic setback affecting people and communities, which in some cases may be unrecoverable. In the transport sector, in the UK restrictions were imposed on all forms of unnecessary travel, with public transport (road and rail services) being shut down, enforcing social distancing and wearing of masks, and offering minimal services for key workers.

The vulnerability of certain groups, communities and sections of society have been highlighted during Covid-19 (discussed later in this chapter). This has included groups who may have been 'invisible' or not traditionally considered as vulnerable, such as health care staff, drivers, shopworkers, those living in multigenerational families or who have to rely on public transport. Many organizations have stepped up or changed their operation to fill gaps in service provision to ensure that the basic needs of vulnerable citizens have been met. It became clear from the beginning that the virus does not affect all individuals in the same way. The first thing we learnt about Covid-19 was that infection was highly risky for older people or people with underlying health conditions. Therefore, local authorities put restrictions in place to shield people from high risk groups first. However, health related risk factors were not the only vulnerabilities that would put citizens at disadvantage. Community health and well-being is associated with socio-economic status and intertwined with other demographic characteristics like race, ethnicity, education, or disabilities. People from lower socio-economic background were not only more exposed to the risk of infection by living in densely congested, poor housing, but they were also more likely to hold low-skilled, temporary, precarious jobs that would be at risk of redundancies in the face of economic fallout. For instance, one of the most vulnerable members of the urban society—women immigrants–were also more likely to work in hospitality and services severely impacted by loss of business. It is estimated that only 1 in 5 countries have specific plans in place to help migrants during and after a crisis[1] despite their forming 14% of key workers in European regions [2]. The needs of migrant workers employed in cities, one of the most vulnerable groups in society do not feature highly in transport related research.

These, and many other examples from across Europe gave ample evidence that economic fallout caused by the pandemic hit disproportionately the most vulnerable members of society.

Cities are formed from businesses reliant on communal service delivery, social gatherings, and hospitality. For historically and culturally attractive cities the economy is reliant on tourism and hospitality. Major cities are also places of learning, swelling their populations by 10,000's of young people each year. In all cases

---

[1] https://migrationdataportal.org/themes/migration-data-relevant-covid-19-pandemic.

increased rates of exposure increases the likelihood of catching the virus. Proximity makes the virus spread faster especially in enclosed and poorly ventilated spaces. Such businesses have been severely affected by the lockdowns causing redundancies or—in case of smaller companies—bankruptcy.

Each city has a different set of challenges determined by its geopolitical landscape and demographic make-up. In most cases the complex systems of operations that create intertwined and interdependent networks have been shaken if not significantly damaged. As has already been stated, the breakages have revealed massive inequalities and inacceptable burdens placed on some of the lowest paid in society.

Covid-19 has also created a chance to reappraise priorities and look for different ways of doing things. Bringing the discussion back to transport, people are more aware of the impact the daily commute had on their lives and the environment and for the first time, there has been a recognition, however short lived, that people working from home can 'get the job done.'

Travel restrictions, bans on social gatherings, total or partial lockdowns, halting non-essential production and operations had a ripple effect on the urban ecosystem. Firstly, a dramatic decrease in car traffic changed the urban environment. Not only air quality improved due to a significant reduction in CO emission, but also noise from traffic decreased. In some touristic cities, like Venice, halting tourism and air travel have improved the overall quality of water due to drastic drop in wastewater; satellite pictures of canals in Venice or citizens enjoying crystal clear air in historically industrial and polluted cities seen for the first time for years were going viral showing the evidence of detrimental impact of human activities on environment. Although sustainable strategies were in plan for years, for many local authorities the pandemic and forced travel restrictions offered ample evidence that—with well synchronized interventions—we can build more sustainable and liveable cities with the resources we already have.

Banning social gatherings has led to transferring non-essential business and services to online operation. Even if the change was just temporary it has transformed the way we think about rigid office and school hours. Flexibility in working pattern or working from home turned out to be very much needed by families with young children, women, persons with disabilities or anyone for whom daily commuting was inconvenient and time-consuming. Traditional working patterns create surges of demand across transport and physical infrastructure and in energy use. With more flexible working patterns, digitalization and remote operations, cities might never again be the unquestionable hubs for employment as many employers are now expected to embrace a hybrid model of working and give employees a choice to work from home if they wish to. Remote work piloted across Europe resulted in a slashing of office space rents. If the trend persists it will affect the business parks and city centres that will no longer be bastions of traditional, office based corporate culture. A key issue going forward is the extent to which people will want to return to a pre-Covid situation, who controls the rhetoric and steers the direction of our cities. Many experts assume that individual mobility will experience a new high. But how can we prevent people from switching from public transport back to private cars—to consider a long-term threat, rather than the more imminent threats to their health? How can you prepare the bike infrastructure (pop-up bike lanes) for expected increase? What role will sharing concepts play?

At the time of writing (April 2021) it has been over a year since the World Health Organization declared Covid-19 a global pandemic. It is perhaps too early to acknowledge the full impact the pandemic has had on cities. However, the early lesson that emerged from the crisis was that resilient and sustainable cities around Europe were able to handle the pandemic better. For many local municipalities' lockdowns tested their strengths and weaknesses. It has set the scene for implementation of holistic action towards inclusive, sustainable development. Many examples on how to build resilient and sustainable cities emerged from good practises across Europe (100 Intelligent Cities Challenge[2]) and highlighted how working together with stakeholders and citizens can enhance preparedness for future obstacles.

17.3 Local Authorities

In the European Region, local governments and other local organizations have become key responders and are still at the forefront of epidemic containment. They are responsible for communicating and implementing government mandates, regulations and guidelines to reduce the risk of infection. Acting as advisory bodies, local authorities also provide access to the local population and can provide supportive guidance in cushioning the worst long-term economic and social consequences of social distancing measures. They are the operational partners of national governments: the central actors in national preparedness and response planning, essential service providers and play a central role in creating a sustainable future. They also serve as a central point for reaching and involving people as part of the solution: through targeted risk communication and public transport and other services, or through guidelines from national governments.

17.4 Effects of Covid-19 on SUITS Cities

By Spring 2021, UK partners and many members of the consortium have worked from, or are still working from home, experiencing the effects of a series of lockdowns and the effects of different national approaches (e.g. herd immunity, mass vaccination). UK academics are still working from home as they have been doing since March 2020. It is unlikely that universities will return to anything like business as usual until autumn 2021, with online learning, restrictions to the number of people who can physically attend lectures, and travel restrictions. It is unlikely that we will return to pre-Covid ways of working, having demonstrated that we can work

---

[2] The European Commission's Intelligent Cities Challenge, Covid19 Good City Practices, accessed March 2021: https://www.intelligentcitieschallenge.eu/covid-19-good-practices.

effectively from home, are sufficiently familiar with technology and have started to enjoy a higher quality of life–which had been eroded by the stress and time spent in commuting to and from the office. In local authorities most employees were expected to work from their office. Again, they have proved that they can work from home, and may not return to former working patterns after Covid-19 lockdown ends.

If the ability to work from home is supported across a lot of industries, this will have a profound effect on mobility services, and local economies as many businesses have grown up to support traditional working patterns and service the needs of the commuter.

In the following sections, we concentrate discussion on how the cities in SUITS responded to the challenges brought about in the pandemic and how European cities can learn from the last months for their future mobility strategies. Special attention has been paid to the employees in the local authorities, as they are the ones who have had to deal with current mobility challenges whilst at the same time engaging on delivery and planning of more sustainable measures. This section is organized along four key topics: (1) Organizational change, (1) Digitalization; (3) New mobility services and (4) Data driven approaches.

(1) **Organizational change**

When the UK government responded to the first wave of the pandemic a tight lockdown was enforced, which required staff to work from home, with very little period of adjustment. This meant that all staff within the LA were working from home along with members of their family who were not classified as key workers, and also looking after children (the system of home schooling took longer to establish).

SUITS LA partners adopted remote working as the new standard with employees working together across departments, changing communication flows, working patterns, hierarchical structure and taking on new tasks—from contact tracing to manning municipal Covid-19 hotlines, to being part of the 'Covid patrol' of the municipal security services.

The immediacy of lockdown left little time for planning new forms of communication or working structures. Teleconferencing quickly became ubiquitous although bandwidth availability from home, making do with existing home computers and laptops, system incompatibilities, lag time[3] and lack of familiarity with different systems remain daily problems. Moving all work on line has also led to general problems such as too many meetings schedules with little time for breaks, lack of control over invites, constant telepresence, and the need to rapidly switch from one group/topic to another has resulted in cognitive overload, 'zoom fatigue[4]', (n)etiquette[5] issues and difficulties in performing essential non screen based tasks. We are working harder, for longer, in home spaces which may be shared by family members (e.g. partners and those who are home schooling), and not designed as full-time offices, without physical social contact.

---

[3] https://www.bbc.com/worklife/article/20200421-why-zoom-video-chats-are-so-exhausting.

[4] https://www.cnbc.com/2021/02/25/zoom-fatigue-why-we-have-it-how-to-fix-it.html.

[5] https://money.howstuffworks.com/business-communications/teleconferencing-etiquette.htm.

Covid-19 made LAs aware of the need for greater flexibility and resilience trust, better communication and technological know-how (from e-financing, e-forms, use of social media to document management systems) and change management in dealing with colleagues, stakeholders, partners and citizens. The adaptability and resilience they have shown has given LAs confidence in their ability to change.

SUITS LAs who had, by the time of the pandemic, built up capacity through attended training events offered as part of the Capacity Building Programme, and worked through most of the organizational change process, were in a strong position to adapt to the crisis. For example, they understand the importance of trust and citizen engagement, had built up networks, and set up communication protocols.

For example, the **transport innovation team at WMCA** had already used teamware, set up communication protocols in their group, were used to sharing information and data with other departments and holding virtual meetings with the directorate. The innovation team increased from 5 people to more than 25 members. They implemented 'learning organization' principles in response to Covid-19 challenges, due to the recent challenges of the pandemic. Additionally, a representative from the innovation team is now part of all major boards, raising the profile of the team and the projects around digitalization.

Organizational change is built on, and requires technical equipment and behavioural change. The cities which were familiar with organizational change processes and had built up their technology know-how were in a better position to cope with the challenges of the last year. For example, end-to-end digital administration is only conceivable with e-files, a document management system and resilient network infrastructures. To use the words of **Rome's** mobility head: 'Smart working and digitalization: it's a structural revolution and not only a passenger phenomena!' The city of Rome intends to redesign its infrastructure according to the requirements of the recent pandemic and preferences to work smartly and remotely.

The Covid-19 crisis is revolutionizing classic work structures as summarized in the term 'home office'. What is important is what work will look like in the future, which skills will be important and how employees will help to shape this. Vocational education and training is central here, and close cooperation between municipalities, states, federal government, academia, educational institutions and trade unions is needed. The SUITS cities became learning organizations through the close cooperation with academic and educational institutions.

To make those changes last, cities will need to develop new strategies for their future. SUITS highlighted the need for 'a vision' and presented different tools to develop and communicate a new vision. Although it was recommended that visions have some longevity, post pandemic reflections may lead to new city visions, shaped by data and through engagement with citizens.

(2) **Digitalization**

This is taken as referring to both the move towards online services and technology-enabled working. Over the course of the project SUITS' LAs were exposed to, and shared different ways of working and smart mobility solutions. They were therefore in a strong place to start using their experience in small, but significant ways during the pandemic. Self-auditing and reflection enabled them to identify situational needs and requirements [3] and how they could respond to the new challenges. For example:

- Many cities (e.g. in the **West Midlands, Torino, Valencia, Stuttgart or Dachau**) introduced digital administrative processes such as appointment booking systems, applying for parking permits, reporting infrastructure problems via portals.
- Citizens found that the administration not only has emails and contact forms, but actually answers them. Interactive platforms had been discussed before the pandemic as an opportunity to enhance citizens' engagement. **Valencia** and **Dachau** presented interactive webpages to receive complaints and requirements from citizens in relation to their mobility needs, e.g. need for a cycle route; more traffic lights or more security around places frequented by children such as playgrounds, nurseries or schools. To keep such portals alive, citizens expect immediate answers. During the pandemic, the cities built on their experiences with such interactive platform spaces to interact with their citizens on health and wellbeing topics, such as needs for medicine or food in particular to elderly citizens.

(3) **New mobility services**

LAs have quickly adapted to working during Covid-19, implementing both planned mobility measures and new ones in response to Covid-19. In this section we review some of the new sustainable transport measures which have been developed and rapidly implemented in response to changing mobility patters and the need for social distancing. For example,

- During Covid-19, the number of taxi bookings has reduced by around 80 percent. To compensate for this, taxis have had to expand their service offering. **Palanga** helped to bring together the city's taxi companies to deliver prepaid groceries and medicines, using e-commerce principals. Palanga has demonstrated that it has the authority and trust to form new partnerships which meet the changing needs of its most vulnerable citizens.
- **Rome** became aware of the need to better use city resources and to avoid 'rush hours' and has promoted 'active modes' of transport such as bikes or scooters. It is making renewed efforts to limit the predominance of cars and support alternatives for mobility and has become more aware of the need for better lifestyle for its citizens and is therefore recovering local spaces to save the environment.
- **Valencia** was able to set up important sustainable transport measures to cope with Covid-19. The most significant of these include

  (i)   reducing bus capacity and keeping the service running with safety measures in order to reduce the Covid-19 risk and keep on providing quality service,
  (ii) a plan to improve public space for pedestrian and bikers, which will improve social distancing, reduce the number of accidents and air pollution and improve the quality of life,
  (iii) collection of proposals for Covid-19 in Valencia through Agencia de la Bici (Bike Agency), with effective impacts to be confirmed,

(iv) RAL COVID-19 I: Protocol for the prevention of Covid-19 in Logistics and Transport Operations, as recommended in the guidelines elaborated by the Ministry

(v) recommendations for the prevention of Covid-19 infections in warehouses and logistics centres.

(vi) millions of Euros have been ring-fenced to create bike lanes and pavements

- **Kalamaria** has set up a series of measures to prevent the spread of Covid-19. The most important ones relate to

(i) increase of car-free zones (pilot program), having a direct impact on green mobility, less pollution, innovation, and better quality of life in urban spaces,

(ii) decrease in the use of PT due to fewer itineraries and decreased occupancy of public transport vehicles (50%), that generated less passenger interaction, more safety for travellers and better public health,

(iii) increase in freight transit over short periods based on increased citizens' needs especially for shopping in supermarkets, generating less congestion due to decreased car use, less air pollution due to optimized vehicle routes, as well as the positive economic impact for companies.

- Covid-19 has had many effects on mobility in **Turin**, with public transport suffering the most because not travelling with a full load does not achieve acceptable levels of cost-effectiveness. Taxi services are also suffering especially for business users who have reduced their travel between cities and towns to a minimum. E-commerce for both goods and food has increased resulting in a rapid rise in freight transport. With the introduction of 'smart working', journeys to and from work have collapsed with a significant reduction in the number of kilometres travelled by private vehicles and a considerable reduction in C02 and Nox.

  In order to promote individual, but at the same time sustainable mobility at the national level, processes of legislative changes have been triggered for the implementation of soft mobility infrastructure. Incentives have also been introduced for the purchase of bicycles and e-scooters. The latter measure was very successful, and funds were quickly exhausted.

  Turin and the main cities in Italy have applied new regulations and immediately implemented soft measures to increase the number of kilometres dedicated to cycling and soft mobility. As a result, roads with a speed limit of 20 km/h were built or traffic bollards were installed to prevent traffic.

In general, all SUITS city partners underlined the importance of bicycles for their cities. During Covid-19, the bicycle has been and is a useful means of transport for unavoidable trips. It is a good alternative to public transport and more environmentally friendly than the car. Also the World Health Organization has underlined the benefits of cycling and walking as a means of transport as they both allow for physical distancing and enable exercise.

- The City of **Stuttgart** for example has supported cyclists since April 2020 with free access to the Bike Citizens navigation app. The Bike Citizens app offers map displays and route planning especially for cyclists. Maps can be downloaded, and routes can be announced using voice control. Cyclists can download the Bike Citizens app in Stuttgart for one year free of charge; there are no additional costs for longer-term use.
- A similar development can be seen in Valencia. Whilst Valencia was already very active to promote cycling in their city, the recent pandemic increased these efforts dramatically.

(4) **Data driven approaches**

Based on comprehensive data collected over the period of four years via several workshops and semi-structured interviews with seven public authorities in Europe, we were able to demonstrate that one of the main obstacles to data sharing was the public employees who distrust online-platforms, their providers and users [4].

The Covid-19 pandemic has acted as a catalyst for data sharing. All cities in SUITS became aware of the need to use and integrate passenger and freight data to inform mobility planning. SUITS supported mobility transformation and decision making for local authorities (LAs) through the gathering and processing of crowd-sourced floating car data (FCD) of citizen and freight mobility, from which integrated transport planning can take place, [5, 6].

This data driven approach showed that with proper incentivization, citizens, taxis and logistics companies can provide accurate and adequate data for monitoring real-time traffic conditions of a city's entire road network. By segregating various traffic sources, freight traffic can be inter-correlated with other urban mobility patterns. This will enable local authorities and policy makers to effectively evaluate the impact of temporary or permanent mobility measures related to freight, e.g. route changes, delivery timeframes, loading/unloading zones, reduction of the number of freight trucks delivering within congested areas, etc., thereby resulting in the overall reduction of driven distances and travelling time, as well as improving urban accessibility and speed of distribution of goods. Such crowdsourcing could easily be extended and incorporate more information, depending on the willingness of the user to share data.

Whilst the benefits of big data in increasing public authorities' efficiency and effectiveness and their citizens' life is well understood, examples from the public sector that highlight public authorities' engagement in such sharing activities is still missing. The use of SBOING's tools showed the decline in taxi service operations during lockdown For example mobile users could indicate their shopping trips with the purpose of helping vulnerable people, i.e. co-buying food and medical equipment for residents who wish to stay or are forced to stay at home.

17.5   The Role of SUITS

A learning organization is characterized by high levels of agility and resilience. SUITS partner cities had been building capacity as part of their organizational change process throughout the project (see Chap. 6, [3] and [7]) Going through this process has helped during the current crisis because LAs:

- did not have such a steep learning curve as they were familiar with organizational change processes, and through SUITS activities had widened their circle of contacts across departments
- were more aware of legal and regulatory frameworks
- were more familiar with the use of technology, e.g. to support meetings, collect data, engage with the public
- had a wider knowledge of sustainable transport measures and how to implement them, during and after the pandemic
- understood how to make changes long-lasting within the organization

LAs have become very agile and efficient during Covid-19, quickly implementing active transport measures—such as creating additional bicycle lanes. **Rome**, **Palanga** and **WMCA** have already showed examples of how Covid-19 is setting new standards in their cities. Rome wants to continue mobility data monitoring (https://romamobilita.it/it/covid-19-impatto-sulla-mobilita) to provide support for the next phases and respect for the social distancing and also became aware of the role of mobility managers in their city. The mobility managers' role needs to be enhanced and integrated in the mobility system of the city in order to have direct feedback of different measures and to coordinate the efforts. This learning will remain and help to successfully manage future mobility challenges.

After SUITS, the cities are keen to build on close relationships with academia. Several '**cross-learnings sets**' have been set up with the support of Coventry University which are designed as informal communication groups to discuss recent trends, developments and activities. For example Valencia, Coventry and WMCA built a group that want to continue to exchange knowledge on mobility topics. Close cooperation has also been built up between Rome and Valencia. Both examples also foster peer-to-peer learning and co-operative problem solving. Rational learning approaches suggest that actors that face similar problems may turn to their peers in search for suitable and proven solutions. Understanding which solutions worked well in other municipalities reduces costs and efforts for the identification of adequate and effective measures and may avoid potentially costly negative lessons from trial and error.

17.6   Emergent Challenges

The main focus of this chapter has been about Covid-19 related transport challenges and how SUITS cities responded to these. Local authorities have played a key role in containing the pandemic and trying to maintain the health and wellbeing of their citizens. Transport and mobility is tightly bound to the fabric of city life. It cannot be considered in isolation. This section looks very briefly at some of the wider challenges which LAs will have to contend with which have a bearing on transport and mobility.

(1)   Structural inequalities

The Covid-19 pandemic has exposed deeply rooted social inequalities [8] that affect how we experience cities and public spaces. Governments around the globe have been urged to recognize the unequal impact the pandemic is having on different communities; particularly where economic inequalities intersect with racial discrimination [9].

Emerging evidence shows that the Covid-19 pandemic is gendered[6] in its consequences and experiences. Whilst mortality has been higher amongst men [10], early indications show that the economic impact is greater for women. An immediate consequence has been the loss of jobs, with women's jobs 1.8 times more vulnerable to this crisis than men's. Women make-up 39 percent of global employment but account for 54 percent of overall job losses. Secondly, men and women cluster in different occupations. Women are more likely to be employed part time on temporary contracts. These are more likely to be cut in economic downtowns. Thirdly, school closures have increased the level of women's unpaid care work, leaving less time for paid work or leisure[7] and causing them to drop out from the job market.

Lockdowns and restrictions saw the closure of retail, recreational and non-essential services with transport, essential retail and health services remaining open. Emerging data is showing links between occupational risk of exposure to Covid-19 associated with race and ethnicity [11]. In the US, data from the Bureau of Labour Statistics (BLS) Current Population Survey (CPS) shows that people of colour were more likely to be exposed to the virus due to the nature of their employment. Black and Asian workers were most likely to be employed in occupations with both frequent exposures to infections and proximity to others leaving them vulnerable to the virus. In the UK, evidence suggests a markedly higher, but non uniform, mortality rate amongst Black and Asian and Minority Ethnic (BAME) groups, including patients in intensive care units and amongst medical staff and Health and Care Workers. Razaq et al., [12] in their evidence-based review conclude that '*there may be a number of factors that increase BAME COVID-19 risks and vulnerability in the general population, such as overrepresentation of BAME populations in lower socio-economic groups; multi-family and multigenerational households leading to increased risk of transmission due to the lockdown; co-morbidity exposure risks especially for CVD,*

---

[6] https://coronavirusexplained.ukri.org/en/article/cad0007/.

[7] https://www.oxfamamerica.org/explore/research-publications/caring-under-covid-19/.

diabetes, renal conditions and complex multi-morbidities in ICU; and there is dispro-portionate BAME employment in lower band key worker roles who either work in high exposure care environments or are unable to implement safe social distancing due to their roles. In addition, for Health and Care Workers, there are increased health and care setting COVID-19 exposure risks'.

Intersectional effects have been a prominent feature in many spikes in cases, which have been seen in the working class neighbourhoods where people live in multigenerational households and are unable to maintain social distancing due to employment and housing conditions [13]. Many local outbreaks were concentrated around factories, farms or food production facilities in which workers were put at risk of high exposure to the virus. In Germany or in Portugal for example slaughterhouses and meat packing plants have been a major risk for Covid-19 infection throughout the pandemic [14]. Such workplaces fail to provide conditions for physical distancing and have working practises which discourage the reporting symptoms. Many industrial or agricultural sites employ migrant workers on insecure low paid contracts with no guarantee of paid sick leave. Disclosure of symptoms incurs penalties and may result in job loss or economic hardship through loss of wages during recovery and quarantine.

Living in overcrowded accommodation, reliance on buses or shared transport and poor working conditions accelerated the spread of the virus amongst low paid workers who were regarded as key workers, providing food and health care.

(2) **Intersectionality/interdependence of transport and health**

Understanding the structural inequalities which were exacerbated by the pandemic requires applying intersectional lenses. For example, gender can be seen as inter-secting with other factors that can further deepen inequalities. Factors like ethnicity, age, socio-economic status, marital status, sexual orientation, disability and many more characteristics can contribute to the person's vulnerability. Globally women were paying a higher price for the crisis and were carrying a double burden of care responsibilities and home labour. However, not all women were equally affected. Specifically, women who were working in essential services such as food retail and production, and health or social care were at double risk of contracting the virus: at work and on the way to and from work. Public Transport worldwide—being predom-inantly used by women—was the first sector affected by the pandemic. Little research has been done on specific risks on transport, but we can apply what we know so far about general virus transmission. Inability to maintain social distancing, small spaces with highly congested and poorly ventilated areas are the main reasons why trains, aeroplanes and buses were ranked as high-risk exposure to the virus.

Despite disruption in public transport and known risks about transmission rates many women had to use public transportation to get to work or to fulfil their caring responsibilities. This puts women at greater risk of coming into contact with the virus. In countries where restrictions on movement have tightened, public transport has been reduced or even shut down. New research on mobility from Denmark emerged concluding that the impact of the pandemic on mobility has been so far the greatest amongst women, especially those with a lower level of education [15].

Similarly, women from lower socio-economic groups are the largest public transport users as they have less access to private vehicles [16]. Moreover, concentration of females in health care and caring jobs made them more exposed to the virus and imposed heavy workloads on them.

In conclusion, it is important to acknowledge overlapping factors impacting women's livelihood and health during the pandemic to recognize the disproportionate impact of the pandemic on women and girls. Researchers have already sparked a discussion around intersections between Covid-19 and gendered burdens, particularly in frontline work, unpaid care work and community activities [17]. However, more research is needed to explore obstacles and inequalities that affected women and girls and how this can inform LA actions to create a fairer and more equitable society.

(3) **Sustainable mobility**

Women form the majority of public transport users. As such, many have been subjected to higher exposure rates, and were also affected by changes in public transport operations. Risk of infection is increased in poorly ventilated spaces/vehicles, where social or physical distancing is difficult, and many surfaces are touched (by many people).

Several approaches have been developed to reduce transmission of disease on public transport including restricting its operation, reducing its use (to essential journeys only), making face masks mandatory, providing sanitization stations, temperature checks, installation of screens, contactless payments and door openings and regular /thorough cleaning, and temperature testing. Whether these measures will be enough to convince people that public transport is safe is still unclear (also see Chap. 15). LAs and transport operators will need to provide clear messages based on sound evidence to install confidence in the coming months.

Many LAs have also offered new, more sustainable ways of travelling. Citizens have been encouraged to cycle or walk to work and take limited recreational/exercise period. The significant decrease in traffic has temporarily changed the urban landscape: cleaner air and fewer cars encouraged people to do more cycling or walking. Cities have responded to this by designing new cycling lanes or low speed zones to keep areas safer for pedestrians or cyclists. For example, **Turin** brought forward its plans on its 'bike plan' which aims to increase the share of bike travel to 15% by 2023. In a very short period of time the local authorities planned 95 km of cycling lanes and introduced bike hubs and new solutions in traffic system giving priority to bikes on traffic lights or crossroads. As a result, it is not only safer for cyclists to move around the city but they also don't have to stop behind the cars in fumes from their engines.

The need to temporarily transform mobility behaviour has prompted many citizens to re-think how they commute and their health and well-being and their relationship with the city.

Micro-mobility solutions have also been accelerated such as bike sharing schemes, e-bikes, e-scooters and robot delivery of groceries and medicine.[8] Some of these have been accompanied by changes in traffic systems, such as the Low Traffic Neighbourhoods, (LTNs) (with over 200 proposed) in the UK[9]. Here changes to the road traffic system and use of street furniture such as planters block roads in residential areas making estates closed-off to through traffic. Such schemes have been designed to make walking and cycling safer, reduce traffic, noise, and pollution and improve overall quality of life by encouraging active lifestyles and community interactions in quieter, safer neighbourhoods.

Although this would appear to be in line with most people's wishes, the rapid implementation of this and other measures has caused problems, confusion and resentment, due to lack of consultation and engagement with members of the public. LTNs have provoked vandalism, bullying, demonstrations and petitions leading to one council suspending the scheme. Worryingly this has also escalated into a cultural war' as it is seen as privileging those living in 'leafy suburbs'.[10]

All the afore-mentioned changes set the ground for a new, more sustainable vision of cities across Europe. Many citizens who had changed mobility patterns, whether willingly or forced by circumstances, have had the opportunity to re-think travelling habits and increase awareness about traffic and its impact on quality of urban lives. LAs have been quick to act on this and have pushed through schemes designed to support more sustainable transport measures. However, either due to the speed of planning and implementation, Covid-19 restrictions on meetings, or poor use of communication channels, some schemes have caused resentment, with some groups disadvantaged or privileged more than others.

Moving forward, governments and LAs may also be further challenged over their handling of the pandemic, such as the use of spot fines, vaccinations, availability of PPE, mandatory wearing of masks, and closure of certain sectors of the economy, lock downs and restrictions on family visits. Clearly such discussion is both too early and beyond the scope of this volume.

However, going forwards LAs need to look at how they can plan the city and transport services to reduce structural inequalities, how they can better engage and inform citizens, develop the levels of trust, transparency and openness which will enable partnership working and create shared visions.

References

1. Newman, A.O.: Covid, cities and climate: historical precedents and potential transitions for the new economy. Urban Sci. **4**(3), 32 (2020)

---

[8] https://www.bbc.co.uk/news/uk-england-northamptonshire-55076342.

[9] https://www.sustrans.org.uk/our-blog/get-active/2020/in-your-community/what-is-a-low-traffic-neighbourhood/.

[10] https://www.theguardian.com/world/2020/sep/20/the-new-road-rage-bitter-rows-break-out-over-uks-low-traffic-neighbourhoods.

2. Kleine-Rueschkamp, K., Özgüze, C.: COVID-19 and key workers: the role of migrants across regions and cities. VOX^{Eu} (2020). https://voxeu.org/article/Covid-19-and-key-workers-role-migrants-across-regions-and-cities. Accessed Apr 2021

3. Nienaber, A.M., Rudolph, F.: Organizational resilience: how SUITS' local authorities were prepared to cope with the COVID-19 pandemic. Policy Brief 4, SUITS (2020)

4. Nienaber, A.M., Soares, A., Spundflasch, S., Woodcock, A.: Distrust as a Hazard for Future Sustainable Mobility Planning. Rethinking Employees' Vulnerability When Introducing New Information and Communication Technologies in Local Authorities. Int. J. Hum. Comput. Interact. 37(4), 390–401 (2021)

5. Georgiadis, A., Kalaitzis, V., Nienaber, A.M., Woodcock, A.: MyPolisLive. net: a tool and a methodology for optimizing urban freight mobility through crowdsourcing. In: 5th Conference on Sustainable Mobility (June, 2020)

6. Pirra, M., Diana, M.: Integrating mobility data sources to define and quantify a vehicle-level congestion indicator: an application for the city of Turin. Eur. Transp. Res. Rev. 11(1), 1–11 (2019)

7. POLIS, and Rupprecht Consult-Forschung & Beratung GmbH (eds.): Topic Guide: Planning for more resilient and robust urban mobility (2021)

8. Alon, T., Dopke, M., Olmstead-Rumsey, J., Tertilt, M.: The impact of Covid-19 on gender equality. NBER Working Paper Series. https://www.nber.org/system/files/working_papers/w26947/w26947.pdf. Accessed April 2021

9. COVID-19 Global Humanitarian Response Plan (2020). https://reliefweb.int/sites/reliefweb.int/files/resources/PRESS%20RELEASE%20GHRP2%20.pdf. Accessed July 2020

10. Jian-Min, J. et al.: Gender differences in patients with COVID-19: focus on severity and mortality. Front Public Health 8 (2020)

11. Hawkins, D.: Differential occupational risk for COVID-19 and other infection exposure according to race and ethnicity. American Journal of Industrial Medicine (2020)

12. Razaq, A., Harrison, D., Karunanithi, S., Barr, B., Asaria, M., Routen, A., Khunti, K.: BAME COVID-19 deaths–What do we know? Raid data and evidence review. The Centre for Evidence Based Medicine (2020). https://www.cebm.net/Covid-19/bame-Covid-19-deaths-what-do-we-know-rapid-data-evidence-review/. Accessed April 2021

13. Bibby, J., Everest, G., Abbs, I.: Will COVID-19 be a watershed moment for health inequalities. The Health Foundation (2020)

14. Middleton, J., Reintjes, R., Lopes. H.: Meat plants—a new front line in the Covid-19 pandemic. BMJ. Jul 9;370:m2716. https://doi.org/10.1136/bmj.m2716. PMID: 32646892 (2020)

15. van der Kloof, A., Kensmil, J.: Effects of COVID-19 Measures on Mobility of Men and Women (Mobycon.com, 2020) (2020). https://mobycon.com/wp-content/uploads/2020/06/Effect-of-covid-19-measures.pdf. Accessed April 2021

16. European Commission (2014) Continuity of passenger mobility following disruption of the transport system. Commission staff working document, SWD (2014) 155 final. http://ec.europa.eu/transport/themes/passengers/doc/swd(2014)155.pdf

17. McLaren, H.J., Wong, K.R., Nguyen, K.N., Mahamadachchi, K.N.D.: Covid-19 and women's triple burden: vignettes from Sri Lanka, Malaysia, Vietnam and Australia. Soc. Sci. 9, 87 (2020)

Chapter 18
The Local Authority Perspective on EU–Funded Collaborative Projects

Andree Woodcock, Ann-Marie Nienaber, Janet Saunders, Sebastian Spundflasch, Sunil Budhdeo, Keelan Fadden Hopper, and Guiseppe Estivio

Abstract The EU, through various funding streams, continues to invest heavily on sustainable mobility to reduce congestion, carbon emissions, improve inclusivity, health and well-being, etc. This is leading to new methods, tools and processes, service propositions and technical innovations (in engineering and computing). Local authorities are responsible for the design and delivery of transport services in their cities. As such they may be regarded as one of the end users of the vast amount of material that is produced by projects such as SUITS. They can also receive funding to participate in some capacity, in many of the programmes, e.g. to run technology trials, to attend training events. This chapter takes a reflective approach to understanding the experience of local authority partners, what were their expectations, what were the costs and benefits, and what recommendations would they make for others to ensure they received maximum benefit if they were invited to participate in future projects. These insights are useful for the designers of research programmes, project managers and local authorities who might be invited into a consortium.

A. Woodcock (✉) · A.-M. Nienaber · J. Saunders
Coventry University, Coventry, UK
e-mail: A.Woodcock@coventry.ac.uk

S. Spundflasch
Technische Universität Ilmenau, Ilmenau, Germany

S. Budhdeo
Coventry City Council, Coventry, UK

Keelan Fadden Hopper
Transport for West Midlands, Birmingham, UK

G. Estivio
Comune di Torino, Turin, Italy

© Transport for West Midlands 2023
A. Woodcock et al. (eds.), *Capacity Building in Local Authorities for Sustainable Transport Planning*, Smart Innovation, Systems and Technologies 319,
https://doi.org/10.1007/978-981-19-6962-1_18

18.1 Introduction from the PI

In line with the rest of this section, this chapter is reflective. It has provided the project team and the LAs with an opportunity to reflect on their experiences in a less formal manner than the required impact assessment. As such, this chapter may be of interest to local authorities or other stakeholders who might be unsure about taking part in large European projects – especially with people and organisations they are unfamilar with. Also, it is important for Principal Investigators to understand the experience of consortium members, to better manage future projects and report back to the commission on issues which might affect success of future projects.

EU projects can appear daunting to those working on them for the first time. A consortium brings together a wide group of partners who work together on a common goal, set by the funding body. Partners (known more formally as beneficiaries) bring their own insights, interests, expertise and networks to a project. The partners may be largely unknown to each other, will be from different countries, cultures and organisational types. They have different expectations, and reasons for joining. In SUITS, many partners had worked together on an earlier project METPEX [1] looking at an inclusive way of measuring end to end multimodal passenger experiences, and are also carrying their association forward into a third project, TInnGO,[1] but others were totally new to European funding. Ideally, those who have worked on the initial proposal will form key personnel and pillars in delivery. However, with long lead times, in most cases, at least over a year (from writing, to negotiation, approval and project start up) and high levels of organisational churn it is often the case that those who finally work on the project may know very little about what their organisation has signed up to. Indeed, it is not unusual for the person who was interested in the project to have left an organisation, or for organisational priorities to have changed. Those whose job it will be to deliver the project may therefore find themselves working with strangers who have different priorities, expressed in a different language, using different management processes, terminologies and ways of expressing arguments. Initial meetings can be a source of great anxiety, as there can be many misunderstandings and operational adjustments as the delivery teams explore ways of aligning the 'Description of Work' (as specified in the contract) with their own 'everyday' work, knowledge, abilities and interests.

Holding even a moderately sized consortium together, from a project management perspective is challenging, with face-to-face meetings held at 6 monthly intervals, supported by regular online meetings. Trust is gradually established as the team work co-cooperatively and collaboratively to fulfil objectives, understand each other's perspectives, and share experiences. From a PI pespective it is reassuring, encouraging and sometimes even surprising, when work is carried out as planned! For the duration of the project my EU partners become my closest colleagues.

Local Authorities may have significant challenges in contributing to EU transport projects. They are lean organisations, sometimes operating in 'fire fighting' mode, as we saw in SUITS when they had to change working practices and develop new

---

[1] https://www.tinngo.eu/ looking at gender and diversity issues in transport.

services to deal with the Covid epidemic and travel restrictions. The smaller LAs may not have large departments, or enough capacity to fully engage and benefit from EU projects. They also work at a different pace needing to provide immediate solutions to current problems and be accountable to their citizens and national governments. Projects of long duration may not seem to provide them with tools to solve immediate problems and can be an unwelcome distraction.

18.2   Methods

In order to discover more about this, we firstly asked Transport for West Midlands/Coventry City Council to share their experiences of working on SUITS and similar EU projects. They represent one of the larger and more experienced local authorities in the consortium, familiar with and experienced in attracting research and innovation awards to support sustainable transport innovations, from national and EU awards. This was then followed by a short survey of all our LAs in SUITS to understand more about the tangible and intangible costs and benefits they had experienced and what advice they would offer to other LAs.

18.3   Case Study of Coventry City Council and Transport for West Midlands (UK)

Coventry City Council was invited to join the SUITS consortium as the UK, Local Authority, partly because of its close association with the university but also because of its prior experience and success across a series of European/UK intelligent mobility projects, where it offered itself as a live testbed for new transport technologies. These have included integrated ticketing, CAV, intelligent parking and route planning, two previous EU projects (in Framework 7) projects where valuable best practices had been learned.

Shortly after the start of the project, a UK-wide reorganisation of local authorities brought Coventry into the West Midlands Combined Authority (WMCA) along with Birmingham, Dudley, Solihull, Sandwell and Wolverhampton. WMCA, through its department. This mean that 'Transport for West Midlands' (TfWM) took on the role of the UK Local authority within the SUITS project.

The aims of the project aligned well to the ambitions of TfWM, which sees innovation as being fundamental to its success, with smart mobility being 'the glue which binds the strategy together'. As such the organisation has a broader remit than just transport, creating challenges in ensuring that there is a joined-up approach and good communication across the organisation.

However, the organisational changes, did for a time, have a detrimental effect on the consortium with new and junior members of staff in TfWM being seconded on to the project, and required to immediately deliver against objectives. Additionally, the project administration and financial reporting had to be realigned to departments in a new, virtual and distributed organisation,

The new, combined authority was required to work in new ways, Fortunately this meant that such the organisational change process of SUITS became significant in fostering many new ways of working, new models of learning as a team, and collaborating with the private sector. At the time of the project, TfWM did not have a SUMP, but worked to a Transport Master Plan and are well regarded as innovators and early adopters of sustainabel transport measures.

In more detail, for TfWM the experience of working with teams from across the EU, is seen as helping LAs manage their own projects better. EU projects are characterised by very robust structures of project management and reporting, besides strong financial management—all key best practises that LAs can take onward into future projects and locally funded initiatives. Internal teams such as Finance, Delivery, Growth, Infrastructure—all benefit through working within the EU project processes, learning a structured way to use funding, deliver a project and engage with internal stakeholders and external business partners. Working on a series of EU projects has led to upskilling of staff and transference of knowledge and best practice across the organisation.

For Coventry these learnings have provided the opportunity to become one of the leading authorities in creating a Smart City. This in turn, attracts more research and development funding and external investment into the city, thereby increasing its economy and (inter)national reputation. Working with EU partners provides extisting projects with added exploitation opportunities, not just within Europe but globally.

In the UK, one of the challenges faced by local government is how to reap the benefits of local devolution whilst also benefiting from economies of scale. Local control of public service delivery leads to better outcomes for citizens, since local government generally knows its citizens and the peculiarities of their own local authority best. These local challenges are the reason why LAs exist, and they often campaign for greater powers to be devolved to local level. However, LAs sometimes have a reputation for being too inward-looking or not learning from the outside, leading to duplication of work.

Projects like SUITS offer an opportunity for LAs to discuss the issues that they are facing and learn from others. Staff often have few opportunities to do this: anecdotally, we have recently found an instance where staff in neighbouring authorities working on an area of highways work had never spoken to one another. This causes operational challenges, but also extends to more strategic issues, such as developing new strategies for helping them do their work. Too often, LAs' opportunities to share information comes only through their suppliers, meaning that it is difficult to get an unbiased and complete picture of others' experiences. All LAs share a common purpose of serving citizens in their areas and so when staff can be brought together, knowledge sharing occurs fairly easily.

Local devolution should not be about LAs existing in isolation, cut off from anyone else – but instead, about having the independence to seek out the very best of what others have done and apply it to their local context – and innovate where suitable solutions do not exist. The SUITS project is one way in which WMCA has been able to learn from elsewhere. This is of value whether we feel we are 'ahead' or 'behind' another local authority.

For example, the visits and workshops showed that other LAs face very similar issues to TfWM. For instance, when visiting Stuttgart, it was valuable to hear about the challenges around public acceptability of measures that the city was facing, in particular in relation to clean air restrictions. Stuttgart is in many respects similar to the West Midlands, with a strong car industry, so it was interesting to see that they had faced similar issues around public acceptability and what they had come to understand. This was reassuring and helped them to both realise that they were not the only ones facing this challenge. In other cases, a LA may be facing far greater upheavals and may benefit from learning about how another LA has, for example, instituted organisational structures that lead to greater success in projects and programmes. In other cases, TfWM learned from other cities, such as Rome's work on citizen engagement for their Sustainable Urban Mobility Plan. TfWM were able to draw inspiration from this to inform the development of a Market Research Online Community to enhance the way they interacted with their citizens.

Ultimately, being a part of the SUITS project has helped TfWM to realise that cities are not simply 'ahead' or 'behind' others—rather, each has individual techniques and practises for working, which can be applied to different projects as appropriate. For example, a discussion was held on the merits of PRINCE2 methodology compared to Agile for a range of innovative projects—leading to a better understanding of the relative merits of each.

Learning from the SUITS project is not simply about learning from other local authorities – we have also been able to learn from others across different sectors, such as academic institutions and consultancies. For instance, having the support from Coventry University's team in organisational change has supported us in being able to make changes within our own organisation, leaving us in a better position for future projects. It takes time for this to come to fruition, but it does make a difference. In turn, this is forging relationships that would not otherwise have existed. This has been of continuing importance for TfWM as we strengthen our relationships with private sector partners that we work with, for example in our Connected and Autonomous Vehicles programme.

Projects which Coventry City has been involved in and their outcomes have depended on engaging with the politicians and senior management, who are willing to support the projects. This experience and the process was shared with partners on the SUITS project. Partners like Kalamaria, Palanga and Valencia all benefited from this process and the long-term support from their politicians and senior management. This in turn helped them to understand how their support can help improve their cities and bring in inward investment and create new jobs by creating a smart intelligent transport City.

As the project developed, various Workshops were held, where city partners gave presentations alongside academic and business partners in the project. These proved a valuable learning experience, as individual cities' experience can in turn be exploited by other cities. Examples of this are presentations about CAV, intelligent parking, traffic management modules, smart crossings, all first trialled in Coventry and shared through project workshops.

Kalamaria acted as a testbed for trailing the MyPolisLive innovative technology tool in the Municipality's vehicle fleet, and was able to share this experience through a Workshop presentation to the whole team, in addition to several presentations from SUITS partner SBOING, the tool's developer. Trialling the tool has helped to improve the municipality's effort and know-how in designing and implementing urban sustainable mobility measures and issues around handling and using mobility data (especially in light of GDPR).

Valencia presented their measures to improve the cycling infrastructure and cycling culture in the city, including setting up the 'bike agency' as part of the local authority, and user-friendly communication campaign, which was followed up as an idea and then exploited by Alba Iulia.

Alba Iulia's experience with intelligent procurement was presented and shared with the other cities, for them to use and exploit. This was amplified by the three Guidelines and 'Decision Support Tool' developed to assist procurement for local authorities.

TfWM shared their experience with holding weekly team meetings. This practice has now been implemented with all city partners. TfWM also gained much from using MS Teams© software to share project developments within the Future Mobility team and later around the organisation. This in turn was a good practise that was shared with other cities.

Being involved in a European project leaves a legacy within the local authority which persists long after the end of the project. New ways of working and problem-solving can become an integral part of the ways teams work. Where proved to be successful, this learning can be extended to other parts of the local authority with wider benefits.

18.4 Views from Other Local Authorities in SUITS

Unfortunately, the project team was not able to meet face to face during it's last year, due to Covid restrictions. In order to collect feedback from the LAs we sent them each five short questions, which could be anwered in about 30 minutes. The results were collated, summarised, thematically analysed and anonymised where appropriate. These are presented in the following sections, with quotes from the LAs where appropriate.

18.4.1 Expectations of LAs at the Start of the Project

Knowledge sharing and exchange

All the LAs expected that the project would provide them with an opportunity to share knowledge, experience, and best practices with other LAs. This expectation was achieved through their participation in ongoing workshops and discussions conducted throughout the project, and 'closed' city partners sessions in the 6 monthly meetings etc.

'SUITS was a very good opportunity to learn quickly from technical partners and partner cities and to access to high level knowledge'.

Training opportunities

Additionally, LAs looked forward to the more formal training and capacity building provided in the webinars, and the CBP specifically in relation to sustainable mobility.

Increasing organisational effectiveness

Key areas where LAs hoped SUITS would contribute included project management, decision-making and validated processes. Partners mentioned the need to update transport departments in terms of human resources. For example, *'The department was composed (and still is) by «old» people, with a professional background mainly focussed on infrastructures and transports than on sustainable mobility solutions… there was the need of bring new concepts into the mindset of the department, in order to be able to update the SUMP.'*

Some LAs expressed a hope that they could spread the tools developed by SUITS through their organisation, i.e. beyond the transport/mobility departments. Indeed, the organisational change process can be applied to any organisation/or part. To our knowledge, not only has SUITS been the first instance of Kotter's model being applied to local authorities, it is one of the few examples that (1) has shown a success, (2) has been fully documented. It was commented that 'behavioural change in organisations promoting the exchange of ideas and the multidisciplinary work between different teams avoiding silos is my favourite one'.

Opportunities to be innovative

Following from the point above, being part of a project such as SUITS gives LAs an opportunity and permission to think in different ways and try out new ideas. This provides a break from traditional modes of thinking. Talking to colleagues from different LAs can show what could be possible and leads to more innovative and exciting thinking.

Opportunities to share

Whilst some LAs stressed that SUITS gave them an opportunity to learn from others, an equal emphasis was given by more experienced LAs of the need to show others their experiences.

For the follower cities, expectations related to enhancing networks, knowledge and reputation of the city staying up to date on EU developments in the mobility sector. They were also interested in finding out about how LAs in other countries were organised, what roles and resources were available to transport planners.

Consideration of the tangible and intangible costs of involvement

Time and levels of commitment were major costs, especially at the start of the project, where staff were new to each other, and in many cases were unfamiliar with the way in which EU projects were structured. For example, projects generate a lot documents—mostly in English—which have to be read, understood and translated into operational details. This is a very big burden for small LAs, those unfamiliar with EU projects, or who do have large administrative teams. As the project progresses more documents are required in terms of administration and financial reporting. The overheads in terms of learning new systems and completing these forms may be underestimated and can be substantial for smaller LAs. Additionally project outputs (e.g. in the form of deliverables) may not easily map on to the 'everyday' challenges and work of LAs.

LAs with little involvement in research and innovation elements spent a larger proportion of their overall time on project administration.

With high levels of organisational churn, getting the right level of sign offs became a great problem, especially when institutional and tacit knowledge is lost. This was especially problematic during reporting periods, when only 'named' people with certain levels of authority are required to sign off documents. Typically such people are far removed from project delivery.

Although LAs were fully supportive of the project, in some cases they did not understand the implications of it (e.g. in relation to organisational change). Getting the right level of buy-in was difficult—with a lot of attention having to be placed on recruiting change agents to lead the organisational change process in the LAs that had sufficient influence with senior management. Many local meetings and discussions were needed to convince departments to make changes, or release colleagues to work on the project. This reduced the speed of delivery and also the initial enthusiasm for making changes or innovations. Being able to share issues with other cities going through the organisational change process enabled strategies to be developed on how to overcome institutional barriers.

Additionally, in many cases staff had to manage the expectations of the project as well as fulfil their usual workloads. For example, *'The mobility department is often overwhelmed by the ordinary workflow (and the situation worsened further during the covid emergency) so that it is not always easy to have internal exchanges on the organisational change, for example, or on internal process'* and *'For many of the staff members this was a very novel type of project which required extra time for implementation besides what they were already allocated'*.

At the start of the project time and resilience were the main costs, *'until you open the right door and identify the curious minds to collaborate in the project. Some of these people have curiosity, open minds and are generous'*.

18.4.2 Discussion of the Tangible and Intangible Benefits

The LAs were able to point to a number of tangible and intangible benefits.

Tangible benefits included:

- Knowledge sharing and learning for LAs. This was the most frequently mentioned point. Being on the project provided LAs with opportunities to talk to their peers and understand where others are facing similar problems. For very small transport departments, this is a big advantage, e.g. Palanga commented on the advantage of gaining insight into innovative organisation and funding schemes for implementation of their sustainable urban mobility plan. Additionally, the project provided access to ('*great*') professionals and experts in different fields, who are of similar mindsets – sharing the same interests and approach. These could be the basis for continued work or networking, e.g. Valencia with Coventry and TfWM, and Palanga with Smart continent
- Use of outputs. Partners reported that the materials, including the guidelines on innovative financing, were useful as prompts to consider new ways of approaching project funding, e.g. Rome used SUITS 'innovative financial tools' in developing their SUMP.
- Access to a great collection of best practices and manuals in different fields from finance innovation, mobility topic and behavioural change.
- Costs savings, e.g. by drawing on the experience of others, using already available hardware and software developed by the project (at no cost to the LA), for example my PolisLive, net and ultranavi systems.

With regard to sustainable transport measures, specifically:

The 3 guidelines (innovative financing, innovative procurement and new business models) developed through SUITS have directly contributed to improve public transport. For example, AIM developed numerous European projects on mobility (mounting to over 30 million Euros) which were approved during the SUITS implementation and which will change the face of transportation in Alba Iulia. This can also be considered a tangible result/benefit.

Kalamaria emphasized that SUITS had directly contributed to:

- the implementation of various sustainable measures such as new pedestrian crossings or the pedestrian touch devices as additional equipment in traffic lights with pedestrian affected phases and for use by visually impaired people.
- more staff from technical department (3 member of technical department's staff) engaged with the project and its core ideas,
- increased understanding of sustainability within the organisation.
- Enhanced user experience and engagement with sustainability issues, for example by using a municipality's online platform '4mycity'. The citizens were more and more engaged with traffic and mobility issues through the platform. For example

they were asked to defining problematic points in traffic or suggesting solutions and updating new problematic areas that need restoration.

For Turin, one of the main problems the municipality had to face in implementing freight monitoring measures was the lack of specialised staff for data analysis inside its department—a common issue for smaller cities. This absence was covered through cooperation with the project partners, mostly with Politecnico di Torino. Participation in European projects always provides the added value of interacting with other entities that can give valuable support in activities that otherwise, would not be possible.

Intangible benefits included:

- the increase of satisfied citizens using new smart pedestrian crossings or the pedestrian touch devices as additional equipment in traffic lights with pedestrian affected phases and for use by visually impaired people, measures implemented during SUITS period
- Improve collaboration culture in the organisation increased LA satisfaction, e.g. consolidation of the local authority human capacity. Attendance of the project's workshops provided project members with the confidence to create new connections between people within their organisation, sometimes even from the same department.
- Gaining of new skills and competencies by team members, e.g. '*I have acquired new competences, not only on technical matters, but also, if not especially, on organisational change. I try to use this knowledge in my work, now. It its very useful also when I talk to external organisations, which are more flexible then the municipal administration. Or when I have to involve other city departments (environment, innovation, etc.)*'

These are especially important as they clearly evidence that SUITS has had a direct effect on the capacity of project members and trainees to engage with external organisations to the future benefit of their cities.

18.4.3   What Actions Could Be Taken to Maximise Tangible and Intangible Benefits and Minimise Their Costs

On being invited on to a project, LAs should consider:

- Ways to maximise the benefits of the outputs of the project, e.g. by expanding on recommendations, and making use of the other LA's experiences (which in this case were mentioned in the CBP or introduced during workshops).
- Using the project as an opportunity to engage in inter and intra departmental working. Teams saw this as an '*opportunity to have a better understanding of each other's work and how they can collaborate if there is room and time windows for initiating the non-formal collaboration between colleagues*'.
- Incorporating a 'European' dimension into job descriptions.

- Find ways of breaking down the silos. '*Most private and public organisations have specific habits and rules, most of them look written in stone when we look at the official organigram in a public institution. Nevertheless, if you have the chance to skip the formalities and bring the people into a workshop or an informal talk far from their desks, then communication and collaboration between people pops-up with some guidance. If I have to provide some insights from my experience in SUITS are:*

  – *Build or improve trustful relationships with your colleagues, adding value and positive vision.*
  – *Get the people far from their desks, if it is possible in a different building to avoid email and calls.*
  – *Get some endorsement from management as Mobility departments are over-loaded with daily tasks most of the time.*'

In terms of design of future projects, Principal Investigators should:

- Create opportunities for small wins at the start of the project, so that 'ideas' can be more easily sold to senior management and the organisation. SUITS required a lot of input from LAs at the start of the project, with the benefits (in terms of project outputs) only rolling out after Year 2.
- Demonstrate to senior and departmental managers the added value of involvement in EU projects at individual, organisational and city levels.

18.4.4 What Advice Would You Give to LAs Who Are Thinking About Joining an EU Project

Common threads emerged in terms of advice to LAs contemplating working on an EU project.

- A high level of commitment is required in relation to administration and reporting. This can be sometimes overlooked but may be a significant burden with tight deadlines, multiple sign offs and unfamiliar processes.
- Projects require a significant level of commitment. LAs need to plan resource and staffing required to participate fully.
- Local administration's human resources are limited (and they have little chance to recruit new employees) and are often already fully allocated. EU projects bring new workload and managers have to deal with this by relying on external support or implementing organisational changes. Therefore, they should evaluate very carefully how they will deal with project activities in case the project is funded. Technical partners may be brought in, but the ownership of the project must remain of the municipality.
- LA's should ensure that their own aims for the project are clear and they are clear about the benefits they want to get out of the project. This should be done with

senior management so that staff other than those directly involved in the project can participate in the project activities, including training.

The smaller follower cities warned that:

- Involvement in some procedures, such as participation in surveys, are time-consuming and do not provide short-term benefits.
- Involvement may require cooperation with other internal departments, and in some cases it can be challenging to get them to participate.
- You should think carefully about what you want to get out of the project and communicate this clearly
- Be aware that what is worked on in the project and the results may not fit the challenges you currently face
- Ensure that staff are paid for the work they do on the project

On the plus side, being in an EU project is seen as:

- A great opportunity to learn from other local authorities and project partners that may be facing similar challenges and have novel ideas as to how those challenges can be addressed.
- Sharing the experience for other EU partners has been a major benefit in terms of improved processes, new technology and new designs in infrastructure.'
- For a LA being in a European Project can influence policies, ensuring a better quality life for each city and region.
- Participating in European projects providing a unique opportunity for LAs with limited budgets to experiment innovation, share information, experience, knowledge, to travel, meet experts, and have access to specific documentation and data.
- Joining an EU project is a way of exchanging and using new ideas, widening staff's way of thinking, and using new technology. Also, in a new situation as 2020's pandemic crisis, EU projects could help LAs avoiding barriers and minimising costs investing in new and innovative sustainable mobility measures, arising from EU projects research.
- Additionally, the projects provide 'another way people from different counties and cultures to collaborate and enhance know- how. Cohesion between different European areas, in economic, cultural and sociable terms is enhanced and improved'

Advice was:

- 'join as many EU projects that you can as long as it meets your LA's transport policies'
- 'Don't think twice! You will learn a lot of new things and will get the chance to develop new cooperation projects with other peers at European level!'
- 'Even if a proposal is not successful, being in a project provides a unique exercise to analyse and audit your needs and challenges as an organisation, which is already a great achievement. If the proposal succeeds you must 'be ready and open to take the most of it as a unique experience to learn, improve and share with

colleagues from different countries with a shared goal and different expertise and experience'. It will allow you to have access to the best European experts with latest developments and you will learn in the process new methodologies, technologies and best practises with new colleagues which will be of great help in this project and in the coming ones. The project results will improve the quality of life of the people in your city, so there is nothing else to say.'

18.5   Conclusions

Looking at the feedback, cities and especially smaller cities should be aware that their participation involves a considerable administrative effort, which should not be underestimated. Therefore, it is important to check whether the technical and administrative capacity exists to support involvement. This should not be too much of a deterrent, as the project leaders and experienced partners will offer guidance and support.

LAs should also be aware that in certain project phases a lot of effort is required, e.g. by participating in interviews, surveys and evaluations, which does not necessarily result in an immediate visible benefit. On the other hand, workshops and meetings provide access to knowledge for which one would have to pay a lot of money elsewhere, for example for consulting services. In any case, there must be a willingness from the city side to play with open cards and to make data and information on plans and projects available to the consortium and to participate actively in project development. It is crucial, that city partners think carefully about what they want to get out of the project and communicate this clearly. Only in this way their expectations will be met, and they can be sure of the full support of the project.

Participating in an EU project provides opportunities for LAs to further develop their own knowledge and capabilities. Establishing contacts with other cities, to learn from their challenges and experiences and to reflect the own activities through mutual exchanges in workshops and conferences was also valued, as many do not have opportunities to look outside of their own departments. Participating in an EU project means working with a wide variety of project partners from science, research and consulting, benefiting from their competencies and expertise and receiving input on a wide variety of topics. These different impulses can make decisive contributions and sustainably change the attitude and the capacity of city administrations.

In addition, cities may be able to implement projects within the framework of an EU project, often with an innovative and experimental character and scientific support, which otherwise would not have been implemented due to a lack financial resources or political support. Under the guise of a research project, unexpected opportunities may arise. Moreover, visibility and reputation of a city is enhanced from participation in a European project and the dissemination that comes with it.

Reference

1. Tovey, M., Woodcock, A., Osmond, J. (eds.): Designing Mobility and Transport Services: Developing Traveller Experience tools, 1st ed. Routledge (2016). https://doi.org/10.4324/9781315587295

Chapter 19
Conclusions

Andree Woodcock

19.1 Introduction

This volume has documented the main results from the H2020 CIVITAS SUITS project, provided insights into the structure of the project and the complexities of developing integrated sustainable transport measures. In this closing chapter, I will discuss some of the wider issues which have been touched on by SUITS and put them in the context of climate change and the COVID-19 pandemic.

SUITS was funded as one of three projects, along with PROSPERITY and SUMPS-UP to develop the capacity of SM LAs to develop SUMPS or to have greater awareness of sustainability issues as part of a wider EU push to become carbon neutral by 2050.

As part of this initiative in December 2020, the EU agreed to a net domestic reduction of at least 55% in greenhouse gas emissions (GHG) by 2030 compared to 1990. This would also spur economic growth, create jobs, deliver health and environmental benefits for EU citizens and contribute to the EU economy by promoting green innovation. Meeting these targets requires multisectoral actions including:

- investing in environmentally friendly technologies
- supporting industry to innovate
- rolling out cleaner, cheaper and healthier forms of private and public transport
- decarbonising the energy sector
- ensuring buildings are more energy efficient
- working with international partners to improve global environmental standards

A. Woodcock (✉)
Coventry University, Coventry, UK
e-mail: A.Woodcock@coventry.ac.uk

© Transport for West Midlands 2023
A. Woodcock et al. (eds.), *Capacity Building in Local Authorities for Sustainable Transport Planning*, Smart Innovation, Systems and Technologies 319,
https://doi.org/10.1007/978-981-19-6962-1_19

Transport accounts for 25% of the EU's GHG emissions, with road transport contributing just under 72% of all transport emissions. The Green Deal[1] seeks a 90% reduction in these by 2050. Relying on technological innovation, implementation and uptake, proposed solutions included greater digitalisation through, for example, automated mobility, smart traffic management, and smart applications such as MaaS.

Technological innovations need to go hand-in-hand with upskilling and training of those ultimately responsible for implementation and planning and fit into a shared vision for the future. As the EU recovers from the covid pandemic and makes new commitments towards carbon neutrality, it is worth considering what barriers remain, how these may be overcome and the position of transport in wider city visons.

19.2 Data and Implementation of Transport Measures

SUITS demonstrated a wide gap in knowledge and ability regarding digitalisation in small-to-medium local authorities. The rate of technological innovation and data acquisition has outstripped people's understanding and capacity to understand how to use this. As a project, we provided training material and resources to allow LAs to independently and easily acquire mobility data through crowdsourcing.

Mobility planning needs to be grounded in experience, knowledge and be evidence based. During the lifetime of the project, the LAs struggled to improve their capacity to handle data. Barriers which prevented them acquiring and using data included: service operators being unwilling to share data as it is part of their USP; tightening up of data protection legislation made it difficult to repurpose existing data [1]; data was not collected in a standard way making it difficult to combine, and additional processing was needed to ensure anonymity and privacy.

Data simply may not be available, accessible or in a usable format [2]. For example, data will be collected at a certain time and location, for a specific reason, for example to support changes to road layout, or to collect passenger satisfaction. Such data may not be shareable (owing to GDPR), or not in standardised format to allow reuse and combination with other datasets. It may not have been well documented or have all the elements essential for new work. For example, there is still a lack of sex-disaggregated mobility data.

Hence, a lot of potentially useful data and knowledge may be distributed round the organisation but is not discoverable. Or it may be collected routinely by service providers but is not shared with LAs [1]. While some data is collected on vehicle movement, multimodal mobility data relating to everyday travel of citizens is only starting to be captured e.g. from mobile phones. Freight data, which is commercially sensitive, is not shared despite a high proportion of freight vehicles in urban areas. This situation improved slightly over the 4 years of the project. Obviously, without a full picture of movements within the city, transport planning cannot be inclusive or optimised.

---

[1] https://ec.europa.eu/info/strategy/priorities-2019-2024/european-green-deal_en.

To address these issues, we would recommend:

- an EU wide sectorial skills audit to address gaps in knowledge
- structured course development and accredited professional qualifications
- standardisation of data collection to enable data to be merged
- development of integrated data hubs in all municipalities, enabling them to process, integrate, receive and service data requirements to enable smart city development (as instigated by TfWM)
- Contractual agreements with fleet operators to provide LAs with freight data (as in Torino).

We also found that LAs were concerned about the introduction of new service offers (e.g. MaaS) or providers, who disrupt transport provision in multiple and unforeseen ways. Cities are being used as living labs—to test out new transport innovations-such as autonomous vehicles, drone deliveries and e-scooters. This can disrupt fragile transport business ecosystems and put existing operators out of business. New start-ups may take passengers off public transport and serve only a small part of the population, leading to greater transport inequities and inequalities.

LAs are responsible for the equal provision of services to all their citizens. Managing new, disruptive market entrants is an added and complex responsibility especially for small transport departments who have to maintain the city transport network and monitor the effects of new operators, e.g. to consider social impact, guarantee employment rights, reduce negative consequences, deal with disputes and re-establish transport systems when trials are finished. For example, some nationwide e-scooter trials were halted in the UK (in 2019) due to misuse, dangerous driving, and reckless abandonment of scooters on pavements. Clearly, older people and those with poor mobility and reduced vision were being made more vulnerable during the trials. As a result, some trials were discontinued with LAs being held responsible for implementation problems by both the service operators and citizens. This creates extra work for the LA, reputational damage and a loss of trust and confidence.

Enabling LAs to be open to innovation was a central theme of the organisational change programme, which sought to transform traditional and hierarchical organisations into ones that were flexible, resilient and innovative [3–5]. With many opportunities to become involved in technology trials and pilots developing out of commercial and research funded projects, there may be a lack of transparency and consultation both with internal and external partners. This can be the case when there is a lag between submitting proposals and funding, when technology needs to be developed or when funding is rushed through. In such cases, there can be multiple breakdowns in communication, with those commissioning the initial project most likely to not be the ones involved in administration and implementation. Employees co-opted on to the project may not be fully briefed and part of the organisation may simply not be informed about activities which may have a profound effect on their work. There needs to be set of ethical standards and codes of conduct in place for consultations, to manage expectations, limit risks, ensure transparency and that all parties benefit from trials which involve members. This enables the build-up of trust within an organisation and between partners (including citizens).

SUITS widened the LAs' familiarity with technology innovations and created opportunities to share experiences with colleagues in similar positions in other cities-so they could learn from each other. We also provided, in consultation with the LAs, a series of business model templates to act as a catalyst for more robust business plans.

19.3 Organisational Trust

The initial audit of cities showed that LAs faced similar internal challenges which reduced their ability to work flexibly and apply new ideas. All cities were able to make small changes and see the benefits of these within the project, becoming models for other departments in the municipality, for example TfWM created newsletters and shared small victories and news of their project with the wider organisation.

The project documented clearly that trust matters in LAs when trying to enhance their capacity building and to become more resilient and agile. It was recognised that a LA or its management can initiate a change but the implementation of the change is carried out by the employees and that employees' commitment to change is one of the decisive factors to allow organisational change [6]. This commitment is primarily driven by employees' trust in their local authority and its top management [7]. Trust is a willingness to be vulnerable to the actions of another party based on the positive expectations that the other will act beneficially, or at least not inflict harm, irrespective of any monitoring or control mechanism [8, 9].

LAs with high trust have higher performance levels, as well as higher levels of reputation. The more positive experiences employees gain with or within the organisation, lead to internal and external recommendations and endorsements. Employees working in an organisation that shows high levels of trust have usually greater job satisfaction and are more co-operative with each other [10] in terms of knowledge sharing, providing innovative/creative ideas, supporting each other and the organisation, have higher levels of intrinsic motivation and will go the extra mile to achieve an organisational target. Finally, trust in the local authority reduces counterproductive working behaviour to a minimum and mistakes or errors are honestly communicated.

Overall, it was found that LAs have to be a trusted organisation as this is the basis for implementing organisational change successfully in the local authority. The positive effects of trust are needed to keep the employees on track during the long journey of change and to keep them motivated. Also, as failures and errors are very likely when new technologies, processes or structures are implemented, a local authority will benefit a lot from high levels of knowledge sharing. Trust allows a local authority to enhance their capacity and to become a learning organisation, being open towards innovative ideas and solutions, reflecting their own culture and processes and being more resilient, all of which are mandatory for sustainable urban mobility planning.

Interestingly the LAs worked on trust issues within their own organisation as a prelude to building up trust with the citizens.

19.4 Partnership, Leadership and Trust

SUITS and other projects have evidenced that there has been a lack of knowledge and shortcomings in understanding and buy-in to sustainable agendas at the LA level. A plethora of project outputs, processes, guidelines and good practises, is reducing existing knowledge gaps around planning and implementation within LAs. This has also included steps to encourage better engagement with citizens (e.g. through participatory planning). Substantial funding is now available through different funding mechanisms to enable cities to implement sustainable transport measures. Research findings have also evidenced the role of transport in climate change and its negative consequences on health and wellbeing.

As a community, the EU needs to consider what other efforts can be made to speed up the transition towards a more sustainable, climate neutral future. Before more regulatory instruments are employed more efforts should be made towards developing trust and partnership between LAs, transport operators, large city employees, so that they work together towards agreed goals. More openness and collaboration in planning has been effective in smaller SUITS cities such as Dachau and Palanga. Such efforts can also be seen at city level when roads are closed to traffic on certain days, or when pollution levels are too high.

Developing new transport services is not only more sustainable, their design also provides an opportunity to assess problems and inequities in current transport provision. Many cities are trying to adapt systems, services and infrastructures which prioritise mobility of people and goods, rather than the efficient movement of traffic. Streets and city centres are being reclaimed and made more accessible. A wider understanding of transport has also led to a far greater understanding of the negative effects on human factors-health, wellbeing, mental health and stress, all of which are affected by the by-products of transport-noise, pollution, lack of green spaces, time lost in traffic jams.

This requires a partnership between technology providers (and their funding bodies), technology users (in this case transport providers), governments and legislators (politicians), local authorities and members of the public. This needs to be built on transparency, openness, and trust. In SUITS, we found that this relationship was lacking and that there were misunderstandings/lack of explanations about decision-making. A balance needs to be created so that everyone understands why certain actions need to be prioritised, buys into the shared vision and supports it. For example, politicians many steer away from commitments that disproportionately disadvantage car drivers (e.g. by introducing parking levies and taxation), green spaces may be eroded by revenue generating land use, city wide vision for creating more healthy cities may be poorly communicated. The need for intersectoral activity and environmental considerations to be built into all policy areas is known as 'climate mainstreaming'.

19.5 Building Up Trust: Modal Shifts and Behavioural Changes

At the heart of all sustainable transport implementations is the core requirement that they will be used and adopted by sufficient members of the public to ensure their longevity. Many pilots, technology trials and new services fail because they are not used by members of the public, or not used in the ways expected. Modal choice can be influenced by many factors, economic, social demographic, psychological, cultural and practical issues [11]. Understanding these can help to shape strategies to nudge behaviour in certain ways, e.g. through incentivization, advertising campaigns and use of key influencers or disincentives.

However, currently uptake of sustainable transport measures is patchy. SUITS LAs recognised the need for better communication and engagement with citizens and developed action plans to address this including poster campaigns and new departments.

Although many people and their cities are benefiting from using active forms of transport, policies are needed that increase trust in the safety of sustainable transport. This could include:

- prioritisation of investment in sustainable infrastructure such as bicycle lanes, cycle superhighways, crash helmet hire, quality of end of trip facilities (to encourage use by underrepresented groups), reducing speed limits, prioritising bikes and pedestrians at junctions. For example, Paris plans to build 650 km of cycleways connecting the city centre with the suburbs—including temporary 'corona bike lanes'.[2]
- Measures to ensure safety and build confidence in public transport systems, e.g. increasing the frequency of services and publishing real-time updates on public transport congestion, compulsory wearing of facemasks, hand sanitizers, more frequent cleaning of vehicles, automatic door openings, contactless payment.
- Pricing and regulatory policies which discourage car use (e.g. road and vehicle tax, parking levies, congestion charging) and encourage mass and active forms of transport. For example, in the Netherlands, cyclists can claim €0.19 for every kilometre cycled to work, and France has introduced tax free incentives for employees who can prove use of sustainable transport modes such as car-sharing and cycling.
- Holistic approach to cost models and financing. There has been a systematic underreporting of the economic and quality of life benefits of sustainable modes of transport. This is being partly addressed through new innovative financing and procurement guidelines, but economic spillover effects are not yet emphasised. For example, the effects of long-term job creation by investment in public transport and cycling, increased footfall and nicer environments created in pedestrianised areas which lead to increase in retail opportunities.

---

[2] https://www.forbes.com/sites/carltonreid/2020/04/22/paris-to-create-650-kilometers-of-pop-up-corona-cycleways-for-post-lockdown-travel/.

- Considering the positive impact on public health budgets. For example, some estimates suggest that if every Londoner walked or cycled for 20 min a day, the UK's public health system could save GBP 1.7 billion in treatment costs[3] over the next 25 years. Others estimate that for each car driver switching to cycling for a daily commute of 5 km (one way), the health benefit from the physical activity is worth about EUR 1300 per year [12]. Recent evidence suggests that cities with higher levels of pollution may have been more adversely affected by the COVID-19 pandemic [13].

19.6 Building Up Trust: The COVID-19 Pandemic Effect

During the Covid pandemic there was a short term, drastic shift in mobility behaviour. Global road transport activity was almost 50% below the 2019 average by the end of March 2020 and commercial flight activity almost 75% below 2019 by mid-April 2020. Public transport has also been affected. For example, the strict lockdown imposed in the UK in March 2020 led to a 95% decrease in underground journeys in London. This is supported by data from one popular transport planning smartphone app showing that trips are down by over 90% since the crisis began in many of the world's major cities.[4]

Shifts to more active forms of transport or less travel were made due to social/physical distancing, government requests for voluntary self-restrictions on travel (backed up by law enforcement in some cases), social influence and the perception of risk (dread behaviour) and closure of venues (e.g. shops, schools, leisure facilities). These could become permanent switches with incentivizations and changes to transport infrastructure (such as more bicycle lanes).

'Dread behaviour' [14] influences mobility behaviour in a number of ways, even when the risk is smaller than other common dangers; for example, fear of terrorist attacks and bombings reduced use of public and air transport; parents fear letting their children walk or bicycle to school or other destinations due to exaggerated fears of kidnapping even though they may be supported at a global level, climate change and sustainability policies; public transport is perceived as unsafe even though road rage and traffic accidents lead to more deaths.[5] Clearly, media coverage has an enormous effect on public perception. Planners and policy advisors have a responsibility to help decision-makers and the general public put risks into perspective by sharing accurate information and leading by example.

---

[3] https://www.london.gov.uk/press-releases/mayoral/setting-out-a-vision-for-getting-londoners-active#.

[4] https://www.iea.org/articles/changes-in-transport-behaviour-during-the-covid-19-crisis.

[5] https://www.planetizen.com/node/51290.

It might have been predicted that post covid-19, dread behaviour and risk aversion would continue to play significant roles in the future of public and shared transport as close contact in confined shared spaces is a major contributor to the spread of the disease. IBM's survey in US[6] showed that along with a shift to less travel, there was a 17–28% swing away from public and shared forms of transport, with 1 in 4 saying that they will use private vehicles exclusively going forward, attributed in part to dread behaviour. This is mirrored in increases in car sales in Korea and rises in morning peak hour traffic in Beijing. Where private cars were a dominant form of transport before the pandemic, an analysis of 'dread behaviour' following previous crises suggests that there will be a switching away from public and shared transport to private vehicles. This is accompanied by a push to 'kick start' economies which, unless carefully managed may result in a return to pre-covid levels of congestion and pollution, with gains made in greening cities quickly forgotten.

There will be a lot of lessons that can be learnt from individual, city and national responses to the COVID-19 pandemic which will be directly relevant to the transport sector. This was a global crisis that had immediate, significant, visible and long-term effects. This contrasts with the climate crisis where a sense of urgency is still lacking.

Governments and local authorities have responded differently to the COVID-19 pandemic, using different strategies and measures to reduce its spread. A key element has been the role of local authorities and the relationship between the authority/governing bodies and the citizens. Many recommendations will come out in the following months; however, the recommendations of [15] are relevant to many of the issues of SUITS. The researchers, working in the context of the COVID-19 pandemic in Japan, where there were no legal mechanisms to enforce a "city lockdown." found that *non-binding requests, soft measures such as campaigns to promote a reduction of non-essential travel might be more effective if they (i) properly convey the severity of the threat posed by COVID-19 as well as its coping mechanisms, and (ii) appeal to the group, rather than the individual, emphasising the behaviour (or at least the perception of behaviour) of others'*. Importantly the agency doing the persuasion has responsibility for guaranteeing that fears are legitimate, that coping strategies are effective, and that support is available to minimize the cost of coping behaviour incurred by businesses and individuals.

It may be predicted that the widening understanding of mobility, as opposed to transport per se, along with a post-covid rising interest in the local natural environment agenda (centred around permaculture, biophilic urbanism and nature based city planning) will set new agenda and strategies, forcing a rethink in priorities. This will create opportunities for climate mainstreaming in smart city development to achieve carbon neutrality through a mixture of methods, across different sectors. Already researchers in urbanism, such as Newman [16] are looking beyond carbon neutral scenarios to regenerative development in which 'not only will power and transport be zero-carbon, but all industrial processes, natural system processes and the whole metabolism of cities and settlements will be like an ecosystem that regenerates itself.'

---

[6] https://newsroom.ibm.com/2020-05-01-IBM-Study-COVID-19-Is-Significantly-Altering-U-S-Consumer-Behavior-and-Plans-Post-Crisis.

In this scenario, cities (or LAs) will begin to work collegiately, across departments and in new partnerships (e.g. with the creative industries) to 'create new centres of zero carbon–zero poverty urbanism' [16].

19.7 Going Forward; Building on SUITS' Legacy

SUITS has had a direct impact on the LAs involved in the project, and has produced a set of material, in the Capacity Building Toolbox which can be used freely by other cites, to help their move towards greater transport sustainability, carbon neutrality and the development of SUMPs. This reduces some of the knowledge capacity gaps acknowledged in 2015/2016.

However, as indicated, future efforts need to be directed towards:

- Increasing ability to collect and use data to inform urban planning by training, standardisation, open access data and collection of all data relevant to movement in the city
- Integration of data across departmental boundaries, through the creation of data hubs which can drive smart city development which creates places that are healthy and satisfying
- The technological push needs to be balanced with greater understanding of human factors and their effects on socio technical systems. Key here is trust, engagement with citizens and partnership working-to deliver systems that are inclusive and work towards eradicating inequalities.
- Protecting and enabling opportunities for knowledge exchange between urban planners, at regional, sectorial, national and EU wide level
- Transforming local authorities and their employees into agencies of change and leadership, which are cooperative, trusting and trustworthy

Lastly, the lessons learnt from the COVID-19 pandemic, need to be considered and shape future EU thinking, strategies and road maps in a coherent and cohesive way, looking beyond sustainability, to resilience and regeneration. Efforts to establish and embed sustainability as a key goal should be maintained. However, more intersectional activity and holistic approaches are needed so that all functions and sectors of society work together to create cities that support citizens equally through the transport system, to enable them to access employment, leisure, education and all the other opportunities that city living provides. Transport is a key part of the sustainability jigsaw, and its importance has been thrown into greater relief through the COVID-19 pandemic. The pandemic has also incidentally demonstrated the capacity for citizens to change some of their behaviours, and shown municipal authorities that they can take action to achieve benefits for all.

Andree Woodcock.

Principal Investigator of SUITS.

The views expressed in this chapter are those of the author.

References

1. Nienaber, A.M.I., Woodcock, A., Liotopoulos, F.K.: Sharing data-not with us! distrust as decisive obstacle for public authorities to benefit from sharing economy. Front. Psychol. **11**, 1–13 (2021)
2. Georgiadis, A., Kalaitzis, V., Nienaber, A.M., & Woodcock, A.: MyPolisLive. net: a tool and a methodology for optimizing urban freight mobility through crowdsourcing. In: 5th Conference on Sustainable Mobility (June 2020)
3. Nienaber, A.M., Rudolph, F.: Organizational resilience: how SUITS'local authorities were prepared to cope with the COVID-19 pandemic. SUITS Policy Brief 4 (2020)
4. Nienaber, A.M., Spundflasch, S.: Managing organisational change in local authorities. A Managers' Manual (2020)
5. Nienaber, A., Spundflasch, S., Soares, A.: Sustainable urban mobility in Europe: implementation needs behavioural change. SUITS Policy brief 3 (2020)
6. Choi, M.: Employees' attitudes toward organizational change: a literature review. Hum. Resour. Manage. **50**(4), 479–500 (2011)
7. Nienaber, A.M., Schewe, G.: Enhancing trust or reducing perceived risk, what matters more when launching a new product? Int. J. Innov. Manag. **18**(1), 1450005 (2014)
8. Mayer, R.C., Davis, J.H., Schoorman, F.D.: An integrative model of organizational trust. Acad. Manag. Rev. **20**(3), 709–734 (1995)
9. Rousseau, D.M., Sitkin, S.B., Burt, R.S., Camerer, C.: Not so different after all: a cross-discipline view of trust. Acad. Manag. Rev. **23**(3), 393–404 (1998)
10. Nienaber, A.M., Romeike, P.D., Searle, R., Schewe, G.: A qualitative meta-analysis of trust in supervisor-subordinate relationships. J. Manag. Psychol. **30**, 507–534 (2015)
11. De Witte, A., Hollevoet, J., Dobruszkes, F., Hubert, M., Macharis, C.: Linking modal choice to motility: a comprehensive review. Transp. Res. Part A: Policy Pract. **49**, 329–341 (2013)
12. Rabl, A., de Nazelle, A.: Benefits of shift from car to active transport. Transp. Policy 19(1), 121–131 (2012). ISSN 0967-1070
13. Ogen, Y.: Assessing nitrogen dioxide (NO2) levels as a contributing factor to coronavirus (COVID-19) fatality. Sci. Total Environ. 726 (2020)
14. Percoco, M.: Environmental consequences of dread behavior: a note on 2005 London bombings. Res. Transp. Econ. **73**, 83–88 (2019)
15. Parady, G., Taniguchi, A., Takami , K.: Travel behavior changes during the COVID-19 pandemic in Japan: analyzing the effects of risk perception and social influence on going-out self-restriction. Transp. Res. Interdiscip. Perspect. 7 (2020)
16. Newman, A.O.: Covid, cities and climate: historical precedents and potential transitions for the new economy. Urban Sci. **4**(3), 32 (2020)